The Earth's Biosphere: Evolution and Dynamics

The Earth's Biosphere: Evolution and Dynamics

Editor: Christopher Ackles

www.callistoreference.com

Callisto Reference,
118-35 Queens Blvd., Suite 400,
Forest Hills, NY 11375, USA

Visit us on the World Wide Web at:
www.callistoreference.com

ISBN: 978-1-64116-284-5 (Hardback)

Printed in China.

Cataloging-in-Publication Data

The earth's biosphere : evolution and dynamics / edited by Christopher Ackles.
p. cm.
Includes bibliographical references and index.
ISBN 978-1-64116-284-5
1. Biosphere. 2. Earth (Planet). 3. Geobiology. I. Ackles, Christopher.
QH343.4 .E27 2020
577--dc23

Table of Contents

Preface

The biosphere, which is also known as the ecosphere, is the global sum of all ecosystems. It can also be characterized as the closed system on earth which is the zone of life, and a system which is largely self-regulating. Earth's biosphere is an ecological system that integrates all living beings and their relationships. It also includes their interactions with the elements of the lithosphere, geosphere, hydrosphere and atmosphere. The biosphere is divided into a variety of biomes on the basis of latitudes. All biomes are inhabited by diverse species of flora and fauna. This book unravels the recent studies on the Earth's biosphere. It presents researches and studies performed by experts across the globe. This book aims to equip students and experts with the advanced topics and upcoming concepts in this area.

This book is the end result of constructive efforts and intensive research done by experts in this field. The aim of this book is to enlighten the readers with recent information in this area of research. The information provided in this profound book would serve as a valuable reference to students and researchers in this field.

At the end, I would like to thank all the authors for devoting their precious time and providing their valuable contribution to this book. I would also like to express my gratitude to my fellow colleagues who encouraged me throughout the process.

Editor

1

Spatial and temporal patterns of plantation forests in the United States since the 1930s: an annual and gridded data set for regional Earth system modeling

Guangsheng Chen[1], Shufen Pan[1], Daniel J. Hayes[2], and Hanqin Tian[1]

[1]International Center for Climate and Global Change Research and School of Forestry and Wildlife Sciences, Auburn University, Auburn, AL, USA

[2]School of Forest Resources, University of Maine, Orono, ME, USA

Correspondence to: Shufen Pan (panshuf@auburn.edu)

Abstract. Plantation forest area in the conterminous United States (CONUS) ranked second among the world's nations in the land area apportioned to forest plantation. As compared to the naturally regenerated forests, plantation forests demonstrate significant differences in biophysical characteristics, and biogeochemical and hydrological cycles as a result of more intensive management practices. Inventory data have been reported for multiple time periods on plot, state, and regional scales across the CONUS, but the requisite annual and spatially explicit plantation data set over a long-term period for analysis of the role of plantation management on regional or national scales is lacking. Through synthesis of multiple inventory data sources, this study developed methods to spatialize the time series plantation forest and tree species distribution data for the CONUS over the 1928–2012 time period. According to this new data set, plantation forest area increased from near zero in the 1930s to 268.27 thousand km^2 in 2012, accounting for 8.65 % of the total forestland area in the CONUS. Regionally, the South contained the highest proportion of plantation forests, accounting for about 19.34 % of total forestland area in 2012. This time series and gridded data set developed here can be readily applied in regional Earth system modeling frameworks for assessing the impacts of plantation management practices on forest productivity, carbon and nitrogen stocks, and greenhouse gases (e.g., CO_2, CH_4, and N_2O) and water fluxes on regional or national scales. The gridded plantation distribution and tree species maps, and the interpolated state-level annual tree planting area and plantation area during 1928–2012.

1 Introduction

A forest plantation is defined as an area of introduced or native tree species established through planting or seeding for wood and/or non-wood forest products (i.e., industrial forests) or the provision of other ecosystem services (i.e., protective forests; FAO, 2005). In the conterminous United States (CONUS), all plantation forests are used for forest products (FAO, 2005, 2015). The United States is ranked as the second largest country in the world for plantation forest, accounting for about 8.5 % of the total national forest area (or 12.72 % of timberland) (Oswalt et al., 2014; FAO, 2015). Plantation forests in the CONUS are generally intensively managed, including practices such as the use of genetically improved seedlings, site preparation, nitrogen (N) and phosphorus (P) fertilization, and pre-commercial thinning. These plantation forests could significantly reduce the pressure on natural forests to meet fiber and other wood product demands (Sedjo, 2001). Upper estimates suggest that world demand for wood could be met by harvesting 10 % of the global forest area under intensive management (Oliver, 1999). In the CONUS, dependence on forest plantations to supply wood and non-wood products is increasing (Stanturf and Zhang, 2003). Plantation forests play a major role in current and anticipated future supplies of timber because of their high growth rates, easy operability, and intensive management (USDA Forest Service, 2011).

Due to intensive interventions of human activities, plantation forests are distinct from naturally regenerated forests. Plantation forests have more uniform stand structure characterized by even-aged stands, single or low diversity of tree species, and less understory vegetation. The simple stand structure is also often characterized by fixed spaces among planted trees, which could significantly reduce the competition for resources by more even allocation of nutrients, water, and light among dominant trees. At present, most of the planted tree seedlings in the CONUS are genetically improved through either selection of the best seed sources or seed orchard breeding (Fox et al., 2007). Site preparation (e.g., root excavation, soil disking and bedding, slash burning, herbicide and insecticide application, fertilizer use) is commonly used before and during plantation forest establishment (Fox et al., 2004; Jokela et al., 2010; Allen et al., 2005). During tree growth, forest plantations are often managed with fertilization, mid-rotation thinning, and weed control. In addition, plantation forests are more frequently harvested at a younger stand age as compared to naturally regenerated forests. These contrasting management practices relative to naturally regenerated forests significantly alter biogeochemical and hydrological cycles in plantation forests (Allen et al., 2005; Sun et al., 2006; Jokela et al., 2010; Achat et al., 2015a).

It is still a challenge to assess the feedbacks between human and natural systems due to the complexity of both systems (Chen et al., 2012). With increasing human interventions and the uniform ecosystem structure, plantation forests are an ideal managed ecosystem to characterize the coupling effects of human activities and natural environmental factors on biogeochemical and hydrological cycling on a large scale. Previous studies have reported the distinct, local-scale carbon, nitrogen, and water cycles in plantation forests as compared to naturally regenerated forests (e.g., Albaugh et al., 2012, 2015; Sun and Vose, 2016; Gyawali and Burkhart, 2015; Vose et al., 2012; Hoover et al., 2014). Although the importance of plantation forests has been recognized, a representation of plantation management practices in current Earth system models is lacking (e.g., Hayes et al., 2012; Tian et al., 2014; Pan et al., 2015), mainly due to few established relationships between management practices and ecosystem biogeochemical and hydrological cycling, as well as no available long-term and high-spatial-resolution gridded plantation maps at regional and national scales (Escalante Fernandez et al., 2002). In the CONUS, the effects of tree planting and management practices on forest productivity, carbon sequestration, and greenhouse gas emissions are being monitored through various ongoing field experiments and measurement programs, including the USDA Forest Service Forest Inventory and Analysis (FIA) program, Forest Productivity Cooperative (FPC, http://forestproductivitycoop.net/), Plantation Management Research Cooperative (PMRC, http://pmrc.uga.edu/), Forest Modeling Research Cooperative (FMRC, http://www.fmrc.frec.vt.edu/), Forest Biology Research Cooperative (FBRC, http://www.sfrc.ufl.edu/fbrc/), and the PINEMAP observation network (http://pinemap.org/). These field observations build a solid basis for extending field- or local-level studies to regional or national scales through remote sensing, modeling, or statistical extrapolation methods. Such scaling-up studies rely on a series of spatially explicit and long-term regional data sets including various management practices, plantation distribution maps, and information on environmental conditions. The critical first step is to generate long-term and spatially explicit plantation distribution maps. Therefore, in this study, we aim to develop long-term (1928–2012) forest plantation area and spatial distribution data for the CONUS through a synthesis of various inventory data sets across multiple scales. This data set can be used for ecosystem modeling and statistical extrapolations of productivity, carbon storage, greenhouse gas fluxes, and hydrological cycling in plantation forests, which will improve the estimation accuracy of greenhouse gas balance in the CONUS as well as advance our understanding of how intensive land management modulates climate–ecosystem feedbacks.

2 Data and methods

2.1 The workflow

Figure 1 shows the data sets collected in this study and the workflow for overall processes. At first, the initial collections of various inventory data (in gray color boxes, Fig. 1) on plot, state, subregional, and regional scales were used to develop the middle products (in black boxes), including gridded plantation forest fraction map and the state-level annual plantation area data. Then, these inventory data and middle products were integrated into the mechanistic program (in the circle; see Sect. 2.7 for more details) to determine spatial distributions of plantation forest area and tree species (our final products) during 1928–2012.

2.2 Divisions of study area

In this study, we collected and synthesized various data from different scales organized by division of the study area into several spatial units, as described here. The FIA reports (e.g., Smith et al., 2009) commonly divide the CONUS into eight ecological subregions (Fig. 2), which are further grouped into three regions (South, West, and North). These eight subregions are South Central, Southeast, North Central, Great Plains Intermountain, Pacific Northwest, and Pacific Southwest.

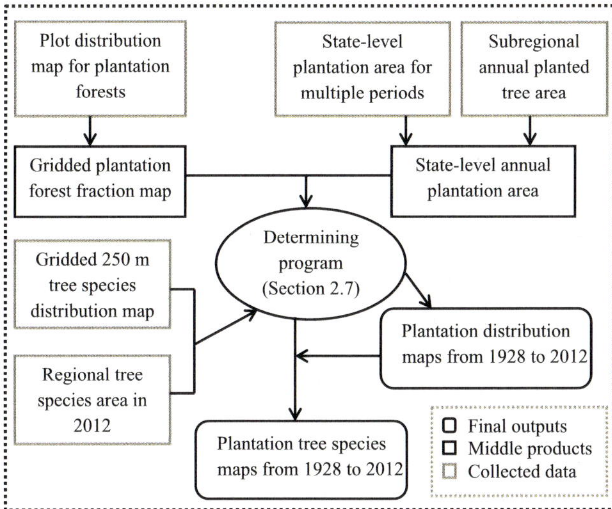

Figure 1. The workflow of overall processes in the generation of gridded (8 km × 8 km) plantation distribution and tree species maps from 1928–2012.

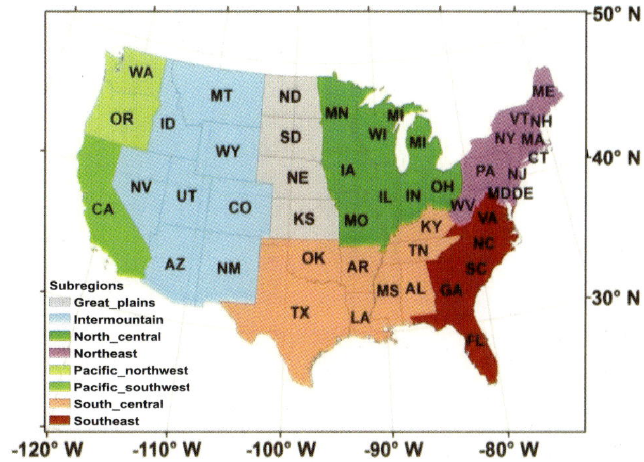

Figure 2. The division of the CONUS into eight subregions for data synthesis and analysis in this study. Note that three regions are further grouped in some reports, i.e., South (South Central and Southeast), North (Northeast and North Central), and West (Great Plains, Intermountain, Pacific Northwest, and Pacific Southwest). Data source: Smith et al. (2009).

2.3 FIA plot-scale data and processing

Due to the difficulty in distinguishing the optical reflectance of plantation forests from naturally regenerated forests, remote sensing products are not currently available to directly identify spatial locations of plantation forests across landscape scales. However, owing to thousands of FIA plots and the plantation forest records, here we are able to roughly determine the spatial locations, despite some inaccuracy due to assumptions and extrapolations.

We collected the USDA Forest Service FIA plot-level stand origin data (the variable is referred to as STDORGCD in the stand condition table of FIA data; https://www.fia.fs.fed.us/) for generating the spatial locations of plantation forests. The earliest available FIA plot data were collected in the mid-1980s. However, due to the inconsistent inventory time periods and missing observations of forest origin for some states, we chose only plot-level inventory data for the 5 years (2000–2004) when most of the states have records for forest origin. We used these data to represent the distribution of plantation forests in 2001, consisting of 16 677 plots in total with plantation forest records (Fig. 3). According to FIA privacy policy, the geographic coordinates (i.e., latitude and longitude) of plots are swapped with nearby (within 0.8–1.6 km), ecologically similar plots and thus do not represent the exact locations at the fixed latitude and longitude (O'Connell et al., 2017). Our spatial units for grid cells in this study are either 1 or 8 km; thus, these deviations in spatial locations may not significantly influence our accuracy for assigned grid cell locations of plantation forests.

Based on the collected plots for plantation forests, we calculated the gridded fraction data using the method

$$F = N \times A/B + \varepsilon, \quad (1)$$

in which F is the fraction of plots with plantation forests within each grid cell; N shows the plot numbers with plantation forest in each grid cell; A is the represented area (0.8 km × 0.8 km) of each FIA plot; B is the grid cell area (8 km × 8 km); ε is a residue, which is used to add a small fraction (at the 0.01 % scale) to the grid cells with the same plot numbers (N) and calculated based on the percentage of forest (% × 0.0001) from NLCD2001 land cover data (http://www.mrlc.gov/nlcd01_data.php). The finally calculated F of each grid cell will be a unique value, which is shown in Fig. 4.

2.4 County-, state-, and region-scale inventory data

The inventory-based plantation forest area data on three spatial scales were collected to generate the gridded data set. First, county-level data from 2007 were collected to evaluate the performance of the generated grid-scale plantation forest area for counties. Second, state-level inventory data of plantation forest area for eight time periods (i.e., 1952, 1962, 1970, 1982, 1989, 1999, 2007, and 2012) for the states in the South Central and Southeast were collected from the southern forest resource assessment report (Wear and Greis, 2002). Due to a lack of available historical data, our data set includes only years 2007 and 2012 for other states in the CONUS, as collected from USDA Forest Service reports (Smith et al., 2009; Oswalt et al., 2014). Third, the subregional (Fig. 1) annual planted tree area data from 1928 to 2011 were collected from Oswalt et al. (2014), in which the data from 2004 to 2011 were not available for most subregions except for the Southeast and South Central (Fig. 5). Annual tree planting area in the Southeast and South Central exhibited two quickly

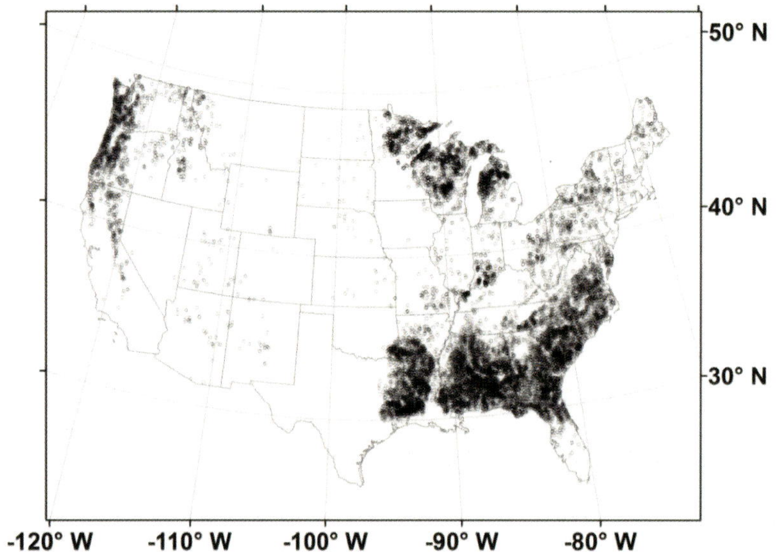

Figure 3. FIA plot distributions (16 677 plots in total) with plantation forest in the conterminous US during 2000–2004.

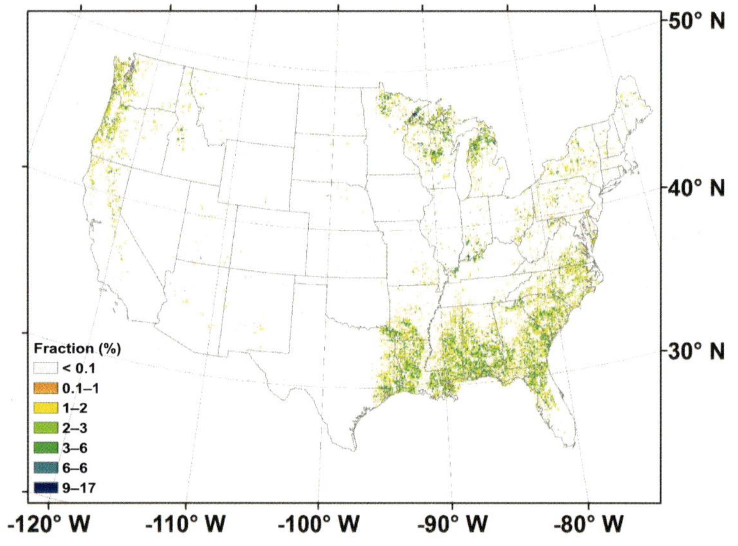

Figure 4. Fraction of the plots with plantation forests in the grid cells at a spatial resolution of 8 km × 8 km.

increasing periods during 1945–1960 and 1966–1989 and no obvious tendency after 1990.

2.5 Forest species data

We collected forest-cover-type data on a spatial scale of 250 m generated by the USDA Forest Service FIA program and Remote Sensing Applications Center (https://www.fia.fs.fed.us/library/maps/). In total, 113 major tree species are divided in this data set. According to the plantation forest species area data for the three regions (i.e., South, North, and West) in Oswalt et al. (2014), we identified the major plantation forest species in the CONUS and further regrouped into 11 major tree species groups, i.e., loblolly–shortleaf pine, longleaf–slash pine, Douglas fir, white–red–jack pine, ponderosa pine, spruce–fir–larch–hemlock, oak–hickory–gum–cypress, elm–ash–cottonwood, maple–beech–birch–aspen, other hardwoods (including juniper, palm, mangrove, and others), and other pine species (including redwood, sand pine, western white pine, lodgepole pine, and others). Using the aggregation method in ArcGIS, the 250 m forest-type data were then aggregated to continuous values representing the fraction of each species group per 8 km grid cell. Based on the same methods in Sect. 2.7, we generated a map with the Boolean (0, 1) data for each forest-type group, with 1 representing the grid cells occupied by this forest

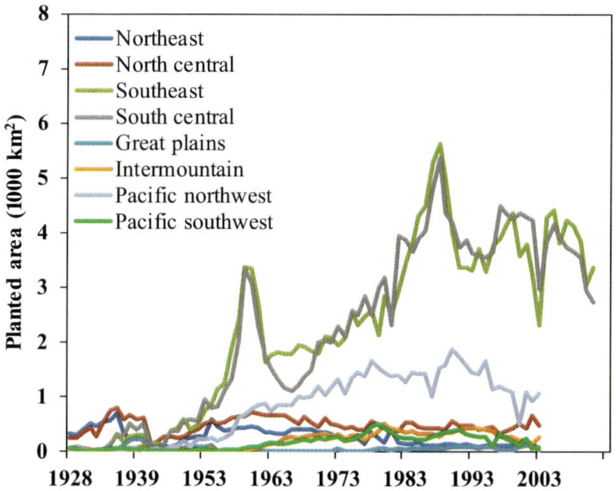

Figure 5. Interannual variations in planted tree area ($1000\,km^2\,yr^{-1}$) for eight subregions in the CONUS during 1928–2011 (data source: Oswalt et al., 2014). Note that the data for the Southeast and South Central are extended to 2011, while the continuous inventory data end in 2003 and resume in 2011 for other subregions.

type. The forest-type data were then overlaid with our generated plantation maps (Sect. 2.7) to obtain the 8 km resolution plantation-forest-type information. In the report of Oswalt et al. (2014), there is a plantation group of non-stocked forest type (3.88 thousand km^2 in total), which mainly includes young plantation stands and seedling orchards that have yet to reach a crown density of 10 % (https://www.fs.fed.us/ne/fia/methodology/def_ip.htm). We were unable to directly assign it to the regrouped 11 plantation types; instead, we compared the fractions of all 11 plantation types in the grid cells with non-stocked, and assigned the plantation types with the highest fractions within these grid cells.

The USDA forest-type map was also generated based on the FIA plot data. Furthermore, the majority of the trees in plantation forests of the CONUS are native species (Escalante Fernandez et al., 2002; FAO, 2005), which can help reduce the pixel contamination due to the neighboring grid cells. Therefore, the forest-type map matches well with our generated plantation distribution data. Using the loblolly–shortleaf species group as an example, Fig. 6 illustrates the generation of plantation tree species groups based on the fractional data and regional inventory area of each tree species group.

2.6 Generation methods for state-level plantation area

We have collected state-level plantation area data for eight periods: 1952, 1962, 1970, 1982, 1989, 1999, 2007, and 2012, but we lack data to capture interannual patterns within these periods. To make the state-level data consistent among all periods, we post-processed these inventory data. In this study, we assumed that the plantation forest area did not decrease with time for each state; thus, if the data from the previous period (e.g., 2007) were less than the data in the present period (e.g., 2012), the data in the present period (e.g., 2012) were then replaced by the previous ones (e.g., 2007). We assumed that the data in 2007 are the actual plantation area (i.e., assume the inventory data in this year are accurate) to control the post-processing. Therefore, the plantation area in other periods could not be exactly the same with the collected inventory data. The annual tree planting area in 1928 was used as the control of initial plantation area (A_0), and the other eight time periods for the South and Southeast were assigned as A_1 to A_8. The other states that had data for only two periods (2007 and 2012) were assigned as A_1 and A_2. We integrated annual planted tree area data for eight subregions with state-level plantation area data to linearly interpolate distribution pattern during 1928–2012 for each state. The interpolation method is as follows:

$$C_{\text{sum}} = \sum_{j=1}^{N} C_j \tag{2}$$

$$\text{TA}_i = A_p + \left(A_{p+1} - A_p\right) \times \frac{C_j}{C_{\text{sum}}}, \tag{3}$$

in which i is the year (1928–2012), TA_i is the generated targeted plantation area in year i, p is the time periods (0–8 for the South and Southeast states and 0–2 for other states), A_i is the inventory plantation area at year i, A_{p+1} is the inventory plantation area at time period $p+1$, N is the total numbers of years during inventory period p to $p+1$, j is the number of year (0–N) during inventory period p to $p+1$ for plantation forest area, C_{sum} is the total planted tree area during inventory period p to $p+1$, and C_j is the planted tree area at time j during period p to $p+1$.

2.7 Methods for spatialization of gridded plantation area and tree species data

Boolean (0, 1) plantation data were developed at 8 km × 8 km spatial resolution (125 718 grid cells), with 0 denoting naturally regenerated forest and 1 denoting plantation forest. Considering the limited plot numbers, we chose this coarse spatial resolution for plantation data. In addition, the Boolean data at a moderate (8 km) spatial resolution might be adequate to apply in future modeling or statistical studies since many other available environmental data, such as climate, land use, and atmospheric composition, are coarser or similar to 8 km. We only have one single time point of plantation forest fraction (Sect. 2.3). To track back the historical spatial patterns based on the state and subregional inventory data, we have to assume that the plantation forests will not be converted back to naturally regenerated forests, i.e., if this grid cell was identified as plantation in 1928, it will always have been plantation since then.

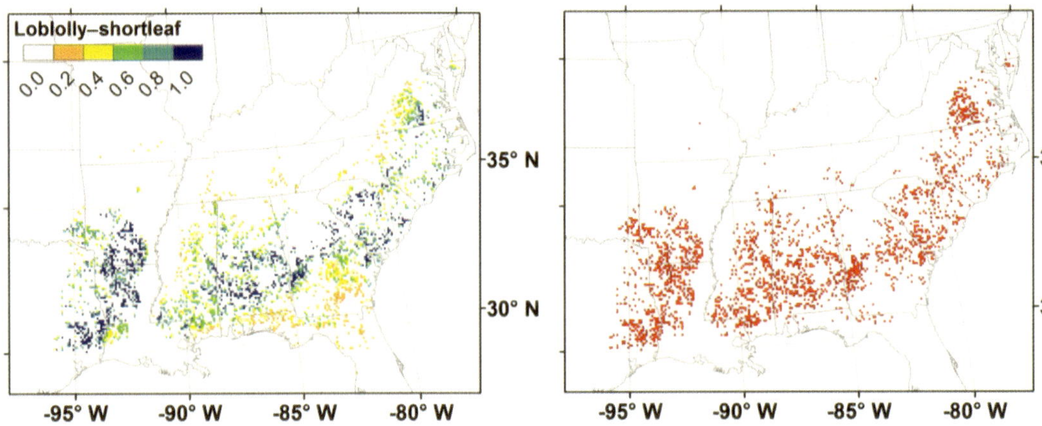

Figure 6. Illustration of the generation of spatial distribution maps for tree species groups in terms of fractional data and regional inventory area data using loblolly–shortleaf pine as an example. Left panel: fraction of loblolly–shortleaf pine species group in each grid cell; right: identified final grid cells with loblolly–shortleaf pine.

Figure 7 describes the procedure to produce the spatial distribution maps of plantation forests. The state-level plantation forest area data set (TA_i) generated in Sect. 2.4 is the targeted plantation area for this specific state i. To determine if a grid cell is plantation forest, the fraction data set for grid cell j in state i (F_{ij}) generated in Sect. 2.3 is used. The principle is to progressively narrow down the fraction threshold ranges ($T_{i,\min}$ and $T_{i,\max}$) to a fixed threshold value (T_i), and based on this determined threshold, we ultimately reach the targeted plantation area for state i. At the first-round run of the program, a minimum threshold 0 and maximum threshold 1 are assigned. The T_i is calculated as the average of $T_{i,\max}$ and $T_{i,\min}$. Based on this T_i value, we run a program to check if the fraction data (F_{ij}) is higher than T_i for each grid cell within the specific state. If yes, then this grid cell is assigned as a value of Boolean 1 ($B_{ij} = 1$); otherwise, it is assigned as 0 ($B_{ij} = 0$). The B_{ij} values for all grid cells within this state are added to calculate the total plantation area (A_i). If the total area is smaller than TA_i, the program will assign $T_{i,\max} = T_i$; if the total area is larger than TA_i, the program will assign $T_{i,\min} = T_i$. Based on the new $T_{i,\max}$ and $T_{i,\min}$, the program will go to the second-round run and repeat all processes above. After the second-round run, if the A_i is still not equal to $TA_i \pm 1\,\text{km}^2$, the program will run more rounds until $A_i = TA_i \pm 1\,\text{km}^2$. Under this condition, the generated state-level plantation area is very close to the targeted plantation area at the end. Finally, the B_{ij} maps (0 and 1 Boolean values) represent the spatial distributions of plantation forests in this specific state i. This program was run for all the CONUS states and eventually resulted in the spatially explicit plantation forest distribution maps from 1928 to 2012.

Based on the regional inventory data and gridded fractional data for individual plantation tree species groups (Sect. 2.5), we also applied the methods above to generate the plantation tree species group maps during 1928–2012.

3 Results and discussion

3.1 Plantation forest area and temporal variations

Plantation forest area in the CONUS showed a continuous increase from 1928 to 2012, with the largest increasing rates during the 1950s (176 % per decade) and 1960s (86 % per decade) and the least during the 1970s (Fig. 8). Plantation forest area was 268.27 thousand km^2 in 2012, accounting for 8.65 % of CONUS forestland area and 2.93 % of the total land area. The global plantation area was reported to account for about 6.95 % of the total forestland area (FAO, 2015), which is lower than the fraction in the CONUS. The increasing rate showed a slight leveling-off trend during recent decades; however, the total plantation area still increased by 36.81 % from 2000 to 2012, with this time period having the largest absolute increase (+72.16 thousand km^2) in plantation area. The West region had the largest forest area (1.40 million km^2; Oswalt et al., 2014) as compared to the North (0.71 million km^2) and South (0.99 million km^2); however, the South has had the highest plantation forest area since 1950, followed by the West since 1976. In 2012, the plantation forest area in the South, North, and West was 191.78, 25.90, and 50.55 thousand km^2, respectively. The plantation forest area accounted for 19.34 % of the total forest area in the South, while only about 3.62 % in both the North and West. Over the earlier time periods (1928–1950), the North had the highest plantation area. The West had the smallest plantation forest area before 1976, but it increased faster than the North and overpassed after 1976. The plantation area in the South has increased the fastest since 1950 as compared to the other two regions. The plantation area in the South and North maintained increasing rates in recent decades while the rate of increase in the West slowed down.

The smaller proportion of plantation forests in the West does not imply a greater potential for increasing plantation

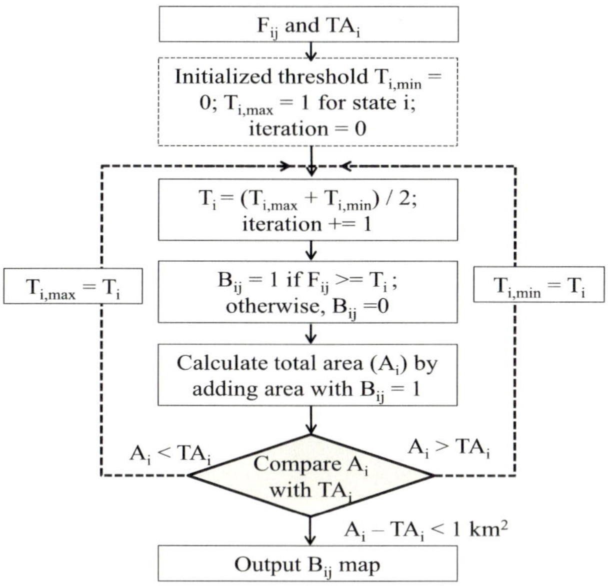

Figure 7. The procedure to identify the spatial distribution maps of plantation forests for each state based on grid cell fractional data and state-level inventory data, in which i: state, j: grid cell ID, F_{ij}: fraction of plantation forest for grid cell j in state i, T_i: calculated threshold fraction for state i, $T_{i,\max}$: identified maximum fraction threshold, $T_{i,\min}$: identified minimum threshold, B_{ij}: plantation distribution represented by Boolean values (0, 1) for grid cell j in state i, A_i: calculated plantation area in state i, and TA_i: targeted plantation area in state i.

forest area in this region in the future because the mountainous terrains and relatively dry climate (the southern and central portions) are not suitable for tree planting and management. In addition, most of the forest area in the West belongs to public land (USDA Forest Service, 2014), which is managed for multiple uses and generally not managed as intensively for forest product yields as privately owned, profit-oriented forest properties. The North region has a far smaller fraction of public forest than the West; however, the cooler climate may result in less productivity and thus restrain its potential in the wide spread of plantation forest area in the future (Escalante Fernandez et al., 2002). In contrast, although the South has a very high fraction of plantation forest and provides most of the wood and/or non-wood forest products for the CONUS, it still has a large potential for increasing plantation forest area, as also predicted by Wear and Greis (2002), in which they have projected increasing rates in plantation forest area with varied extents from about 14 to 57 % from 2010 to 2040 under four future scenarios. Their predictions for the period from 1995 to 2010 ranged from about 9 to 19 %, which were far lower than the realistic increasing trend (45 %) as indicated by our study, implying an underestimated projection of future plantation forest area in the South. Wear and Greis (2012) updated their predictions and indicated increasing rates of 20 and 65 % under the lowest

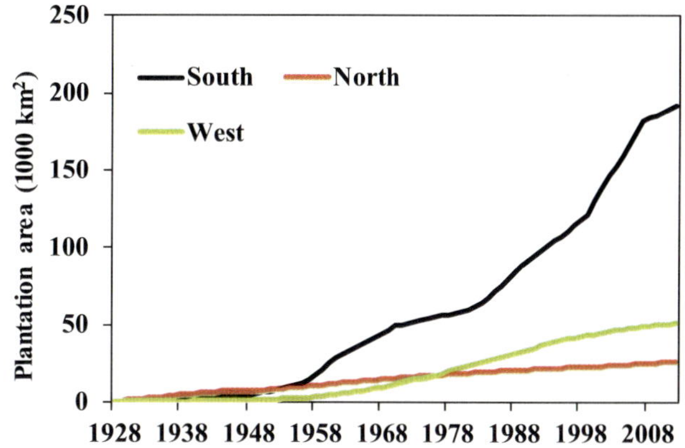

Figure 8. Total plantation forest area (1000 km^2) for different regions in the CONUS during 1928–2012.

and highest scenarios, respectively, from 2010 to 2060 in the South. Based on our data in 2010 (187.8 thousand km^2), the highest estimation of total plantation area in 2060 would be 309.9 thousand km^2, which accounts for about 31.30 % of total forestland area in this region. The temporal and spatial changing patterns in the CONUS were primarily determined by the demands for wood products, the changes in policy and incentives, and food and bioenergy price and demands (Wear and Greis, 2012). The development of plantations was divided into three stages: initial stage (before 1946), acceleration (1946–1976), and steady growth (1977–present) (Zhang, 2004). During the initial stage, the major incentives were the government policies, such as the 1928 McSweeney–McNary Act, 1924 Clarke–McNary Reforestation Act, and Agricultural Conservation Program (ACP, 1936). The acceleration stage was primarily driven by increasing wood product demand and price and decreasing food commodity price. The Soil Bank Program (SBP), which encourages agricultural abandonment to environmental conservation, was the most important government incentive during this period. During the steady growth stage, the declining cost of plantation establishment was the major incentive. Other incentives include tax reduction through the reforestation tax incentive provision (1980) and Tax Reform Act (1986), ACP (ended in 1995), Forest Incentive Program (FIP, 1974), and Conservation Reserve Program (CRP, 1986).

3.2 Spatial distribution patterns

Before the 1950s, there was only small plantation forest area (230 grid cells), mainly scattered throughout the South, Northeast, Pacific Northwest, and North Central (Fig. 9). The late 1950s were essentially marked as the beginning of extensive pine plantation establishment in the South (Frederick and Sedjo, 1991). The time period 1950–1970 had witnessed the fastest increasing rate of plantation forests; therefore, the plantation forests were widely spread across the South, the

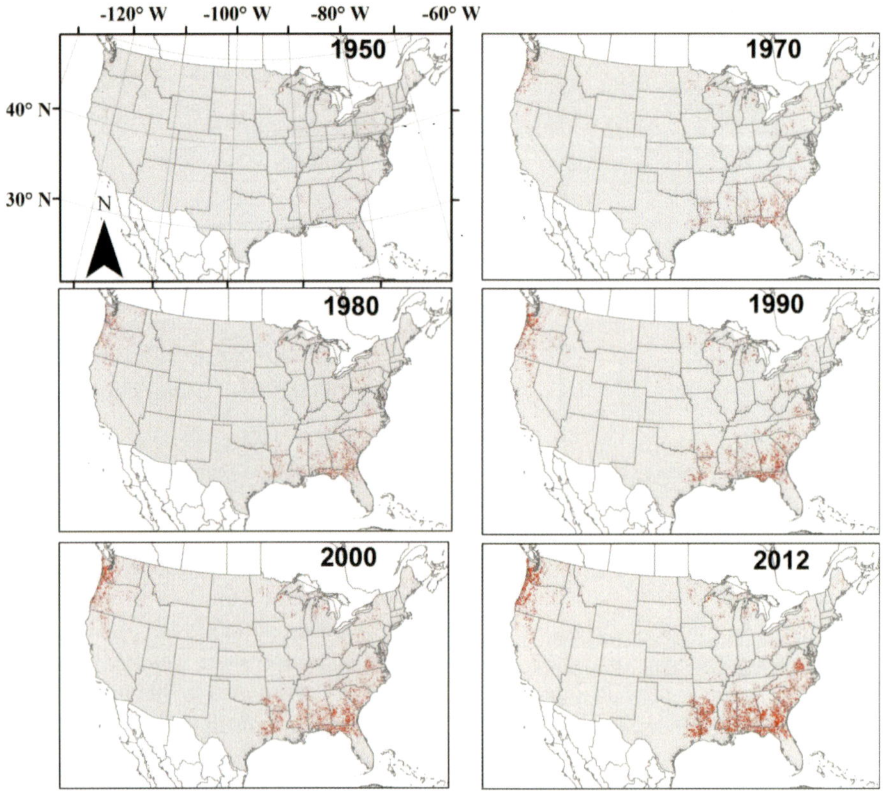

Figure 9. Spatial distributions for plantation forests during 1950, 1970, 1980, 1990, 2000, and 2012 at a spatial resolution of 8 km for the CONUS.

Northeast, and the Pacific Northwest. The spatial distribution patterns of plantation forests were quite similar among the time periods after 1980 and area expansions occurred within these three regions. Further analyses indicated that the 20 states with the largest plantation area accounted for about 96.32 % of the total CONUS plantation area in 2012, and the top 10 states accounted for about 76.62 % of the total (Fig. 9). Among the 20 states, GA had the highest plantation area, followed by AL, OR, and MS, while OK and TN had the smallest area. The plantation forest area accounted for 31.2, 30.6, 30.4, 29.3, and 28.4 % of total forestland area in GA, LA, AL, MS, and FL, respectively. Although LA has lower total forestland area (about 59 % of GA) as compared to the other four states, it had the second largest plantation proportion. Plantation area in these southern states was projected to continue increasing from the present to 2060 (Wear and Greis, 2012). Notably, the Pacific Northwest states of OR and WA had relatively high proportions of plantation forests (20.1 and 19.9 %, respectively), with OR ranked as the third largest state of forestland area in the CONUS, and they might have a greater potential for a continuing increase in plantation area in the future.

During 1990–2012, AL had the largest increase (238.0 %) in plantation area, followed by MS (236.8 %) and LA (191.2 %, Fig. 10). These states had small increasing rates of 13.2, 5.49, and 5.77 %, respectively, during 1950–1990. In contrast, plantation area in GA, FL, and OR showed continuous and stable increasing trends during 1950–2012. Among the top 20 states, the absolute plantation area was the smallest in OK; however, this state exhibited a large increasing rate (137.9 %) during this period. In addition, the states of TX and AR also displayed a relatively high increasing rate. These two states might become major contributors to the increasing plantation area in the CONUS in the future since their forestland area is relatively larger and could sustain more conversions of plantations from naturally regenerated forestland. Conversely, several Northeast states (e.g., WI, MI, NY, and PA) and the Southeast state FL showed the smallest rates of increase.

3.3 Plantation tree species

Tree species is key information to estimate both tree endogenous growth rates as well as the responses of exogenous growth to environmental changes and management practices. To identify the tree species in the plantation forests during 1928–2012, the plantation maps were overlaid with the tree species distribution map in 2012 (Fig. 11). In the CONUS, almost all planted tree species are native species and planted for productive purposes (Escalante Fernandez et al., 2002;

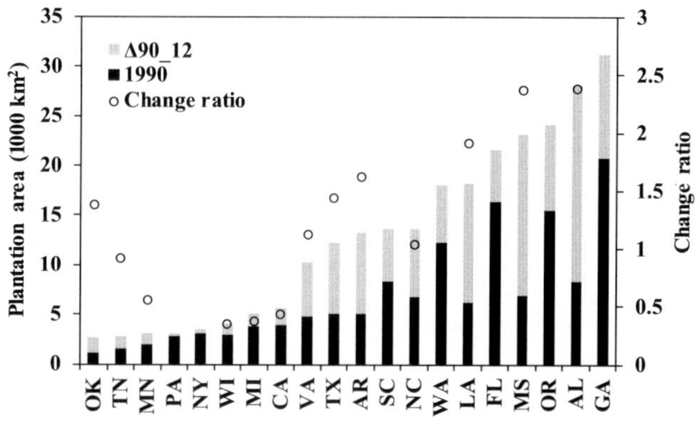

Figure 10. Plantation area in 1990 (dark gray), area change from 1990 to 2012 (light gray), and change ratio (Δ90_12 / 1990; red circle) for the selected top 20 states with the largest plantation area in the CONUS.

FAO, 2005). In the South, over 69.2 % of the planted tree species were loblolly–shortleaf pine, followed by longleaf–slash pine (15.6 %), oak–pine (7.5 %), and oak–hickory (Oswalt et al., 2014). The slash pine forests have less productivity than loblolly pine but generally produce higher-quality wood (Escalante Fernandez et al., 2002). Therefore, this species was widely planted in southern AL, GA, and northern FL. In the North, about 48.8 % of the planted tree species were white–red–jack pine, followed by spruce–fir (11.3 %). The white–red–jack pine types are scattered across the North Central states, while spruce–fir mainly shows up in ME and MN. In the West (primarily Pacific Northwest), Douglas fir accounted for 60.3 % of the planted tree species, followed by Oak–hickory–gum–cypress (11.9 %) and Ponderosa pine (9.4 %). Douglas fir is primarily located along the coastline in WA and OR. On a national scale, loblolly–shortleaf pine accounted for most (49.45 %) of the plantation forest area, followed by longleaf–slash pine (11.04 %) and Douglas fir (11.17 %).

3.4 Plantation management practices and their impacts

The plantation forests in the CONUS are mostly privately owned and about two-thirds of the plantations are timberland (Escalante Fernandez et al., 2002). Therefore, intensive management practices were widely applied to promote productivity, especially after 1990 (Fox et al., 2004, 2007; Stanturf et al., 2003). Generally, management intensity among regions is greatest in the South and lowest in the Northeast (Escalante Fernandez et al., 2002). The major plantation management practices include site preparation (e.g., soil disking, bedding, litter raking, and herbicide use), genetic improvement (e.g., breeding and seed tree selection), fertilization, thinning, prescribed fire, and harvesting (Vance et al., 2010; Fox et al., 2004). The late 1950s was regarded as the beginning of extensive pine plantations in the CONUS (Frederick and Sedjo, 1991; Vance et al., 2010). During the most recent 2 decades (1990–2009), pine plantations were harvested (including partial and clear-cut harvest) at a rate of about 3.15 thousand km^2 per year in the CONUS (Smith et al., 2009). Thinning, site preparation, and slash burning area per year were 1.25, 2.87, and 2.70 thousand km^2, respectively (Smith et al., 2009). About 6.47 thousand km^2 of pine plantations was fertilized in 1999 alone, while about 40.47 thousand km^2 in total has been fertilized in the South since 1969 (Fox et al., 2007).

Vance et al. (2010) synthesized the extent and benefits of multiple intensive management practices and the factors influencing productivity in the different subregions of CONUS. Fox et al. (2004) even indicated that multiple management practices would increase pine volume at harvest by over 4 times in the South. In addition to carbon dynamics, the intensive management practices were reported to significantly change the ecosystem hydrological and nitrogen cycles based on numerous field experiments and observations from various observational networks (e.g., FPC, FIA, FMRC, PMRC, FBRC, PINEMAP, AmeriFlux, and the Long Term Ecological Research Network (LTER)). These studies have addressed the ecological impacts of plantation forestry in terms of tree species and environmental conditions, as well as management regime, intensity, and frequency. Continued observational and experimental evidence of plantation forest function is critical to assess or predict the relationships between environmental changes, plantation management practices, and managed forest carbon, nitrogen, and water cycles. At present, it is highly likely for researchers to scale up the field or local experiments and observations to regional or national scales through remote sensing, modeling, or statistical extrapolation methods.

4 Data availability

The gridded (8 km × 8 km) plantation distribution and tree species maps and state-level tree planting area and plantation forest area during 1928–2012 are available from https://doi.org/10.1594/PANGAEA.873558 (Chen et al., 2017). There are two data formats for gridded data: text/ASCII and ArcGIS GRID formats; an Excel format table is used to organize the annual tree planting area and total plantation area data for the 48 states in the conterminous US during 1928–2012. A supplemental file is added to show the plantation distribution maps in 1952, 1962, 1970, 1982, 1989, 1999, 2007, and 2012.

5 Conclusions and outlooks

This study developed an annual and spatially explicit data set for plantation forests in the CONUS during 1928–2012. The data set showed that plantation forests have increased rapidly since the 1960s. While these increasing rates have stabilized during recent decades, there was still great poten-

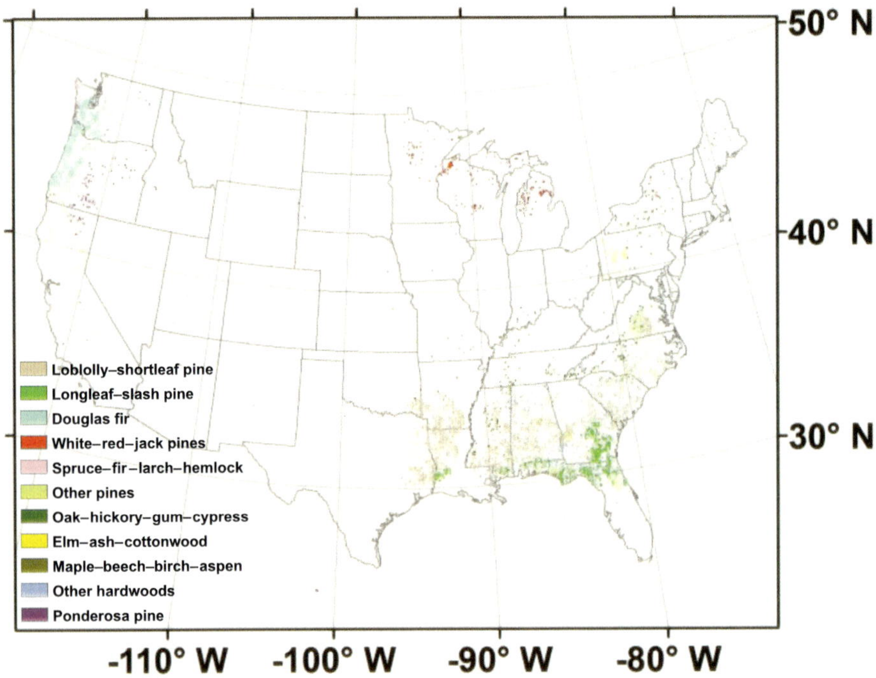

Figure 11. Spatial distribution of plantation tree species in the CONUS in 2012.

tial to increase plantation area in terms of the small fraction of plantation forests (8.65 %) currently existing in the CONUS. With suitable climate and geophysical environmental conditions, the southern US is the major plantation forest base, with plantation forests accounting for 19.34 % of total forestland.

Many short- and long-term field experiments in the CONUS, especially in the South, are ongoing to monitor intensive management practice effects on plantation forests. The large number of available observational data has greatly improved our understanding of the impacts of forest planting and management practices on ecological and socioeconomic services. Scaling up these studies from local-scale observations to regional understanding requires a series of spatially explicit and long-term regional–national data sets that include information on various management practices, plantation distribution, environmental conditions, and vegetation maps. The first and critical step is to generate the long-term plantation distribution maps. Recognizing this, we synthesized various inventory data to generate the gridded plantation distribution and species maps during 1928–2012. There are some aspects of uncertainty in our methods in which the data sets might be unable to track the exact plantation locations; however, our data sets had a relatively high spatial resolution (8 km) as required for terrestrial ecosystem modeling or statistical extrapolations on regional or national scales. The detailed spatiotemporal data for plantation tree species enables future research in simulating and extrapolating the regional- or national-scale carbon, nitrogen, and water dynamics in plantation forests based on species-specific parameters, which could further improve the mechanisms and estimation accuracy of regional Earth system models.

The future plantation area and distribution will be determined by many factors, including wood product markets, bioenergy technology and biofuel prices, food supply and demand, environmental policies, and other socioeconomic factors (Wear and Greis, 2012). The plantation forest area in the South is projected to increase to 26 % of total forestland (high scenario; Wear and Greis, 2012). From a socioeconomic perspective, present plantation forests in the CONUS generate positive economic profits along with providing good environmental services. From a carbon credit perspective, the plantation forests in the South are regarded as a major contributor to carbon sink in the CONUS and North America (Hayes et al., 2012; King et al., 2012; Tian et al., 2012, 2014); however, recent studies (Achat et al., 2015a, b; Nave et al., 2010) suggested that the shorter rotation age and some intensive management practices (e.g., site preparation for soil bedding, slash burning, and harvest residue raking) might reduce soil carbon stocks in plantation forests, implying that plantation forests could be a carbon source. From the hydrological perspective, plantation forests may increase water use and alter the water cycle due to higher productivity and management practices (e.g., short rotation, mechanic site preparation, and drainage), especially in the regions with strong precipitation limitation (Vose et al., 2012). From the perspective of nutrient cycling, plantation management practices could change soil available–total nitrogen, soil nitrous oxide emission, vegetation nitrogen, and nitrogen contents in nearby water bodies. Many past assessments have been conducted at the scale of the individual perspective; however, there is still a lack of a comprehensive assessment of plantation forests' function in mitigating future climate change by considering carbon, nitrogen, and water fluxes across broader regions. Such a comprehensive assessment is critical for determining whether the policymakers or land managers are going to plant more trees and how to best manage the forests in the CONUS (Sun and Vose, 2016).

Author contributions. GSC collected and compiled the inventory data, developed spatialization methods, and led the writing. SFP, DJH, and HQT participated in developing methods and writing the paper.

Competing interests. The authors declare that they have no conflict of interest.

Acknowledgements. This research was supported by the NASA Carbon System Monitoring Program (NX14AO73G and NNX12AP84G), NASA Interdisciplinary Science Program (NNX11AD47G and NNX14AF93G), and Auburn University IGP program.

Edited by: Hannes Grobe
Reviewed by: three anonymous referees

References

Achat, D. L., Fortin, M., Landmann, G., Ringeval, B., and Augusto, L.: Forest soil carbon is threatened by intensive biomass harvesting, Scientific Report, 5, 15991, https://doi.org/10.1038/srep15991, 2015a.

Achat, D. L., Deleuze, C., Landmann, G., Pousse, N., Ranger, J., and Augusto, L.: Quantifying consequences of removing harvesting residues on forest soils and tree growth – A meta-analysis, Forest. Ecol. Manage., 348, 124–141, https://doi.org/10.1016/j.foreco.2015.03.042, 2005b.

Albaugh, T. J., Vance, E. D., Gaudreault, C., Fox, T. R., Allen, H. L., Stape, J. L., and Rubilar, R. A.: Carbon emissions and sequestration from fertilization of pine in the southeastern United States, Forest Sci., 58, 419–429, https://doi.org/10.5849/forsci.11-050, 2012.

Albaugh, T., Alvarez, J., Rubilar, R., Fox, T., Allen, H., Stape, J., and Mardones, O.: Long-term *Pinus radiate* productivity gains from tillage, vegetation control, and fertilization, Forest Sci., 61, 800–808, https://doi.org/10.5849/forsci.14-207, 2015.

Allen, H. L., Fox, T. R., and Campbell, R. G.: What is ahead for intensive pine plantation silviculture in the South? South. J. Appl. For., 29, 62–69, 2005.

Chen, G., Tian, H., Zhang, C., Liu, M., Ren, W., Zhu, W., Chappelka, A., Prior, S. A., and Lockaby, G.: Drought in the southern United States over the last century: Variability and its impacts on terrestrial ecosystem productivity and carbon storage, Climatic Change, 114, 379–397, https://doi.org/10.1007/s10584-012-0410-z, 2012.

Chen, G., Pan, S., Hayes, D. J., and Tian, H.: Spatial and temporal patterns of plantation forests in the United States since the 1930s, links to gridded result files in different formats, PANGAEA, https://doi.org/10.1594/PANGAEA.873558, 2017.

Escalante Fernandez, R., Monreal Rangel, S., Stanturf, J., Arseneau, C., and Nantel, P.: Forest plantations in North America. Report to the North American Forestry Commission, 26 pp., 2002.

FAO: Global Forest Resources Assessment 2005, UN Food and Agriculture Organization, Rome, Italy, available at: http://www.fao.org/forestry/fra/fra2005/en/, last access: 26 July 2017, 2005.

FAO: Global Forest Resources Assessment 2015, UN Food and Agriculture Organization, Rome, Italy, available at: http://www.fao.org/3/a-i4808e.pdf, last access: 26 July 2017, 2015.

Fox, T. R., Jokela, E. J., and Allen, H. L.: The evolution of pine plantation silviculture in the Southern United States, in: Gen. Tech. Rep. SRS–75, U.S. Department of Agriculture, Forest Service, Southern Research Station, Asheville, NC, chap. 8, 63–82, 2004.

Fox, T., Allen, H., Albaugh, T., Rubilar, R., and Carlson, C.: Tree nutrition and forest fertilization of pine plantations in the southern United States, South. J. Appl. For., 31, 5–11, 2007.

Frederick, K. K. and Sedjo, R. A.: America's renewable resources: historical trends and current challenges, Resources for the Future, Washington, DC, 296 pp., 1991.

Gyawali, N. and Burkhart, H. E.: General response functions to silvicultural treatments in loblolly pine plantations, Can. J. Forest Res., 45, 252–265, https://doi.org/10.1139/cjfr-2014-0172, 2015.

Hayes, D. J., Turner, D. P., Stinson, G., McGuire, A. D., Wei, Y., West, T. O., Heath, L. S., Dejong, B., McConkey, B. G., Birdsey, R. A., Kurz, A. W., Jacobson, A. R., Huntzinger, D. N., Pan, Y., Post, W. M., and Cook, R. B.: Reconciling estimates of the contemporary North American carbon balance among terrestrial biosphere models, atmospheric inversions, and a new approach for estimating net ecosystem exchange from inventory-based data, Global Change Biol., 18, 1282–1299, https://doi.org/10.1111/j.1365-2486.2011.02627.x, 2012.

Hoover, C., Birdsey, R., Goines, B., Lahm, P., Marland, G., Nowak, D., Prisley, S., Reinhardt, E., Skog, K., Skole, D., Smith, J., Trettin, C., and Woodall, C.: Chapter 6: Quantifying greenhouse gas sources and sinks in managed forest systems, in: Quantifying greenhouse gas fluxes in agriculture and forestry: Methods for entity-scale inventory, Tech. Bull. 1939, edited by: Eve, M., Pape, D., Flugge, M., Steele, R., Man, D., Riley-Gilbert, M., and Biggar, S., Office of the Chief Economist, U.S. Department of Agriculture, Washington, DC, 6-1-6.114, 2014.

Jokela, E. J., Dougherty, P. M., and Martin, T. A.: Production dynamics of intensively managed loblolly pine stands in the southern United States: a synthesis of seven long-term experiments, Forest Ecol. Manag., 192, 117–130, https://doi.org/10.1016/j.foreco.2004.01.007, 2004.

Jokela, E. J., Martin, T. A., and Vogel, J. G.: Twenty-five years of intensive forest management with southern pines: important lessons learned, J. Forest., 107, 338–347, 2010.

King, A. W., Hayes, D. J., Huntzinger, D. N., West, T. O., and Post, W. M.: North America carbon dioxide sources and sinks: magnitude, attribution, and uncertainty, Front. Ecol. Environ., 10, 512–519, https://doi.org/10.1890/120066, 2012.

Nave, L. E., Vance, E. D., Swanston, C. W., and Curtis, P. S.: Harvest impacts on soil carbon storage in temperate forests, Forest Ecol. Manag., 259, 857–866, https://doi.org/10.1016/j.foreco.2009.12.009, 2010.

O'Connell, B. M., LaPoint, E. B., Turner, J. A., Ridley, T., Pugh, S. A., Wilson, A. M., Waddell, K. L., and Conkling, B. L.: The Forest Inventory and Analysis Database: Database Description and User Guide for Phase 2 (Version 6.0.2), U.S. Department of Agriculture, Forest Service, 748 pp., available at: https://www.fia.fs.fed.us/library/database-documentation, last access: 26 July 2017.

Oliver, C. D.: The future of the forest management industry: Highly mechanized plantations and reserves or a knowledge-intensive integrated approach?, Forest. Chron., 75, 229–245, https://doi.org/10.5558/tfc75229-2, 1999.

Oswalt, S. N., Smith, W. B., Miles, P. D., and Pugh, S. A.: Forest Resources of the United States, 2012: a technical document supporting the Forest Service 2015 update of the RPA Assessment, Gen. Tech. Rep. WO-91, U.S. Department of Agriculture, Forest Service, Washington Office, Washington, DC, 218 pp., available at: https://www.srs.fs.usda.gov/pubs/gtr/gtr_wo091.pdf, last access: 26 July 2017, 2014.

Pan, S., Tian, H., Dangal, S. R. S., Yang, Q., Yang, J., Lu, C., Tao, B., Ren, W., and Ouyang, Z.: Responses of global terrestrial evapotranspiration to climate change and increasing atmospheric CO_2 in the 21st century, Earth's Future, 3, 15–35, https://doi.org/10.1002/2014EF000263, 2015.

Sedjo, R. A.: From foraging to cropping: the transition to plantation forestry, and implications for wood supply and demand, Unasylva 52 (2001/1) No. 204, available at: http://www.fao.org/DOCREP/003/X8820E/x8820e06.htm#P0_0, last access: 26 July 2017, 2001.

Smith, W. B., Miles, P. D., Perry, C. H., and Pugh, S. A.: Forest Resources of the United States, 2007, Gen. Tech. Rep. WO-78, U.S. Department of Agriculture, Forest Service, Washington Office, Washington, DC, 336 pp., available at: https://www.fs.fed.us/nrs/pubs/gtr/gtr_wo78.pdf, last access: 26 July 2017, 2009.

Stanturf, J. A. and Zhang, D.: Plantations forests in the United States of America: past, present and future, XII Word Forestry Congress, Quebec City, Canada, available at:http://www.fao.org/docrep/article/wfc/xii/0325-b1.htm, last access: 26 July 2017, 2003.

Sun, G. and Vose, J. M.: Forest management challenges for sustaining water resources in the Anthropocene, Forests, 7, 68–80, https://doi.org/10.3390/f7030068, 2016.

Sun, G., Zhou, G., Zhang, Z., Wei, X., McNulty, S., and Vose, J. M.: Potential water yield reduction due to forestation across China, J. Hydrol., 328, 548–558, https://doi.org/10.1016/j.jhydrol.2005.12.013, 2006.

Tian, H., Chen, G., Zhang, C., Liu, M., Sun, G., Chappelka, A., Ren, W., Xu, X., Lu, C., Pan, S., Chen, H., Hui, D., McNulty, S., Lockaby, G., and Vance, E.: Century-scale response of ecosystem carbon storage and flux to multifactorial global change in the Southern United States, Ecosystems, 15, 674–694, https://doi.org/10.1007/s10021-012-9539-x, 2012.

Tian, H., Chen, G., Lu, C., Xu, X., Hayes, D., Ren, W., Pan, S., Huntzinger, D., and Wofsy, S.: North American terrestrial CO_2 uptake largely offset by CH_4 and N_2O emissions: toward a full accounting of the greenhouse gas budget, Climatic Change, 129, 413–426, https://doi.org/10.1007/s10584-014-1072-9, 2014.

USDA Forest Service: National Report on Sustainable Forests – 2010, FS-979, USDA Forest Service, Washington, DC, 2011.

Vance, E. D., Maguire, D. A., and Zalesny, R. S.: Research strategies for increasing productivity of intensively managed forest plantations, J. Forest., 108, 183–192, available at: https://www.srs.fs.fed.us/pubs/ja/2010/ja_2010_vance_002.pdf, last access: 26 July 2017, 2010.

Vose, J. M., Ford, C. R., Laseter, S., Dymond, S., Sun, G., Adams, M. B., Sebestyen, S., Campbell, J., Luce, C., Amatya, D., Elder, K. H., and Scalley, T.: Can forest watershed management mitigate climate change effects on water resources, in: Revisiting Experimental Catchment Studies in Forest Hydrology, Proceedings of a Workshop held during the XXV IUGG General Assembly in Melbourne, June–July 2011, edited by: Webb, A. A., Bonell, M., Bren, L., Lane, P. N. J., McGuire, D., Neary, D, G., Nettles, J., Scott, D. F., Stednick, J., and Wang, Y., IAHS Publ. 353, Oxfordshire, UK, 12–25, 2012.

Wear, D. N. and Greis J. G.: Southern Forest Resource Assessment – Technical Report, Gen. Tech. Rep. SRS-53, U.S. Department of Agriculture, Forest Service, Southern Research Station, Asheville, NC, 635 pp., available at: https://www.srs.fs.usda.gov/pubs/gtr/gtr_srs053.pdf, last access: 26 July 2017, 2002.

Wear, D. N. and Greis, J. G.: The Southern Forest Futures Project: summary report, Gen. Tech. Rep. SRS-GTR-168, USDA-Forest Service, Southern Research Station, Asheville, NC, 54 pp., available at: https://www.srs.fs.fed.us/pubs/gtr/gtr_srs168.pdf, last access: 26 July 2017, 2012.

Zhang, D.: Markets, policy incentives and development of forest plantation resources in the United States of America, in: The Role of Incentives in Forest Plantation Development in Asia and the Pacific, Asia-Pacific Forestry Commission, FAO Regional Office for Asia and the Pacific, Bangkok, 237–261, 2004.

2

Diversity II water quality parameters from ENVISAT (2002–2012): a new global information source for lakes

Daniel Odermatt[1,2,3], Olaf Danne[2], Petra Philipson[4], and Carsten Brockmann[2]

[1]Odermatt & Brockmann GmbH, Zurich, 8005, Switzerland
[2]Brockmann Consult GmbH, Geesthacht, 20502, Germany
[3]Eawag, Swiss Federal Institute of Aquatic Science and Technology, Dübendorf, 8600, Switzerland
[4]Brockmann Geomatics Sweden AB, Kista, 164 40, Sweden

Correspondence: Daniel Odermatt (daniel.odermatt@eawag.ch)

Abstract. The use of ground sampled water quality information for global studies is limited due to practical and financial constraints. Remote sensing is a valuable means to overcome such limitations and to provide synoptic views of ambient water quality at appropriate spatio-temporal scales. In past years several large data processing efforts were initiated to provide corresponding data sources. The Diversity II water quality dataset consists of several monthly, yearly and 9-year averaged water quality parameters for 340 lakes worldwide and is based on data from the full ENVISAT MERIS operation period (2002–2012). Existing retrieval methods and datasets were selected after an extensive algorithm intercomparison exercise. Chlorophyll-*a*, total suspended matter, turbidity, coloured dissolved organic matter, lake surface water temperature, cyanobacteria and floating vegetation maps, as well as several auxiliary data layers, provide a generically specified database that can be used for assessing a variety of locally relevant ecosystem properties and environmental problems. For validation and accuracy assessment, we provide matchup comparisons for 24 lakes and a group of reservoirs representing a wide range of bio-optical conditions. Matchup comparisons for chlorophyll-*a* concentrations indicate mean absolute errors and bias in the order of median concentrations for individual lakes, while total suspended matter and turbidity retrieval achieve significantly better performance metrics across several lake-specific datasets. We demonstrate the use of the products by illustrating and discussing remotely sensed evidence of lake-specific processes and prominent regime shifts documented in the literature. The Diversity II data are available from https://doi.pangaea.de/10.1594/PANGAEA.871462, and Python scripts for their analysis and visualization are provided at https://github.com/odermatt/diversity/.

1 Introduction

Freshwater ecosystems have undergone more dramatic changes than any other type of ecosystem (Sectretariat of the Convention on Biological Diversity, 2010). Lakes contain about 87 % of all surface freshwater (Gleick, 1996). The major threats that affect lakes and reservoirs are water-level changes, toxic pollution, salinization, eutrophication, acidification, sediment pollution and invasion of exotic species. Several upstream anthropogenic activities are related to these threats, such as agriculture, forestry, grazing, mining, irrigation, urbanization and dams, hydraulic engineering and industrial development. All these pressures are interconnected and act concurrently to reduce water quality and contribute to the deterioration of the ecosystem, including habitat loss and reduced biodiversity.

The conditions of inland waters vary over a wide range of spatial and temporal scales, leading to important logistic and economic difficulties in monitoring them on a regular basis. Some countries have national or regional lake monitoring

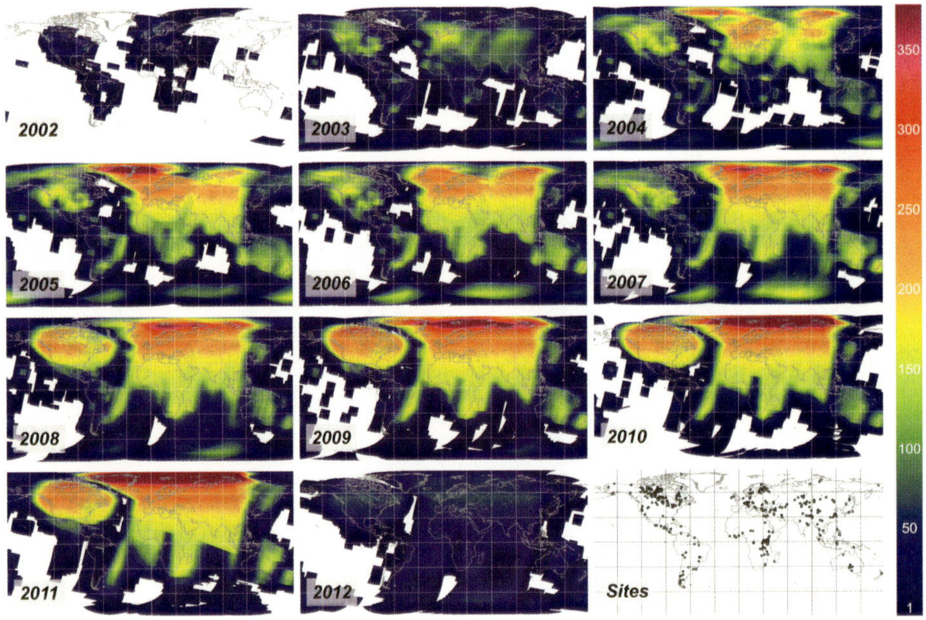

Figure 1. Global density maps for the bulk reprocessed MERIS FR dataset in the years 2002–2012, and distribution of the 340 lakes available in the Diversity II water quality dataset (bottom right).

programmes, which are primarily based on ground surveys. However, ground surveys often fail to sample on appropriate spatial and temporal scales. Other countries do not have monitoring programmes due to a lack of funds. The use of satellite remote sensing is a potentially cost-effective and efficient way to supplement the conventional in situ point sampling surveys. Remotely sensed products for water availability and quality are complementary to in situ data in terms of spatial and temporal coverage. They provide synoptic views of spatial distribution unachievable by other means and are ideally suited to covering the broad range of space scales and timescales associated with inland water applications. However, remote sensing is limited in terms of parameter coverage and depth resolution.

The Medium Resolution Imaging Spectrometer (MERIS) was operated by the European Space Agency (ESA) in 2002–2012 and demonstrated unparalleled capabilities for water quality remote sensing. Extensive reviews of popular retrieval methods revealed a wide range of different algorithms, but the usage of MERIS data prevailed (Matthews, 2011; Odermatt et al., 2012). The first globally representative lake water quality dataset from remote sensing provided a snapshot of chlorophyll-*a* (chl-*a*) concentrations in 80 000 lakes worldwide based on MERIS full-resolution (FR) data acquired in 2011 (Sayers et al., 2015). However, this dataset was compiled with an algorithm optimized for ocean colour remote sensing, whose suitability for inland waters was disavowed on several occasions (see e.g. Mobley et al., 2004; Morel and Prieur, 1977). Therefore we carried out an intercomparison of well-known and publicly available algorithms for the retrieval of chl-*a* and other water quality parameters in optically complex waters with a heterogeneous reference dataset for more than 40 lakes (Odermatt et al., 2015a). The Diversity II water quality dataset is produced with the most suitable retrieval methods identified through these investigations.

In addition to the optimized methodology, the Diversity II water quality dataset excels previous work by covering the full MERIS operation period with monthly, yearly and 9-year product aggregates, and several additional water quality parameters. Hence it provides a generically specified database that can be used for assessing a variety of locally relevant ecosystem properties and environmental problems. Several case studies are available that demonstrate such assessments with lake-specific foci (Odermatt et al., 2015b), but the larger part of the dataset is yet to be exploited.

2 Input data

2.1 Geographical scope

We selected 340 lakes for processing (Fig. 1) based on their biodiversity relevance, size, auxiliary and reference data availability, geographic distribution and particular user requests. Sixty-six of those lakes are at least 50 km^2 large and located within Ramsar Wetlands or listed as LakeNet Biodiversity Priority sites (www.worldlakes.org). The data table (https://doi.pangaea.de/10.1594/PANGAEA.871462?format=html#download) allows for their identification. The dataset includes 250 of the world's 350 largest lakes by extent, whereas size implies regional relevance and favours the feasibility of remote sensing retrievals in general and of lake surface water temperature (LSWT) in particular (Politi et al., 2016). Various contributors provided in situ water quality measurements for 42 lakes, which are used as reference sites for quality assessment (Odermatt et al., 2015a). The largest reservoirs in South America and individual sites in Asia and Australia are included in order to improve the global representativeness. Fifty additional lakes are included due to specific stakeholder requests.

In principle, the Diversity II water quality dataset could be extended to a much larger number of lakes. Size is the most important restriction in this regard, with a contiguous open water surface of roughly 1 km by 1 km being the theoretical minimum, but with certain complications occurring even for larger water bodies. The total number of suitable lakes worldwide is expected to be between the 80 000 demonstrated by Sayers et al. (2015), and, neglecting shape properties, the 350 000 lakes larger than 1 km^2 identified by Verpoorter et al. (2014).

2.2 ENVISAT MERIS L1B FSG imagery

MERIS was operated in 2002–2012 on-board the near-polar orbiting ENVISAT satellite by the European Space Agency (ESA; Rast et al., 1999). It measured reflected solar radiance in 15 narrow spectral bands across visible and near-infrared (NIR) wavelengths. In FR mode, its push-broom charge-coupled device (CCD) arrays sampled the 1150 km wide swath at approximately 260 by 290 m ground resolution in across-track and along-track directions, respectively. MERIS had a nominal revisit time of 2–3 days at the Equator and less at higher latitudes, but FR data were not systematically acquired in the early years until 2005, and in later years they varied slightly due to mission operations, and therefore the availability of usable data varies regionally and temporally (Fig. 1).

We refer to three widely, but not consistently, used satellite image processing levels, in which Level 1 (L1) consists of top-of-atmosphere (TOA) signals, Level 2 (L2) includes derived geophysical quantities and Level 3 (L3) represents spatio-temporally aggregated data. Approximately 300 000 MERIS L1 images were used as input for the production of the Diversity II water quality dataset. The data represent calibrated TOA radiance, also referred to as at-sensor radiance. It emerged from the 2014 bulk reprocessing using MERIS Instrument Processing Facility version 6. Its geo-orthorectification was improved using the Accurate MERIS Ortho-Rectified Geolocation Operational Software (AMORGOS; Bourg and Etanchaud, 2007); thus, the data are referred to as MERIS L1B Full-Swath Geo-corrected (FSG), and have an absolute and relative pixel location accuracy of 77 and 52 m, respectively.

2.3 AATSR ARC-Lake LSWT products

The Diversity II water quality dataset includes LSWT products that were readily provided by the ESA ARC-Lake project as version 3 production in rasters of 0.05° cell size. The LSWT retrieval was performed with an optimal estimation approach (MacCallum and Merchant, 2013, 2012). Lake-specific prior surface temperatures were generated using an iterative scheme that is initiated with the monthly MODIS land and sea surface temperature climatologies. The ARC-Lake processor then uses valid satellite observations, simulations with the FLake model (Mironov, 2008) and data interpolating empirical orthogonal function (DINEOF; Alvera-Azcárate et al., 2005) techniques to iteratively create from this spatially and inter-annually invariant initial guess a field of spatially resolved temperature fields.

Several product types using different processing techniques and spatial and temporal aggregations are available (MacCallum and Merchant, 2014). We selected the DINEOF reconstructed, day- and night-time acquired monthly products in 0.05° spatial resolution, whose filenames are ALIDXXXX_PLREC9D_TS012SR.nc and ALIDXXXX_PLREC9N_TS012SR.nc, respectively, with XXXX being a four-digit lake ID. They are available for 298 out of the 340 lakes considered, and empty LSWT product layers are contained in the remaining 42 lakes. Users requiring actual measurements of LSWT are recommended to download them from ARC-Lake or a comparable database directly.

2.4 Auxiliary data

Each lake's perimeter was defined in a shapefile that resulted from vectorized outlines of the Synthetic Aperture Radar Water Bodies (SAR-WB) map created by Santoro and Wegmüller (2014). These perimeters represent the maximum extent of water available from ENVISAT-ASAR acquisitions between 2002 and 2012, and each polygon's area and circumference were added to an attribute table. The polygons are intersected with the Global Lakes and Wetlands Database (GLWD; Lehner and Döll, 2004) Level 1 dataset, and ambiguities were manually resolved. The merged tabulated attributes are available in a metadata list (https://doi.pangaea.de/10.1594/PANGAEA.871462?format=html#download). Alternative lake names were added to the list at every opportunity, but are neither exhaustive nor tracked.

Lake water surface level data (Crétaux et al., 2011) provided by the Laboratory of Studies on Spatial Geophysics and Oceanography (LEGOS) through their Hydroweb portal (http://www.legos.obs-mip.fr/soa/hydrologie/hydroweb/) were originally distributed with the Diversity II database. This data come as 1-D discrete time samples, as opposed to the 2-D temporal aggregated water quality maps. Furthermore, they are based on independent developments and updates that would require continuous mirroring. Due to these differences, we refrained from adding them to the Pangaea Diversity II repository.

3 Data processing methods

The bulk production of temporally aggregated water quality parameters from L1B and auxiliary data requires a combination of several methods in an unsupervised processing chain. For this purpose we implemented the *CaLimnos* v1 processing chain (Fig. 2) for deployment on ESA's Earth observation data processing cluster *Calvalus* (Fomferra et al., 2012). It is composed of several processors for the ESA BEAM Toolbox (Fomferra and Brockmann, 2005), which has recently evolved into the Sentinel Application Platform (SNAP). The same input and auxiliary data and pre- and post-processing modules were also used to create 10-day aggregates for the investigation of phenological cycles in Lake Balaton (Palmer et al., 2015), and corresponding *CaLimnos* v1 L2 intermediate outputs were used for assessing the spatio-temporal variability of chl-*a* in Lake Geneva (Kiefer et al., 2015).

3.1 Pre-processing

The identification of pure water pixels is an essential preprocessing step, because even sub-pixel signal contributions from land surfaces can strongly affect the retrieval procedures, especially when using band arithmetic algorithms that do not check for input signal compliance at runtime. The Idepix algorithm is an open-source SNAP processor and performs such identification for clouds, cloud shadows, cloud buffers, land, snow/ice, sun glint and mixed pixels (Danne, 2016) based on bottom-of-Rayleigh reflectance (BRR; Santer et al., 1999). BRR is subject to a partial correction of atmospheric effects, representing reflectance at the hypothetical boundary between an infinitesimally small aerosol layer and gaseous air layers above. It is the preferred signal when background reflectance for the estimation of aerosol optical thickness is highly uncertain, and therefore BRR intermediate products are also used for the identification of shallow water areas and as input for the Maximum Peak Height processor (Matthews et al., 2012) according to Fig. 2.

Idepix uses the Shuttle Radar Topography Mission (SRTM) Water Body Dataset (SWBD; Slater et al., 2006) as a static a priori land–water mask, which is a snapshot of global water surface extent between 56° S and 60° N in February 2000. It applies several arithmetic expressions, a spectral unmixing algorithm for mixed pixel identification, and two back-propagation neural networks (NNs) for cloud identification to MERIS FSG L1B and BRR input data (Kirches et al., 2013). Output is a pixel identification flag layer which is much better suited for water constituent retrieval than the original L1B product flags (Ruescas et al., 2014). However, usage with inland waters is subject to two particular challenges. First, Idepix' sea ice identification uses climatological auxiliary data that are not available for lakes, and therefore lake ice identification is less accurate. Second, ephemeral water surfaces that may extend far beyond the SRTM observed extent are always clipped to the latter.

Bottom visibility is a critical and unmastered error source for water quality retrieval, because most algorithms that provide concentrations of water constituents do not account for benthic reflectance contributions in these so-called optically shallow waters. In fact, a pre-condition for them is optically deep water (i.e. no bottom reflection). Sandy or vegetated substrates cause surface-leaving signals that can closely resemble increased suspended sediment and phytoplankton concentrations in the water column, respectively, and thus distort retrievals. Only very few algorithms actually deal with the detection of optically deep water, and none of them applies to inland waters. Based on recommendations for clear coastal waters (Cannizzaro and Carder, 2006) and our own investigations, we defined a band ratio that evaluates the relative elevation of oligotrophic lakes' 555 nm water-leaving reflectance peak, but using BRR in three MERIS bands as input due to the lack of robust automated atmospheric correction algorithms for such conditions (Eq. 1; Odermatt et al., 2015a).

$$\text{IF ratio}_{490} = \frac{\text{BRR}_{\text{band3}} \cdot \text{BRR}_{\text{band7}}}{(\text{BRR}_{\text{band5}})} < \text{Thres, shallow} = \text{TRUE} \quad (1)$$

Due to the ambiguity of certain substrates' shallow water reflectance and deep-water reflectances, this optical signature alone is prone to false positive identifications. It becomes much more robust when applied to temporally aggregated *ratio_490* due to the relative persistence of benthic features as opposed to the dynamically changing water composition. Summer half-year mean averages were selected after evaluation of several statistical aggregation methods. Corresponding aggregates are composed of all cloud-free MERIS observations in May to October 2008 for the Northern Hemisphere, and November 2008 to April 2009 for the Southern Hemisphere. Lakes with constantly high turbidity, such as Lake Balaton, still trigger false positives. Therefore, each *ratio_490* aggregate map was verified with high-resolution satellite imagery and bathymetry information. Considerable shallow water areas in about 30 oligotrophic to mesotrophic lakes were masked using a threshold of 0.65, which, in the case of the Beaver Island Archipelago in Lake Michigan, masks areas that are between 5 and 10 m deep (Fig. 3). Pixels removed in such a manner are indicated in a separate product layer (*shallow*, Fig. 3c). In the Lake Michigan example, there are some patterns in the *chl_fub* product that still resemble bathymetry features, but their concentration levels are within variations for deep-water areas apart from a few individual pixels. Especially in more turbid lakes, lower thresholds are applied to prevent false positives according to the corresponding column in the lakes list (https://doi.pangaea.de/10.1594/PANGAEA.871462?format=html#download).

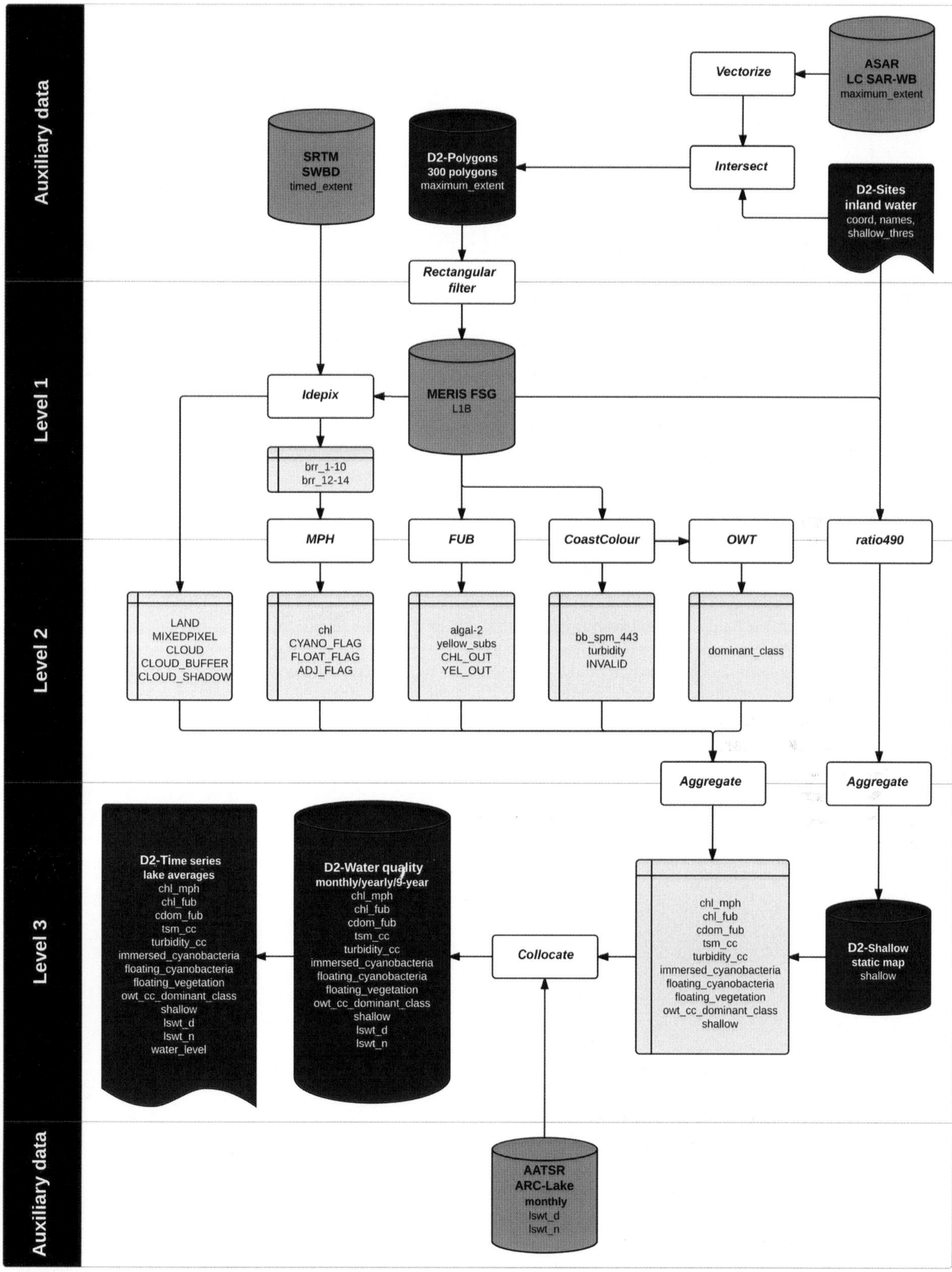

Figure 2. The *CaLimnos* v1 processing chain for inland waters. Colouration indicates algorithms and downstream processes (white), input and auxiliary data (dark grey), intermediate products (light grey) and output products (blue).

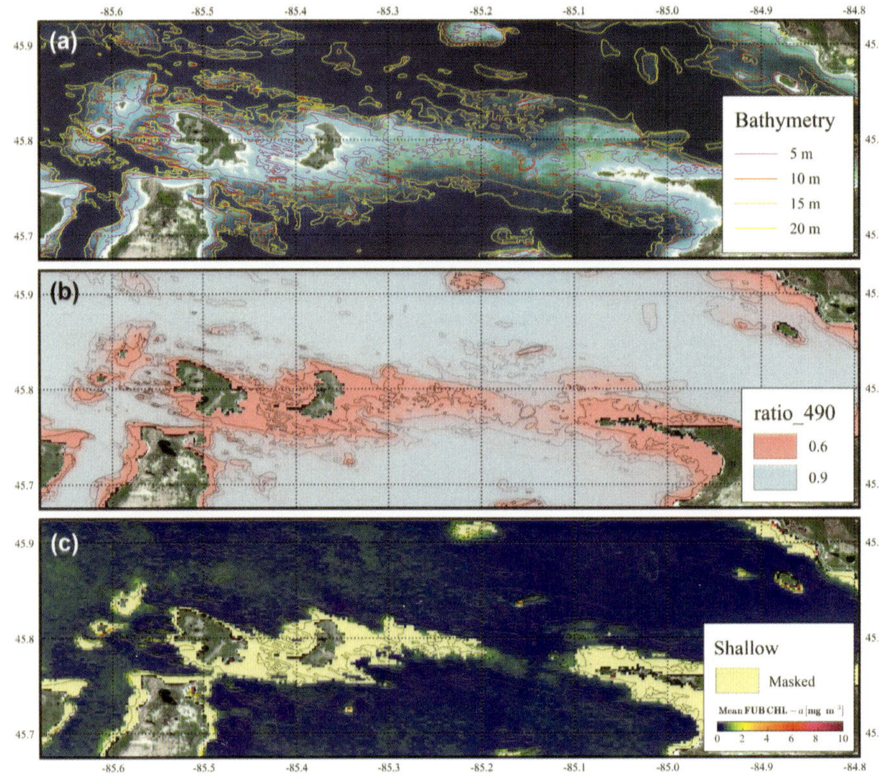

Figure 3. Shallow water flagging for the Beaver Island Archipelago in the north of Lake Michigan. **(a)** Sentinel-2A true colour image, 8 May 2017. **(b)** Ratio_490 from MERIS data, acquired in May–October 2008. **(c)** Shallow water mask for ratio_490 with a threshold of 0.65 on top of the chl layer for October 2011 as contained in product layer *shallow*. Bathymetry data provided by NOAA-NCEI.

Table 1. Input, output and aggregation specifications for the monthly products.

Algorithm	L2 input layer(s)	Aggr.	L3 output layer(s)
MPH	chl: float ($mg\,m^{-3}$)	mean	chl_mph: float ($mg\,m^{-3}$)
FUB	algal-2: float ($mg\,m^{-3}$)	mean	chl_fub: float ($mg\,m^{-3}$)
FUB	yellow_subs: float (m^{-1})	mean	cdom_fub: float (m^{-1})
CoastColour	bb_spm_443	mean	tsm_cc: float ($g\,m^{-3}$)
CoastColour	turbidity: floating numbers (FTU)	mean	turbidity_cc: float (FTU)
MPH	if CYANO_FLAG not FLOAT_FLAG: binary	mean	immersed_cyanobacteria: float (0–1, dL)
MPH	if CYANO_FLAG and FLOAT_FLAG: binary	mean	floating_cyanobacteria: float (0–1, dL)
MPH	if FLOAT_FLAG not CYANO_FLAG: binary	mean	floating_vegetation: float (0–1, dL)
OWT	dominant_class: integer (1–7, dL)	mode	owt_cc_dominant_class: integer (1–7, dL)
ratio_490	See Sect. 4.1	mean	shallow: binary
lswt_d	See Sect. 3.3	none	lswt_d_mean: float (K)
lswt_n	See Sect. 3.3	none	lswt_d_mean: float (K)

3.2 Water quality retrieval

Chl-*a* retrieval in optically complex waters is straightforward when using the secondary reflectance peak at red and near-infrared (NIR) wavelengths (e.g. Gitelson, 1992; Gons, 1999; Gower et al., 1999). However, using MERIS observations this peak is only accessible in moderately productive or turbid waters, while clear and humic waters call for different approaches (Odermatt et al., 2012). Moore et al. (2014) developed an optical water type classification (OWT) framework, which supports the distinction of these different water types, and which is available as a SNAP plugin (Peters, 2016). It assigns water-leaving reflectance spectra to seven end members, which were identified through cluster analysis of in situ measurements. Classes 1–3 represent clear or absorbing waters, classes 4–5 represent high phytoplankton and classes 6–7 represent high suspended mineral contents (Fig. 4). The OWT algorithm depends on the accurate correction of atmospheric effects (Eleveld et al., 2017), which was assessed by classifying 42 matchup pairs of in situ reflectance measurements in 10 diverse lakes and MERIS water-leaving reflectance from various atmospheric corrections. Water-leaving reflectance obtained with the Coast-Colour NN algorithm (description below) achieved the best agreement, in which half the matchup pairs were assigned to the same OWT, and adjacent classes were assigned in 14 cases (Odermatt et al., 2015a). In five out of the remaining seven cases, the optically quite similar classes 1 and 3 are confused. This mismatch due to differences between in situ measured and satellite observed reflectance is quite significant. However, when considering only the separation between classes 1–3 and 4–7, and thus the feasibility of chl-*a* retrieval based on the secondary reflectance peak, the approach becomes very robust, with only 2 out of the 42 pairs being confused. The OWT maps therefore provide a rough but robust indicator for chl-*a* algorithm selection. Most lakes are relatively clearly dominated by either OWT 1–3 or 4–7, which makes the chl-*a* product selection straightforward. In rare cases like Lake Turkana (Fig. 5), such a selection can only be made if either the lower or upper end of the dynamic range is considered more relevant. Otherwise, it is recommended to either split the lake perimeter or merge the chl-*a* products, e.g. by weighting them with turbidity levels.

For the Diversity II production the Maximum Peak Height algorithm (MPH; Matthews et al., 2012) was developed further and implemented in a SNAP operator (Block, 2016) because it outperformed other red-NIR reflectance peak algorithms in the algorithm intercomparison study (Matthews and Odermatt, 2015; Odermatt et al., 2015a). It uses BRR in MERIS bands 6–10 and 14 for the retrieval of the red-NIR reflectance peak height and position, which allow for the identification of cyanobacteria- and eukaryote-dominated pixels, water surface covering by cyanobacteria scum or floating vegetation, and chl-*a* quantification. Dedicated empirically calibrated equations are used for the retrieval of chl-*a* concentrations in eukaryote- and cyanobacteria-dominated waters. Technically, the algorithm is designed to cover the range of 0–1000 mg m^{-3} chl-*a*. However, retrieval accuracy is significantly better for lakes that are predominantly OWT 4–7, namely eutrophic to hypertrophic waters.

The FUB algorithm (Schroeder et al., 2007), named after the Free University of Berlin, is a bundle of dedicated NN algorithms for chl-*a*, total suspended matter (TSM) and coloured dissolved organic matter (CDOM) retrieval from MERIS L1B data, and a fourth NN that computes AOT at four wavelengths (440, 550, 670, 870 nm) and water-leaving reflectance in all bands up to 708 nm, except at 680 nm. The algorithms are trained with radiative transfer simulations using the Matrix Operator Model (MOMO; Fell and Fischer, 2001) covering chl-*a*, TSM and CDOM concentration ranges of 0.05–50 mg m^{-3}, 0.05–50 g m^{-3} and 0.005–1 m^{-1}, respectively, and using MERIS bands 1–7, 9, 10 and 12–14

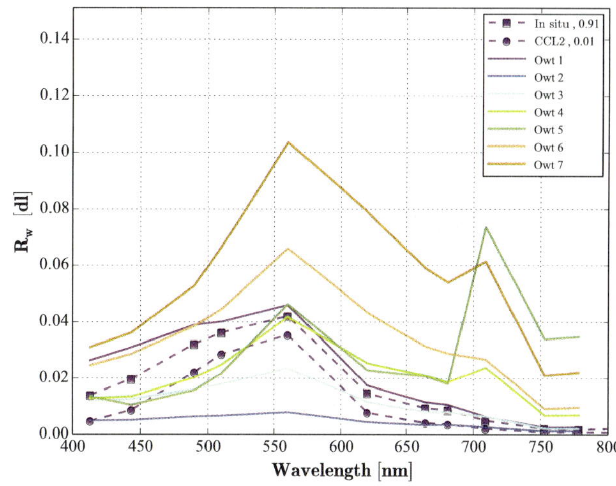

Figure 4. OWT end-member water-leaving reflectance spectra and a OWT 2 retrieval example for an in situ and MERIS CCL2 reflectance pair from Lake Zurich, 15 August 2007. Respective class membership scores are indicated in the legend.

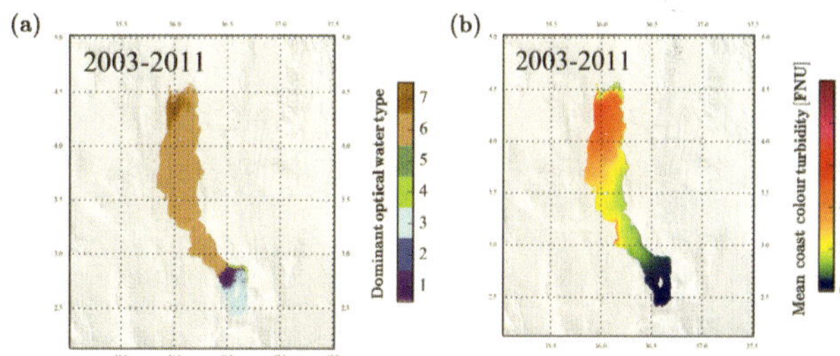

Figure 5. Nine-year aggregated OWT in Lake Turkana (**a**; *owt_cc_dominant_class_mode*), which features a very prominent gradient in turbidity (**b**; *turbidity_cc_mean*). Maximum turbidity and predominantly OWT 7 are observed in the north, where its main tributary, the Omo River, provides about 90 % of the lake's inflow (Beadle, 1981). In contrast, the terminal basin in the south corresponds to OWT 1 and 3, which are second and third lowest in turbidity according to the end members in Fig. 4.

as input. The training ranges are a severe limitation for global usage, but specific retrieval quality flags indicate for each of the four NN algorithms whether the input or output exceeds the training range. However, for oligotrophic to eutrophic and in particular humic lakes, which are commonly identified as OWT 1–3, the FUB algorithm's chl-*a* output outperformed all other candidates in the intercomparison study (Odermatt et al., 2015a). Note that FUB uses shorter wavelengths that reach deeper into the water column than MPH, which means that the two chl-*a* products represent different depths and may not converge at intermediate concentrations (ca. 10–30 mg m^{-3}), where both algorithms produce valid results.

For the retrieval of TSM via particulate backscattering at 443 nm (bb_spm_443 in Fig. 2) and turbidity, as well

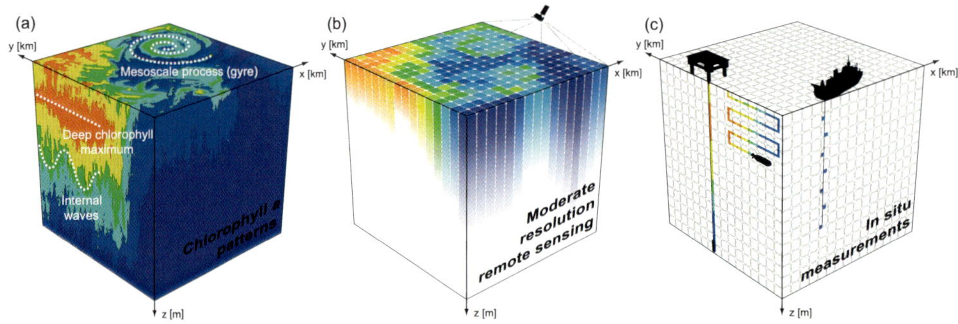

Figure 6. Comparison of how typical chl-a distribution patterns **(a)** are resolved with moderate-resolution satellite sensors **(b)** and in situ measurements **(c)**.

as water-leaving reflectance input for the OWT classification, we used the CoastColour NN algorithm. Its architecture is based on the approach described in Doerffer and Schiller (2007), with two dedicated NN systems performing atmospheric correction and inherent optical property retrieval (Doerffer, 2011; Ruescas et al., 2014). In contrast to earlier NN algorithms, the CoastColour NN was trained with significantly larger concentration ranges, namely 0.03–1000 $g\,m^{-3}$ TSM and 0.03–500 $mg\,m^{-3}$ chl. It was extensively validated with the CoastColour Round Robin dataset (Nechad et al., 2015; available in Pangaea) and lake in situ measurements (Odermatt et al., 2015a).

3.3 Post-processing and auxiliary data

The aggregation of L2 to L3 products (Fig. 2) facilitates temporal binning and collocation in a common coordinate grid with the WGS 84 (EPSG 7030) coordinate system. Monthly aggregates are created using the input, output and aggregation methods listed in Table 1, and the same aggregation methods are used to create yearly and 9-year aggregates from monthly and yearly aggregates, respectively, which ensures that all months inputting aggregate periods are weighted equally even if the numbers of L2 available in these periods may differ strongly. Aggregation of biophysical parameters is done using the mean of all valid nearest-neighbour input pixels for each output pixel, while the OWT L3 output layer consists of the most frequently observed class value across all available input layers, using the lower class in the rare case of a draw.

Monthly, yearly and 9-year aggregates for each lake are saved in individual GeoTIFF files, and compressed in 13 ZIP files representing 11 annual archives for the monthly aggregates, one archive for the yearly aggregate, and the 9-year aggregate. These 13 ZIP archives are zipped again to make each lake available for download in a single file of up to 18.9 GB in size (Caspian Sea).

To extract product statistics and for visualization of the products, a Python package is available at https://github.com/odermatt/diversity. The scripts included in the package allow for creation of spatial and temporal plots such as shown in Sect. 5 (Figs. 5, 12–18). They also feature the use of blacklists, e.g. to exclude all products with scarce lake extent coverage from further analyses.

4 Product assessment

4.1 Performance metrics and reference data

The estimation of remotely sensed water quality product accuracy is complicated by the relevance, consistency and availability of reference measurements, especially across different lakes and countries. Authoritative in situ reference measurements are obtained from lab-analysed samples or probe records; they apply well-known standards and are checked for consistency. But such measurements are generally not optimized for comparability with remote sensing data. They represent volumes in the order of litres, either from discrete samples or along depth profiles, in contrast to the pixel-sized satellite sampling (Fig. 6), whose vertical sensitivity depends on the derivative of the round-trip attenuation in each pixel (Zaneveld et al., 2005). Furthermore, our 10-year satellite mission with at least 3 days revisit time yields only an expectancy value of 20 matchup data pairs with monthly in situ measurements in a location with 50 % cloud probability. Against this background, Seegers et al. (2018) recently suggested a set of performance metrics that consists of error and decision metrics, as well as spatial and temporal maps and decision graphics. We provide these metrics, namely the mean average error (MAE) and bias of log-transformed parameters as defined in Seegers et al. (2018), and corresponding graphics for chl, TSM and turbidity matchup data pairs with datasets from 24 lakes and a group of reservoirs in 10 countries on 4 continents. Correlation coefficients are obtained from ordinary least square linear regression, neglecting inevitable errors of in situ measured reference datasets (McArdle, 1988). The remote sensing estimates used in these comparisons are means and standard deviations for 3×3 pixel kernels at sampling coordinates from L2 products that were acquired on the same day.

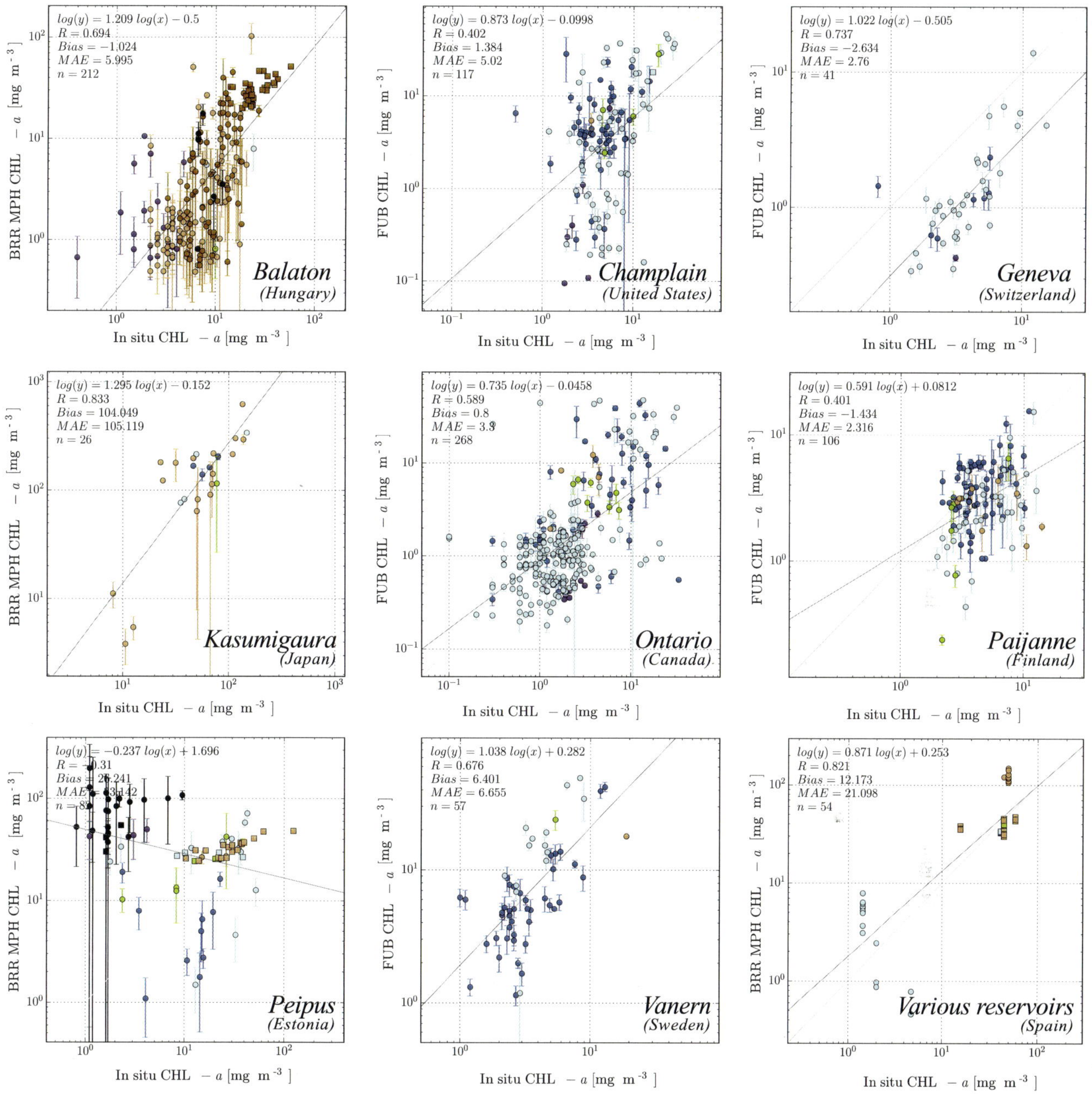

Figure 7. Matchup comparisons between remotely sensed chl-*a* and monitoring measurements. Colours represent the dominant OWT in the nine matchup pixels according to Fig. 2 (black: OWT classification not possible), and error bars indicate their standard deviation.

It should be noted that the accuracy estimates given hereafter refer to L2 products, but for the L3 products distributed with our database, randomly distributed errors are, depending on acquisition frequency (Fig. 1), reduced through averaging.

All reference datasets are from continuous monitoring programmes according to the enclosed acknowledgements, the only exception being the 2001–2005 dataset collected in Spanish reservoirs (Ruiz-Verdú et al., 2008; Simis et al., 2007), which was sampled in reservoirs Almendra, Cuerda del Pozo, Iznajar, Rosarito and Talarn. Due to the small number of samples per reservoir, and for their methodological consistency, the data from Spain are used as if they represented a single water body. The data of the Albufera lagoon are excluded because the reference locations were too far inshore and did not produce meaningful data pairs. We generally use records averaged across the top 5 m of the water column where data represent vertical gradients, e.g. from fluorometers or turbidimeters. In the case of Lake Champlain,

Table 2. Statistical metrics and ancillary information for chl-*a* matchup data and the MPH algorithm. n/Min./Med./Max. describe the in situ measurement counts and ranges represented by the matchups. OWT are indicated as the mode of the 9-year averaged products where available, as well as for the available matchup data for MPH (in brackets).

Lake	Country	OWT	n	Min.	Med.	Max.	R	Bias	MAE
Albert Falls Reservoir	South Africa	– (3)	31	3.1	12.3	32.8	0.229	−9.034	10.594
Lake Balaton	Hungary	7 (7)	212	0.4	7.6	57.0	0.694	−1.024	5.995
Lake Champlain	United States	2 (3)	128	0.5	4.9	30.5	0.193	7.023	10.398
Lake Geneva	Switzerland	3 (3)	41	0.8	3.6	15.2	0.051	−2.603	2.821
Hartbeespoort Reservoir	South Africa	– (3)	29	1.0	35.7	573.9	0.845	−22.994	57.697
Lake Haukivesi	Finland	– (2)	34	1.4	5.9	17.0	0.522	−2.335	3.040
Lake Huron	Canada	3 (1)	16	0.1	0.4	2.2	0.741	2.247	2.247
Lake Kallavesi	Finland	– (6)	19	6.1	9.7	34.0	0.389	0.722	3.802
Lake Kasumigaura	Japan	6 (6)	26	8.1	61.1	149.0	0.833	104.049	105.119
Klipvoor Reservoir	South Africa	– (6)	23	12.6	73.3	497.2	0.434	167.235	211.419
Lake Malaren	Sweden	3 (6)	79	1.3	11.9	35.0	0.463	7.932	14.867
Lake Ontario	Canada	3 (3)	275	0.1	1.8	34.3	0.433	2.213	4.238
Lake Paijanne	Finland	2 (2)	102	2.0	4.0	14.1	0.616	0.480	2.485
Lake Peipus	Estonia	2 (6)	82	0.8	12.2	123.8	−0.310	26.241	23.142
Lake Pielinen	Finland	2 (3)	93	0.9	5.4	17.0	0.206	0.854	3.517
Lake Pyhajarvi	Finland	– (2)	21	2.4	5.8	9.8	0.424	−4.035	4.035
Lake Vanajavesi	Finland	– (2)	86	3.7	11.0	43.0	0.570	−1.956	6.294
Lake Vanern	Sweden	2 (2)	66	1.0	3.4	37.0	0.695	6.198	10.405
Lake Vattern	Sweden	2 (2)	17	0.6	1.0	5.7	0.888	0.848	0.920
Lake Vesijarvi	Finland	– (2)	42	1.8	6.8	25.0	0.575	−5.323	5.439
Lake of the Woods	Canada	6 (2)	23	0.6	4.5	18.5	0.479	1.939	4.421
Lake Zug	Switzerland	3 (3)	17	0.9	1.5	4.3	0.569	2.574	2.994
Lake Zurich	Switzerland	3 (3)	18	1.1	6.5	13.0	0.106	−4.113	4.266
Various reservoirs	Spain	– (6)	54	1.4	44.3	58.3	0.821	12.173	21.098

chl measurements are taken with a vertically integrating hose sampler across twice the measured Secchi depth. The comparability of this approach with remotely sensed concentrations is expected to decrease with increasing Secchi depth, and therefore we removed all matchups sampled across more than 10 m depth (i.e. 5 m Secchi depth). In contrast to the algorithm selection study in Odermatt et al. (2015a), we did not consider any datasets that yielded less than 15 matchups with any of the algorithms, because the statistics resulting from these data are increasingly erratic. Note also that we did not obtain an interpretable number of CDOM measurements, and therefore this parameter remains unvalidated and should be used with caution.

4.2 Chl matchup comparisons

Figure 7 shows the largest chl matchup datasets of each country, whereas FUB or MPH are selected according to the lakes' dominant water type in the 9-year averaged products. Eight out of nine lakes feature correlation coefficients between $R = 0.40$ and 0.83. Larger concentration ranges result usually in higher R; accordingly, eutrophic lakes have generally higher R than oligotrophic and mesotrophic lakes. For most of the latter, i.e. lakes dominated by OWT 1–3 (Champlain, Geneva, Ontario, Paijanne), FUB matchups result in MAE that are close to median concentrations, and lower absolute bias (Table 3), while the comparison for Lake Vanern shows significantly higher concentrations from remote sensing than from in situ measurements. Concerning eutrophic and hypertrophic lakes, MAE and absolute bias are likewise clearly lower than median concentrations for Lake Balaton and the Spanish reservoirs, while a larger bias occurs with the data of Lake Kasumigaura. These successful examples in Fig. 7 are contrasted by Lake Peipsi, whose high CDOM levels and cyanobacteria-rich plankton community are known to create challenging conditions (Kutser et al., 2016). A large number of the matchup samples (black dots) even come with reflectance shapes that do not sufficiently match any of the OWT, which is not observed for any of the other 23 datasets.

Concerning the performance metrics for all 24 datasets, the MPH algorithm (Table 2) achieves $R > 0.4$ for 17 of them, including all datasets with OWT 4–7 apart from Lake Peipus (see above) and Lake Kallavesi ($R = 0.39$). But it does not meet $R > 0.4$ for 5 out of the 16 datasets that are predominantly OWT 1–3, which confirms earlier findings that the algorithm's sensitivity decreases below about $10\,\mathrm{mg\,m^{-3}}$ (Odermatt et al., 2012). In contrast, the FUB algorithm removes all pixels that exceed its training range (see Sect. 4.2); i.e. it does not allow any matchup for max. chl $> 50\,\mathrm{mg\,m^{-3}}$ but for Lake Peipus (chl $= 52.4\,\mathrm{mg\,m^{-3}}$).

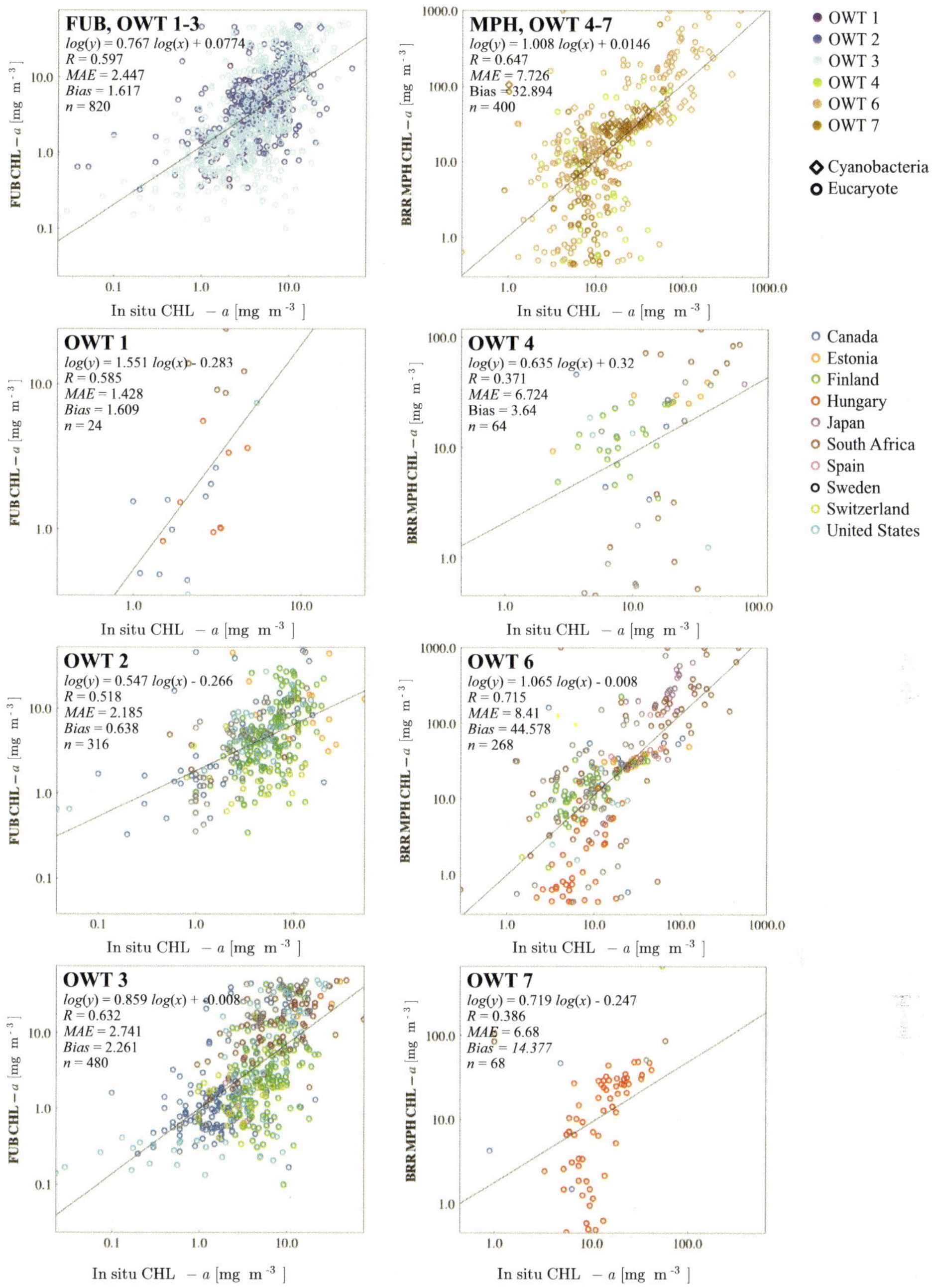

Figure 8. Global chl-*a* matchup scatter plots and statistics for the FUB (left) and MPH (right) algorithms, for all matching OWT (top row) and specifically per OWT (rows 2–4). Note that none of the matchups was identified as OWT 5 (extremely productive water). Marker shapes represent MPH cyanobacteria flagging.

This is reflected by the number of matchups for FUB, n in Table 3, which is strongly reduced for the most productive waters (Hartbeespoort, Kasumigaura, Klipvoor, Malaren, Peipus, Spanish reservoirs). When considering only datasets with OWT 1–3, FUB achieves $R > 0.4$ in 10 out of 15 cases, the remaining 5 being predominantly lakes in Finland (Haukivesi, Pielinen, Pyhajarvi, Vanajavesi), where a combination of high CDOM levels and a somewhat smaller dynamic range than in Lake Peipus complicates retrievals (Attila et al., 2018). As shown already for the examples in Fig. 7, the MAE for lake-specific chl concentrations is within 30 % of the lake median concentration for about three-quarters of the datasets, and the absolute bias is similar to or significantly smaller than the MAE.

Finally, we segment the validation matchups by OWT (Fig. 8), whereas FUB matchups are given for OWT 1–3,

Table 3. Statistical metrics and ancillary information for chl-*a* matchup data and the FUB algorithm. n/Min./Med./Max. represent the in situ measurement counts and ranges represented by the matchups for all $n > 15$. OWT are indicated as the mode of the 9-year averaged products where available, as well as for the available matchup data for FUB (in brackets).

Lake	Country	OWT	n	Min.	Med.	Max.	R	Bias	MAE
Albert Falls Reservoir	South Africa	– (3)	31	3.1	12.3	32.8	0.520	8.182	10.250
Lake Balaton	Hungary	7 (7)	188	0.4	7.2	33.3	0.813	4.493	6.600
Lake Champlain	United States	2 (3)	117	0.5	4.68	27.6	0.402	1.384	5.020
Lake Geneva	Switzerland	3 (3)	41	0.8	3.6	15.2	0.737	−2.634	2.760
Hartbeespoort Reservoir	South Africa	– (3)	15	1.0	9.89	35.72	–	–	–
Lake Haukivesi	Finland	– (2)	36	1.4	5.9	17.0	0.150	−4.314	4.501
Lake Huron	Canada	3 (1)	16	0.1	0.4	2.2	0.406	0.079	0.393
Lake Kallavesi	Finland	– (6)	20	6.1	9.7	34.0	−0.404	−6.169	6.846
Lake Kasumigaura	Japan	6 (6)	4	8.1	24.0	38.4	–	–	–
Klipvoor Reservoir	South Africa	– (4)	1	36.1	36.1	36.1	–	–	–
Lake Malaren	Sweden	3 (3)	37	1.3	9.5	25.7	0.007	18.884	19.880
Lake Ontario	Canada	3 (3)	268	0.1	1.8	32.8	0.589	0.800	3.300
Lake Paijanne	Finland	2 (2)	106	2.0	4.1	14.1	0.401	−1.434	2.316
Lake Peipus	Estonia	2 (2)	28	1.8	14.8	52.4	−0.212	11.462	19.404
Lake Pielinen	Finland	2 (3)	96	0.9	5.5	17.0	0.177	−3.178	4.039
Lake Pyhajarvi	Finland	– (2)	22	2.4	5.8	9.8	0.274	3.889	5.029
Lake Vanajavesi	Finland	– (2)	83	3.7	11.0	43.0	0.284	4.227	10.241
Lake Vanern	Sweden	2 (2)	57	1.0	3.2	19.0	0.676	6.401	6.655
Lake Vattern	Sweden	2 (2)	17	0.6	1.0	5.7	0.525	−0.410	0.487
Lake Vesijarvi	Finland	– (2)	41	1.8	6.8	16.0	0.607	4.012	5.843
Lake of the Woods	Canada	6 (2)	19	0.6	3.9	10.8	0.188	16.892	17.324
Lake Zug	Switzerland	3 (3)	18	0.9	1.5	4.3	0.576	1.575	1.896
Lake Zurich	Switzerland	3 (3)	18	1.1	6.5	13.0	0.156	−4.803	4.891
Various reservoirs	Spain	– (3)	6	2.0	3.4	4.7	–	–	–

and MPH matchups for OWT 4–7. The overall comparisons per algorithm (Fig. 8, top row) show again that the concentration range available in the FUB products is limited to $< 40\,\text{mg}\,\text{m}^{-3}$, but they reflect a large part of in situ measurement variations, especially for the $1–10\,\text{mg}\,\text{m}^{-3}$ range. Contrariwise, the few MPH estimates for OWT 4–7 that fall into this range hardly align with in situ measurements. Altogether, MPH still achieves a relatively high correlation due to a better match in the $10–1000\,\text{mg}\,\text{m}^{-3}$ range. The interpretation of individual classes is somewhat more complicated, because OWT 1, 4 and 7 consist of far fewer samples than OWT 2, 3 and 6, while OWT 5 is missing entirely. This reflects on the one hand the relative scarcity of the extremely clear, productive and turbid OWT 1, 5, and 7, respectively. For example, Lake Balaton and Lake Huron make up about 80 % of the samples in OWT 1, and Lake Balaton alone makes up about 90 % of the OWT 7 data. On the other hand, it is a consequence of the CoastColour atmospheric correction, which facilitates a very robust discrimination of OWT 1–3 versus OWT 4–7 but tends to underestimate the occurrence of OWT 4 and 5 in favour of OWT 6. The correlations are around $R = 0.5$ to 0.7 for large matchup counts, but lower where $n < 100$. But on the whole, we can again conclude that all MAE and biases are in the order of the subsample's median concentration or lower, with a tendency of MPH to overestimate the reference samples.

4.3 TSM and turbidity matchup comparisons

Most monitoring programs provide either for TSM, turbidity or only Secchi depth measurements. TSM and turbidity are, with caveats, directly correlated, while their relationship with Secchi depth is more complicated (Neukermans et al., 2012; Pfannkuche and Schmidt, 2003). Therefore we consider here only lakes where TSM or turbidity is available, which is the case for 3 and 10 lakes, respectively, whereas TSM prevails for the most turbid lakes, and turbidity is of units FNU apart from Lake Geneva (FTU), Lake Michigan and Lake Zug (NTU). In terms of remotely sensed estimates, it should be noticed that TSM (i.e. *tsm_cc_mean*) is estimated from particulate scattering across all visible wavelengths, while turbidity (i.e. *turbidity_cc_mean*) is estimated only from scattering at 865 nm according to ISO 7027-1 (2016), which means that the latter represents a significantly light smaller penetration depth.

The scatter plots in Fig. 9 depict all available datasets but the one for Lake Pyhajarvi (Finland, $R = 0.3$). The correlations depend even more on the value range than for chl. A surprisingly high correlation but a significant bias and

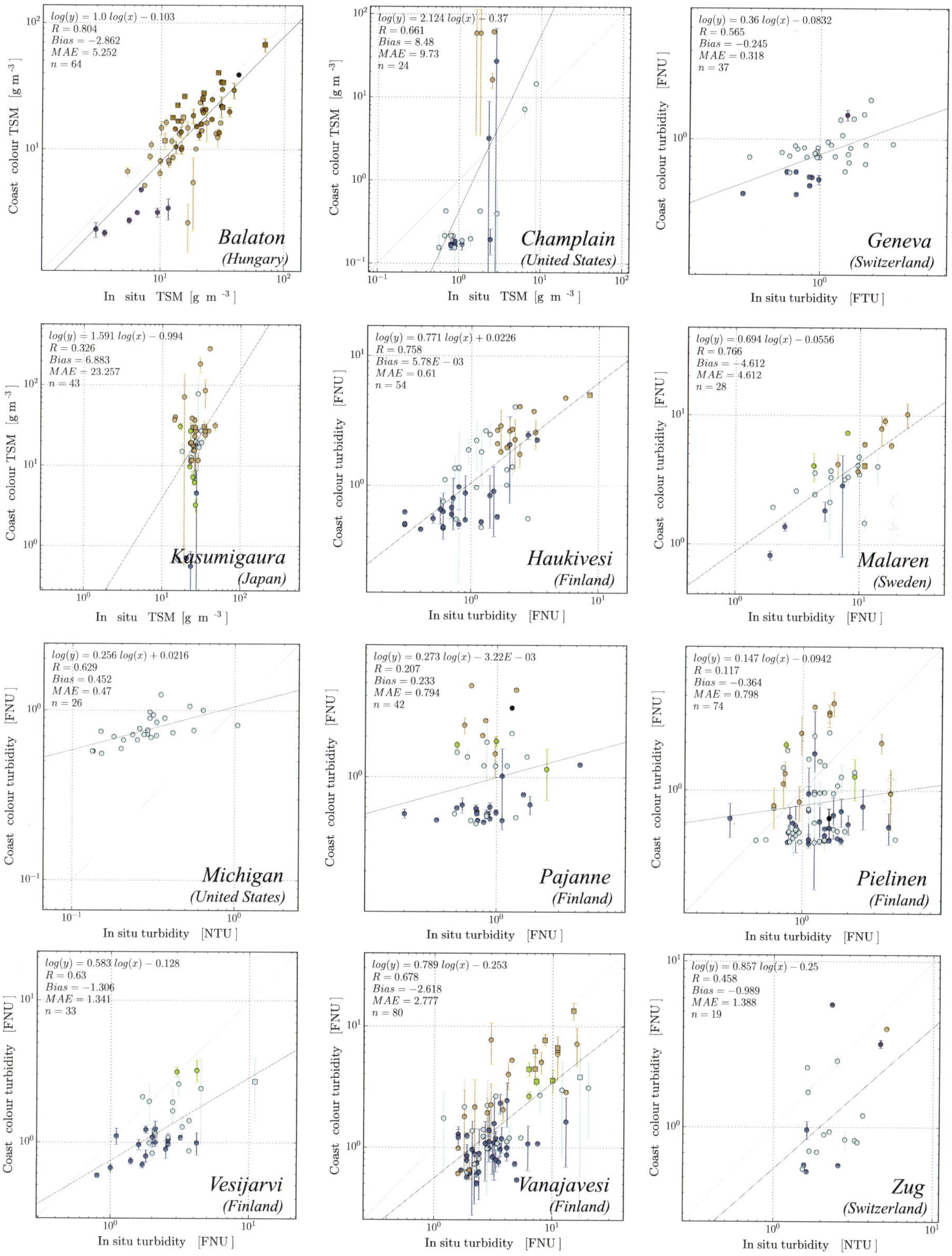

Figure 9. Matchup comparisons between remotely sensed TSM or turbidity, and monitoring measurements. Colours represent the dominant OWT in the nine matchup pixels according to Fig. 2 (black: OWT classification not possible), and error bars indicate their standard deviation.

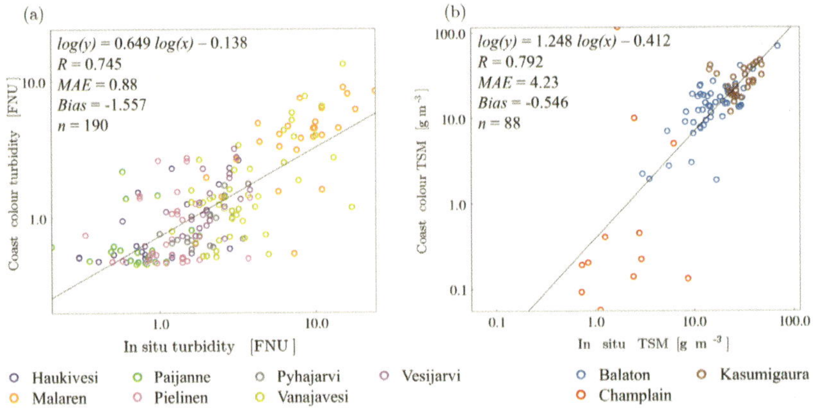

Figure 10. Turbidity **(a)** and TSM **(b)** matchup scatter plots and statistics across different lakes. Note that only turbidity data in units FNU were merged.

MAE are observed for turbidity in Lake Michigan, which might be related to the measurement standard NTU, which does not exceed a maximum of 1.0. The performance for the next least turbid lakes (Paijanne, Pielinen, Pyhajarvi, Geneva, Zug; $R = 0.1$ to 0.5, max. turbidity 3 to 6 FNU/NTU) is significantly worse than for the more turbid cases with $R = 0.6$ to 0.7. We therefore conclude that the lake-specific turbidity estimates tend to be of somewhat better accuracy than the chl estimates, apart from variations below 3 FNU, where the scattering signal at 865 nm decreases towards the noise level, in particular in high CDOM boreal lakes. At least for clear water lakes (e.g. Geneva, Zug), the TSM estimates (i.e. *tsm_cc_mean*) are a promising alternative to the turbidity products due to higher signal levels at visible wavelengths. Unfortunately, TSM reference data are only available for Lake Champlain and two much more turbid lakes, and the results for these three individual datasets are too inconsistent for generalized interpretation.

When merging all three TSM and seven consistent turbidity datasets (Fig. 10), the correlations are just below $R = 0.8$, which is again due to the larger dynamic range, and MAE are only about half the median concentrations. This corresponds to earlier findings in a validation study review (Odermatt et al., 2012), where TSM retrieval is in general subject to smaller uncertainties than chl retrieval. Apart from these performance metrics, it should be noted that there is a cluster of remotely sensed turbidity estimates in the order of 0.3 FNU for Finnish lakes, which are up to 2 FNU according to reference measurements. This cluster is a different view of the sensitivity loss described above.

4.4 Limitations

Most remotely sensed water quality parameters are provided, by definition, in per volume concentration units (chl, TSM, CDOM), which are obtained through generalized relationships between these concentration units and absorption and scattering intensities. This approach has limitations that are similar to those for phytoplankton pigment measurements using fluorometer probes (Huot and Babin, 2010). Both are subject to the superimposition of congeneric signals, such as fluorescence by CDOM or other algal pigments in the case of fluorometer probes, and CDOM and detritus absorption in the case of remotely sensed pigment absorption. In addition, uncertainties in the relationship between concentration units and optical quantities affect the quantification, namely fluorescence quantum yield or weight-specific absorption and scattering. Therefore Huot and Babin (2010) call for repeated calibrations of fluorometers to quantify pigment concentrations. When comparing fluorometer measurements across a large number of different lakes, calibration aspects become a major source of uncertainty. Therefore, it was suggested to remove inconsistencies in absolute magnitude across fluorometer profiles by normalizing each profile to a 0–100 index range (Leach et al., 2017). Likewise, the optically derived constituent concentrations provided in the Diversity II water quality data come without lake-specific calibration. The comparison of absolute chl, TSM and CDOM levels across different lakes should thus be avoided. Instead, we recommend either applying spatial or temporal normalization during data analyses or calibrating the data with lake-specific gains and offsets (e.g. Kiefer et al., 2015). Note that parameters that express solely optical (turbidity, cyanobacteria/floating matter) or thermal (LSWT) parameters are not subject to this limitation.

The second major limitation is related to erroneous pixel identification for melting ice coverage and ephemeral lakes. Areas of melting lake ice are optically very similar to water, and the false retention of a lake ice covered pixel by the Idepix algorithm can lead to highly irregular constituent estimates. Indicators for such cases are the seasonal timing, sharp linear features in water constituent products, indicating lake ice borders or cracks, and very low values in mean LSWT. Auxiliary data from the NOAA/NSDIC Global Lake

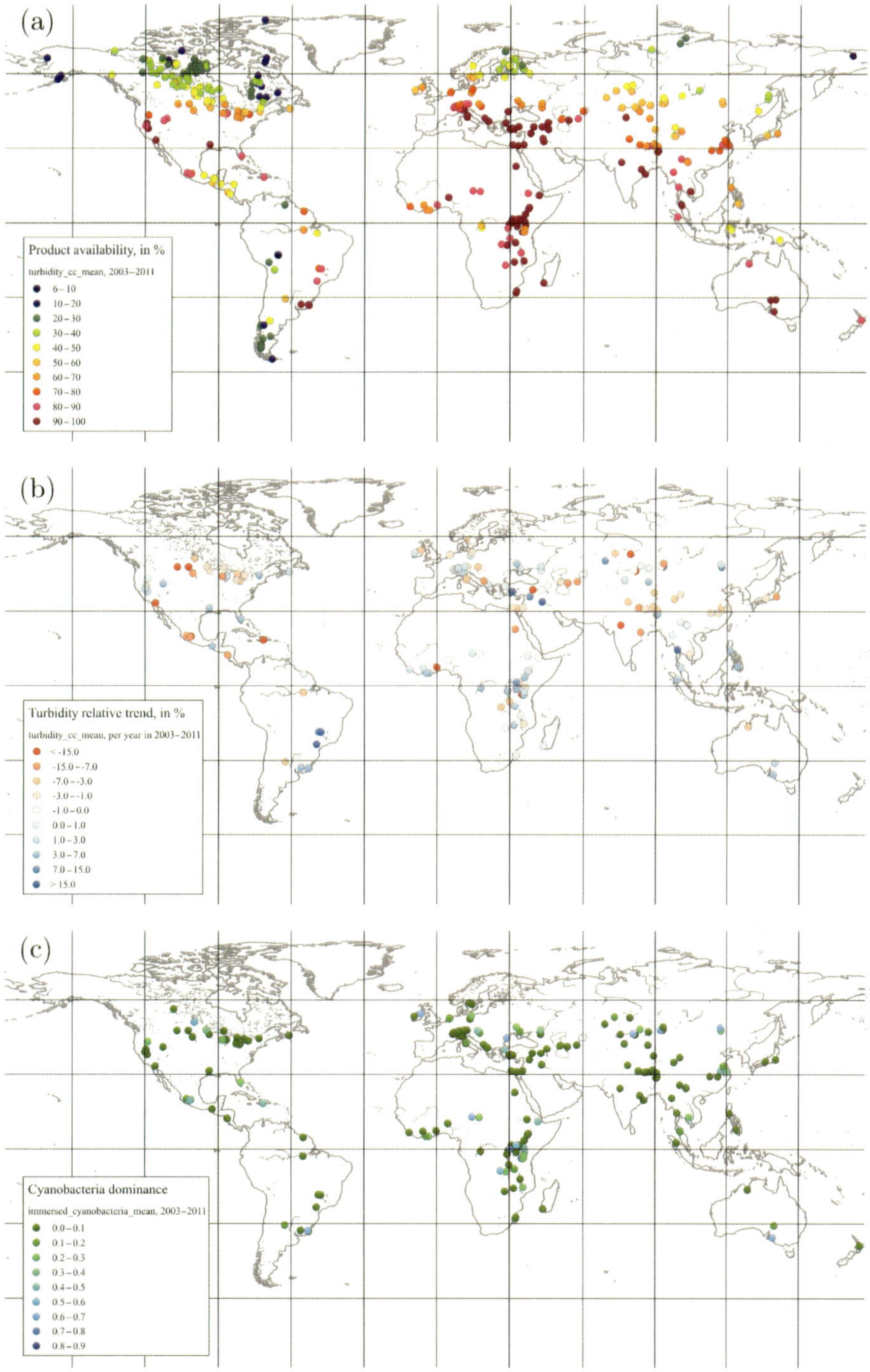

Figure 11. **(a)** Percentage of the 108 monthly *turbidity_cc_mean* products with at least 50 % coverage of the maximum lake area. **(b)** Turbidity trend, relative to median concentrations, for lakes with at least 50 % product availability. **(c)** Mean cyanobacteria dominance as the average of all spatially averaged monthly *immersed_cyanobacteria_mean* values for the same lakes.

and River Ice Phenology Database could also help with the identification. A procedure to fix monthly products that are affected by melting lake ice artefacts using the BEAM L3 binner is described in Odermatt et al. (2015b). Ephemeral (also intermittent or seasonal) lakes as well as other lakes and reservoirs that significantly change their extent over time are usually not well represented because the lakes' areas are clipped to the extent of the SWBD in February 2000. Their change in extent also complicates the identification of shallow water areas in a way that makes our approach based on

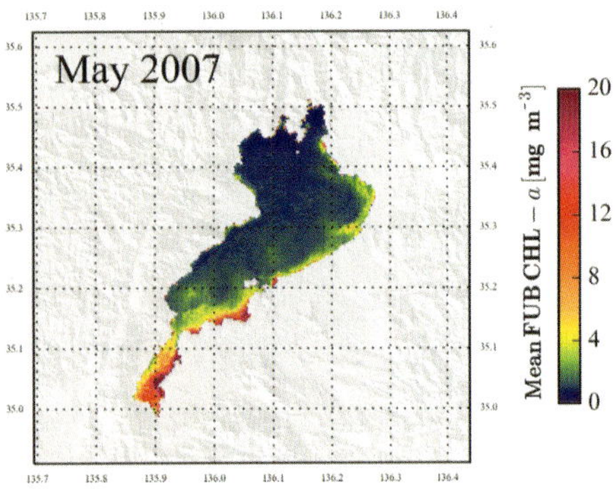

Figure 12. Chl in Lake Biwa, May 2007; L3 aggregate of four cloud-free images and one partly cloudy image (*chl_fub_mean*).

temporally consistent spectroradiometric properties inapplicable. Therefore *ratio_490* thresholds for such lakes were set to 0.0 to disable shallow water identification. Furthermore, many of these lakes are subject to very high salinity levels and other typical habitat properties that favour extraordinary types of water constituents and accordingly bio-optical properties, which can significantly affect the validity of the water quality products. In both cases, melting lake ice and ephemeral lakes, and both the spatial extent of the products and their spatial gradients can significantly deviate from usual conditions. Particular care is thus needed when a lake's product layer extent is significantly smaller than expected. The minimum percentage of water pixels identified in a product relative to the lake's maximum number of water pixels can thus be defined in the blacklisting procedure available in the Diversity II Python library. We recommend setting a minimum of 50 % and checking product integrity visually where appropriate.

Finally, note that the relative abundances in the *floating_vegetation* and *floating_cyanobacteria* layers include only pixels that passed the foregoing Idepix masking. This means that especially very densely covered water pixels are previously identified as land pixels and not counted, resulting in an overall underestimation of abundances.

5 Example data usage

The Diversity II water quality datasets were used for a preliminary global assessment, from which we describe a few example results hereafter. In addition, we performed several lake-specific assessments, most prominently for indicating fish assemblages and status assessments in Lake Vänern (Sweden; Sandström et al., 2016). Other use cases are described as *biodiversity stories* and made available from www.diversity2.info. A summary of three selected examples verifies how the remotely sensed parameters respond to spatio-temporally evident or relatively well-documented biophysical events.

5.1 Global assessments

In order to obtain a robust database for global analyses, we first created blacklists that list all 2003–2011 monthly *turbidity_cc_mean* products that come with a spatial coverage of less than 50 % or the maximum depicted area in the period. In this manner, we eliminate a lot of products that were acquired under suboptimal conditions, namely cloud or ice coverage. Gaps in MERIS FR data acquisition (Fig. 1) are also accounted for in such a manner. By counting the remaining products, we obtain an overview of the percentage of available products per lake (Fig. 11a). Note that high-altitude lakes are often only ice-free between June and September; values as low as 30 % can thus mean almost complete coverage. For a rough trend estimation, we still exclude all lakes with less than 50 % of the 108 months in 2003–2011 covered, and we normalize the annual trend with the 9-year median concentration in order to account for turbidity magnitude differences (Fig. 11c). Note that such normalization can also account for lake-specific water quality parameter underestimation or overestimation, e.g. where phytoplankton pigment to absorption ratios differ. The overview allows the quick identification of regime shifts and extreme events. Concerning the former, we observe a dramatic decrease in turbidity in Lake Enriquillo (Dominican Republic), which apparently accompanied a debated doubling of its surface area during the observation period (Méndez-Tejeda et al., 2016). As a case for the latter, we see a high positive trend of 0.2 FNU / year in Lake Sinakharin (Thailand), which is usually below 1 FNU, but went through several individual months with up to 6 FNU since summer 2009. Finally, using the same product availability constraint, but no normalization, we mapped cyanobacteria abundance (Fig. 11c).

5.2 Lake Biwa

Lake Biwa is a monomictic lake north-east of Kyoto (Japan), up to 104 m deep and with a surface area of 658 km^2 among the smaller lakes in the dataset. The primary productivity of the lake is relatively low, but subject to strong spatial gradients that are related to the distribution of residential and industrial areas, which are concentrated on the south-eastern shore and responsible for increased riverine nitrogen input (Ohte et al., 2010) that increases near-shore phytoplankton growth (Fig. 12). In December 2007 investigations with an AUV (autonomous underwater vehicle) revealed more than 2000 dead organisms on the lake's bottom, mostly endemic Isaza gobi fish and lake prawns. Low dissolved oxygen concentrations of less than 1.0 $mg\,L^{-1}$ near the lake bottom in November were identified as the main cause of an increased exposure of aquatic organisms to heavy metals and the die-off (Itai et al., 2012; Kawanabe et al., 2012). Oxygen supply depends on wintertime vertical mixing, which, aside from wind stress, depends on the vertical density gradients and thus thermal stratification. In situ temperature profiles from

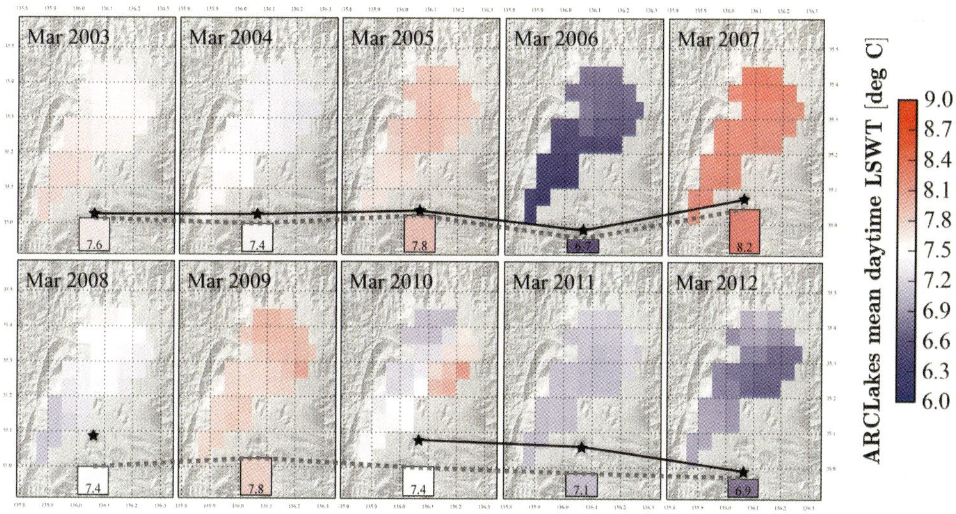

Figure 13. Lake surface water temperature (LSWT) in March in Lake Biwa (*lswt_n_mean*). The March LSWT marks the annual minimum for every year contained in the dataset.

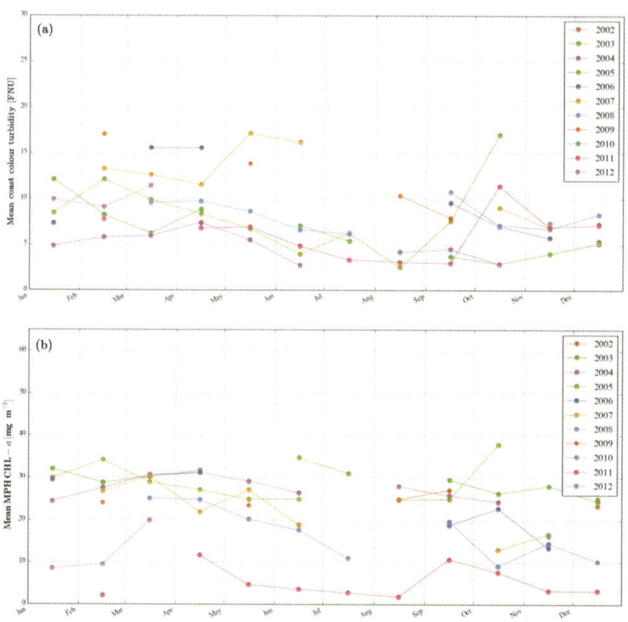

Figure 14. Time series of spatially averaged turbidity **(a)** and chl **(b)** in Lake Nicaragua, with data gaps especially during the rainy season (June–October).

the Lake Biwa Environmental Research Institute's regular limnological survey programme suggest that vertical mixing remained very weak in the winter of 2006/07 (Kawanabe et al., 2012). Even though relating surface to bottom temperatures is not without caveats, significantly higher LSWT with spatially averaged 8.2 °C is observed in March 2007 than in the other years (Fig. 13), suggesting that minimal annual LSWT could be a valuable proxy for vertical mixing in temperate lakes.

5.3 Lake Nicaragua

Lake Nicaragua/Cocibolca is the largest lake in central America, with a surface area of 7851 km^2. It is polymictic, with a maximum and average depth of only 26 and 15 m, respectively. It is subject to prevalent ecological issues such as untreated urban wastewater discharge and immissions from agriculture (soil erosion, fertilizer and pesticide immissions) and aquacultures that introduce non-resident Tilapia species and possibly novel diseases. Moreover, the planned construction of the Nicaragua Canal connecting the Caribbean Sea to the Pacific Ocean would bring about a significant shift in the lake's ecological status, most directly through the excavation of a 27.6 m deep, 520 m wide and 286 km long waterway across the centre of the shallow lake, which will strongly affect light availability within the water (Meyer and Huete-Pérez, 2014). In spite of the limited availability of MERIS FR data over Latin America (Fig. 1), it can contribute to estimating baseline conditions prior to the intervention. Generally maximum and minimum turbidity occur around August and February, respectively (Fig. 14a), within a range between 2 and 20 FNU. Outliers such as in October 2005 and May 2007 can occur when only a small area of the lake is sampled. However, the 2011 turbidity peak in October is related to an extraordinary shift from cyanobacteria to eucaryotic algae (Fig. 15), which comes with significantly lower chl concentrations throughout the year from both the MPH (Fig. 14a) and the FUB algorithm (not shown), but also a second productivity peak. Even though data continuity and in situ measurements are required for further interpretation, the available data suggest that the lake was in a relatively unstable state at the end of the observation period.

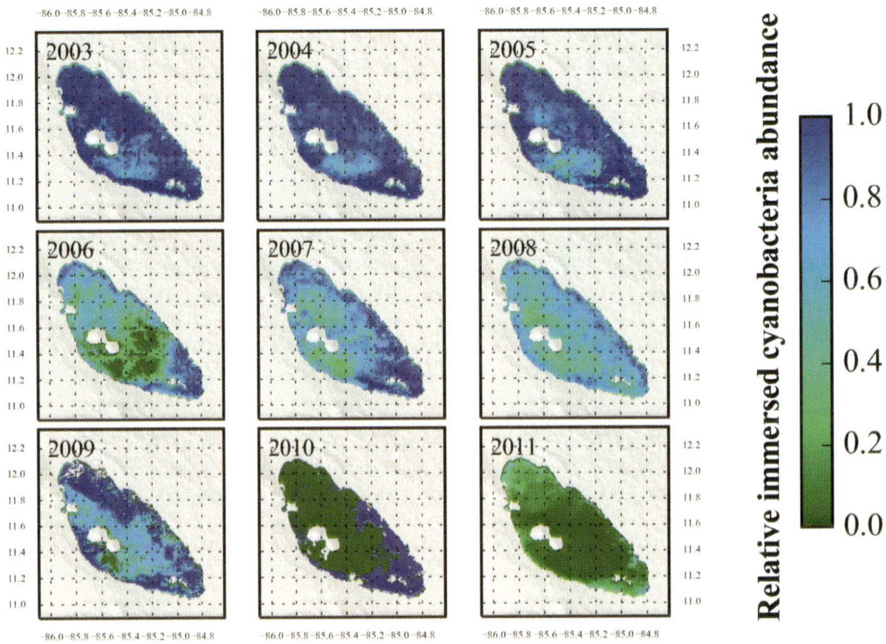

Figure 15. Annual relative cyanobacteria abundance in Lake Nicaragua, 2003–2011 (*immersed_cyanobacteria_mean*). Note that the number and distribution of valid observations across the years are quite unequal (Fig. 16).

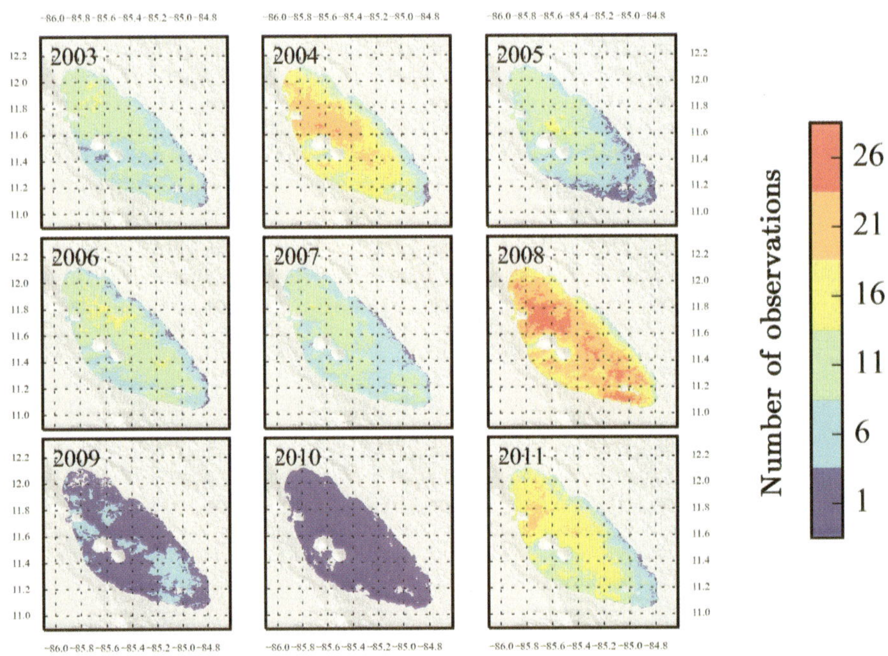

Figure 16. Number of observations for the annual cyanobacteria abundance maps in Fig. 15 (*num_obs*).

5.4 Lake Victoria

Lake Victoria is the second largest freshwater lake in the world and is situated in a shallow depression between the Great Rift Valley and the western Albertine Rift, with a shoreline shared by Kenya, Uganda and Tanzania. It is up to 83 m deep, eutrophic and light-limited (Hecky et al., 2010), and its thermocline is usually at around 30–40 m, with complete mixing occurring once a year (MacIntyre et al., 2014; Payne, 1986). About 80 % of the water input to Lake Victoria is from direct rainfall (Swenson and Wahr, 2009), and atmospheric deposition is the most important phosphorous source in pelagic areas of the main basin (Tamatamah et al., 2005).

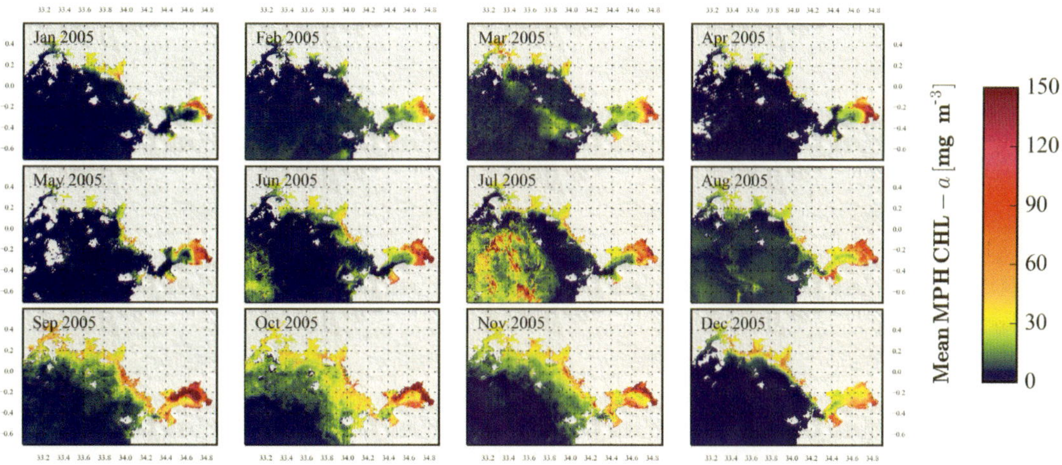

Figure 17. Monthly chl concentrations obtained with the Maximum Peak Height (MPH) algorithm in north-eastern Lake Victoria, 2005 (*chl_mph_mean*).

In contrast, the Nyanza Gulf (also known as Winam or Kavirondo Gulf), the lake's most distinctive morphological feature in the north-east, receives about 10 % of the lake's terrestrial inflow, and it was concluded from ground measurements between March 2005 and March 2006 that the Nyanza Gulf even received phosphorous input from the main basin, in contrast to the paradigm that the gulf is a major contributor to the lake's increasing nutrient enrichment (Gikuma-Njuru et al., 2013). As a matter of fact, MERIS observations confirm that the most productive areas are located in the very east of the gulf throughout 2005 (Fig. 17) and for the first half of 2006 (not shown). During this period, the chl levels in the lake's centre in July 2005 appear extraordinarily high. This observation can be verified through a comparison with the number of available observations (2–7) and the abundance of immersed cyanobacteria (0.4–1) in this area and month. This means that most parts of this cyanobacteria bloom were identified in several observations.

The Nyanza Gulf was also subject to intensive growth of water hyacinths (*Eichhorna crassipes*) in response to El Niño precipitation anomalies in 1998 (Albright et al., 2004) and 2007 (Fusilli et al., 2013). Figure 18 displays the 2007 proliferation event according to Diversity II data. The hyacinths appear in *floating_cyanobacteria_mean* rather than the expected *floating_vegetation_mean*, presumably due to persistent cyanobacteria dominance in the gulf. Given that Idepix is likely to mask completely overgrown pixels as land (see Sect. 4.1), the remaining water pixels counted for the abundance consist partly of immersed cyanobacteria, and partly of floating eucaryotes. Despite these limitations and the fact that monthly aggregates lack the spatial details of individual observations, the extent and course of the proliferation match well with the MODIS observations presented by Fusilli et al. (2013).

6 Data availability

MERIS FR data are the primary input data source for the creation of the Diversity II dataset, and are available from the European Space Agency. LSWT is provided by the ARC Lake project. Auxillary data were collected in the LakeNet and GLWD databases. We use maximum (SAR-WB) and instantaneous (SWBD) lake outlines from ASAR and SRTM data, respectively. In situ data for validation are provided by various public authorities (see Acknowledgements). All input data are available free of charge.

7 Conclusions

The Diversity II dataset is the first globally representative, temporally resolved and methodologically consistent information source for inland water quality dynamics from satellite Earth observations. It includes monthly, yearly and 9-year temporally aggregated geophysical maps of various water quality parameters, which provide unprecedented possibilities for exploitation at global and local scale. Global analyses are yet to be carried out, with caution as to the limitations mentioned hereafter. At local scale, several case studies demonstrated how the data could effectively contribute to traditional investigations of lake-specific processes and events. The Diversity II product user handbook (Odermatt et al., 2015b) helps to improve such interpretation of remotely sensed data by providing background knowledge of acquisition and retrieval methods, and Python scripts are available that facilitate standard information extraction and visualization from the individual GeoTIFF files.

The methods used for producing the Diversity II dataset are widely used and validated by several independent users beyond the end of the ENVISAT era. We complemented these studies with a relatively extensive performance assess-

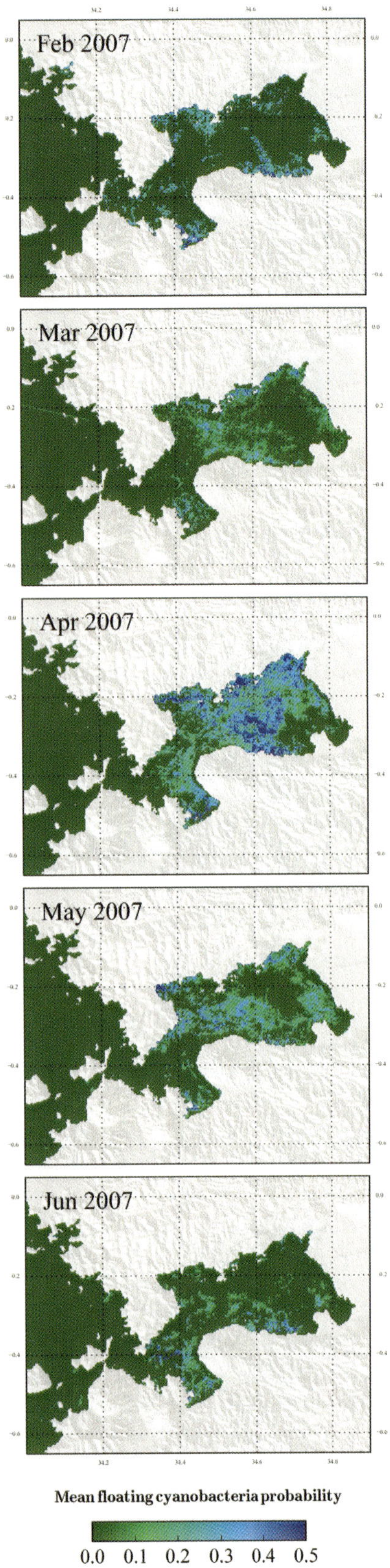

Figure 18. Peak of the 2007 water hyacinth growth in the Nyanza Gulf, Lake Victoria (*floating_cyanobacteria_mean*).

ment, which also refers to the complexity of comparisons between in situ measured and remotely sensed water quality parameters. For the performance indicators given here, R, MAE, bias and graphics, we repeatedly used the median concentration as a reference. We assume that the median represents typically a background concentration, which is several times lower than e.g. the magnitude of seasonal variations. It must also be remembered that the performance assessment refers to L2 observations, while our monthly product's temporal consistency is significantly enhanced where several observations are available in a month.

It is a major asset of Earth observation that productions from L1 observations can be repeated as improved methods become available, so there is no doubt that the methods in use for Diversity II will be improved and overhauled in the future. Several lessons were learned for such repeated productions, some of them involving challenges for future research. For example, the binary identification of optically shallow waters and floating lake ice has received relatively little attention in recent years, but under certain circumstances they have much larger effects on the final product accuracy than retrieval algorithms. Our method for the identification of clouds, land and mixed pixels is more advanced, but it remains critical for all partly cloudy situations. Therefore, the development of such methods should receive much more attention, relative to the number of retrieval algorithms that were developed in recent years. The latest generation of retrieval algorithms will be based on further advanced water types (Spyrakos et al., 2018) and ensemble approaches that account for the selection of multiple algorithms' estimates (as e.g. left to the user with *chl_mph_mean*/*chl_fub_mean*). Finally, new approaches are needed for the consolidation of such increasingly large geospatial datasets, and for the extraction of relevant information, which is still mostly based on lake-specific knowledge.

Author contributions. DO, PP and CB were in charge of overall direction and planning. All authors contributed to the product specifications. OD developed the processing framework and facilitated the production. DO wrote the manuscript with input from all the authors.

Competing interests. The authors declare that they have no conflict of interest.

Acknowledgements. This work was funded by the European Space Agency through DUE project Diversity II, contract number 4O00106622I 12 II-LG, and the Worldbank Global Solutions Group Water, Poverty and the Economy, contracts 7185491 and 7187375. We thank the ESA ARC-Lake project for providing the lake surface water temperature products, and Helga Ganz for her contributions to data analysis and visualization. For in situ reference measurements, we appreciated the support by Antonio

Ruiz-Verdù (data from the Spanish Centre for Hydrographic Studies – CEDEX), Caren Binding (data from Environment Canada), Ghislaine Monet (data from the International Commission for the Protection of Lake Geneva – CIPEL), Mark Matthews (data from the South African Department of Water Affairs, Resource Quality Information Services), Megumi Nakagawa (data from the Japanese National Institute for Environmental Studies – NIES), Oliver Köster (data from Zurich Water Supply – WVZ), Peter Keller (data from Zug's Office for Environmental Protection – AfU), Sampsa Koponen (data from the Finnish Environment Institute – SYKE), Steven Greb (data from the U.S. Environmental Protection Agency – EPA) and Tiit Kutser (data from the Estonian National Monitoring Programme). We thank the Vermont Department of Environmental Conservation for providing Lake Champlain data online and acknowledge the Swedish Agency for Marine and Water Management for data access for Swedish lakes.

Edited by: Falk Huettmann
Reviewed by: three anonymous referees

References

Albright, T., Moorhouse, T. G., and McNabb, T. J.: The Rise and Fall of Water Hyacinth in Lake Victoria and the Kagera River Basin, 1989-2001, J. Aquat. Plant Manage., 42, 73–84, 2004.

Alvera-Azcárate, A., Barth, A., Rixen, M., and Beckers, J. M.: Reconstruction of incomplete oceanographic data sets using empirical orthogonal functions: application to the Adriatic Sea surface temperature, Ocean Model., 9, 325–346, https://doi.org/10.1016/j.ocemod.2004.08.001, 2005.

Attila, J., Kauppila, P., Kallio, K. Y., Alasalmi, H., Keto, V., Bruun, E., and Koponen, S.: Applicability of Earth Observation chlorophyll-*a* data in assessment of water status via MERIS – With implications for the use of OLCI sensors, Remote Sens. Environ., 212, 273–287, https://doi.org/10.1016/j.rse.2018.02.043, 2018.

Beadle, L. C.: The inland waters of tropical Africa. An introduction to tropical limnology, Longman Group Limited, London, England, 1981.

Block, T.: diversity-mph-chl Operator, Brockmann Consult, Geesthacht, Germany, available at: https://github.com/bcdev/beam-diversity-auxdata/tree/master/diversity-mph-chl (last access: 25 July 2018), 2016.

Bourg, L. and Etanchaud, F.: The AMORGOS MERIS CFI (Accurate MERIS Ortho-Rectified Geolocation Operational Software), Software User Manual and Interface Control Document, ACRI-ST, Sophia-Antipolis, France, available at: earth.esa.int/services/amorgos/download/Amorgos_ICD-SUM_3.0a.pdf (last access: 25 July 2018), 2007.

Cannizzaro, J. P. and Carder, K.: Estimating chlorophyll *a* concentrations from remote-sensing reflectance in optically shallow waters, Remote Sens. Environ., 101, 13–24, 2006.

Crétaux, J.-F., Jelinski, W., Calmant, S., Kouraev, A., Vuglinski, V., Bergé-Nguyen, M., Gennero, M.-C., Nino, F., Rio, R. A. D., Cazenave, A., and Maisongrande, P.: SOLS: A lake database to monitor in the Near Real Time water level and storage variations from remote sensing data, Adv. Space Res., 47, 1497–1507, https://doi.org/10.1016/j.asr.2011.01.004, 2011.

Danne, O.: beam-idepix Operator, Brockmann Consult, Geesthacht, Germany, available at: https://github.com/bcdev/beam-idepix (last access: 25 July 2018), 2016.

Doerffer, R.: Alternative Atmospheric Correction Procedure for Case 2 Water Remote Sensing using MERIS, HZG, 27 pp., available at: https://earth.esa.int/documents/700255/2042855/MERIS_ATBD_2.25_v1.0+-+2011.pdf (last access: 25 July 2018), 2011.

Doerffer, R. and Schiller, H.: The MERIS Case 2 water algorithm, Int. J. Remote Sens., 28, 517–535, https://doi.org/10.1080/01431160600821127, 2007.

Eleveld, M. A., Ruescas, A. B., Hommersom, A., Moore, T. S., Peters, S. W. M., and Brockmann, C.: An Optical Classification Tool for Global Lake Waters, Remote Sens., 9, 420,https://doi.org/10.3390/rs9050420, 2017.

Fell, F. and Fischer, J.: Numerical simulation of the light field in the atmosphere–ocean system using the matrix-operator method, J. Quant. Spectrosc. Ra., 69, 351–388, https://doi.org/10.1016/S0022-4073(00)00089-3, 2001.

Fomferra, N. and Brockmann, C.: BEAM – The ENVISAT MERIS and AATSR Toolbox, in: MERIS (A)ATSR Workshop 2005, 597, p. 13.1, 2005.

Fomferra, N., Böttcher, M., Zühlke, M., Brockmann, C., and Kwiatkowska, E.: Calvalus: Full-mission EO cal/val, processing and exploitation services, in: 2012 IEEE International Geoscience and Remote Sensing Symposium, Munich, Germany, 22–27 July 2012, 5278–5281, https://doi.org/10.1109/IGARSS.2012.6352418, 2012.

Fusilli, L., Collins, M. O., Laneve, G., Palombo, A., Pignatti, S., and Santini, F.: Assessment of the abnormal growth of floating macrophytes in Winam Gulf (Kenya) by using MODIS imagery time series, Int. J. Appl. Earth Obs., 20, 33–41, https://doi.org/10.1016/j.jag.2011.09.002, 2013.

Gikuma-Njuru, P., Hecky, R. E., Guildford, S. J., and MacIntyre, S.: Spatial variability of nutrient concentrations, fluxes, and ecosystem metabolism in Nyanza Gulf and Rusinga Channel, Lake Victoria (East Africa), Limnol. Oceanogr., 58, 774–789, https://doi.org/10.4319/lo.2013.58.3.0774, 2013.

Gitelson, A.: The peak near 700 nm on radiance spectra of algae and water: relationships of its magnitude and position with chlorophyll concentration, Int. J. Remote Sens., 13, 3367–3373, 1992.

Gleick, P. H.: Water Resources, in: Encyclopaedia of Climate and Weather, edited by: Schneider, S. H., Oxford University Press, New York, United States, 2, 817–823, 1996.

Gons, H. J.: Optical Teledetection of Chlorophyll *a* in Turbid Inland Waters, Environ. Sci. Technol., 33, 1127–1132, https://doi.org/10.1021/es9809657, 1999.

Gower, J. F. R., Doerffer, R., and Borstad, G. A.: Interpretation of the 685 nm peak in water-leaving radiance spectra in terms of fluorescence, absorption and scattering, and its observation by MERIS, Int. J. Remote Sens., 20, 1771–1786, https://doi.org/10.1080/014311699212470, 1999.

Hecky, R. E., Mugidde, R., Ramlal, P. S., Talbot, M. R., and Kling, G. W.: Multiple stressors cause rapid ecosystem change in Lake Victoria, Freshwater Biol., 55, 19–42, https://doi.org/10.1111/j.1365-2427.2009.02374.x, 2010.

Huot, Y. and Babin, M.: Overview of Fluorescence Protocols: Theory, Basic Concepts, and Practice, in: Chlorophyll a Fluorescence in Aquatic Sciences: Methods and Applications, edited by: Suggett, D. J., Prášil, O., and Borowitzka, M. A., Springer Netherlands, Dordrecht, 31–74, 2010.

ISO 7027-1: Water quality – Determination of turbidity – Part 1: Quantitative methods, available at: https://www.iso.org/standard/62801.html (last access: 25 July 2018), 2016.

Itai, T., Hayase, D., Hyobu, Y., Hirata, S. H., Kumagai, M., and Tanabe, S.: Hypoxia-Induced Exposure of Isaza Fish to Manganese and Arsenic at the Bottom of Lake Biwa, Japan: Experimental and Geochemical Verification, Environ. Sci. Technol., 46, 5789–5797, https://doi.org/10.1021/es300376y, 2012.

Kawanabe, H., Nishino, M., and Maehata, M.: Lake Biwa: Interactions between Nature and People, Springer, Dordrecht, Netherlands, 2012.

Kiefer, I., Odermatt, D., Anneville, O., Wüest, A., and Bouffard, D.: Application of remote sensing for the optimization of in-situ sampling for monitoring of phytoplankton abundance in a large lake, Sci. Total Environ., 527–528, 493–506, https://doi.org/10.1016/j.scitotenv.2015.05.011, 2015.

Kirches, G., Krueger, O., Boettcher, M., Bontemps, S., Lamarche, C., Verheggen, A., Lembrée, C., Radoux, J., and Defourny, P.: Algorithm Theoretical Basis Document v2.3, ESA CCI Project LandCover, available at: https://www.esa-landcover-cci.org/?q=documents# (last access: 25 July 2018), 2013.

Kutser, T., Paavel, B., Verpoorter, C., Ligi, M., Soomets, T., Toming, K., and Casal, G.: Remote Sensing of Black Lakes and Using 810 nm Reflectance Peak for Retrieving Water Quality Parameters of Optically Complex Waters, Remote Sens., 8, 497, https://doi.org/10.3390/rs8060497, 2016.

Leach, T. H., Beisner, B. E., Carey, C. C., Pernica, P., Rose, K. C., Huot, Y., Brentrup, J., Domaizon, I., Grossart, H.-P., Ibelings, B. W., Jacquet, S., Kelly, P. T., Rusak, J. A., Stockwell, J. D., Straile, D., and Verburg, P.: Patterns and drivers of deep chlorophyll maxima structure in 100 lakes: The relative importance of light and thermal stratification, Limnol. Oceanogr., 63, 628–646, https://doi.org/10.1002/lno.10656, 2017.

Lehner, B. and Döll, P.: Development and validation of a global database of lakes, reservoirs and wetlands, J. Hydrol., 296, 1–22, 2004.

MacCallum, S. and Merchant, C.: ATSR Reprocessing for Climate Lake Surface Temperature: ARC-Lake Algorithm Theoretical Basis Document, Univ. of Edinburgh, available at: http://www.laketemp.net/home/ARC-Lake-ATBD-v1.4.pdf (last access: 25 July 2018), 2013.

MacCallum, S. and Merchant, C.: ARC-Lake Data Product Description, University of Edinburgh, Edinburgh, Scotland, available at: http://www.laketemp.net/home/ARCLake_DPD_v3.pdf (last access: 25 July 2018), 2014.

MacCallum, S. N. and Merchant, C. J.: Surface water temperature observations of large lakes by optimal estimation, Can. J. Remote Sens., 38, 25–45, https://doi.org/10.5589/m12-010, 2012.

MacIntyre, S., Romero, J. R., Silsbe, G. M., and Emery, B. M.: Stratification and horizontal exchange in Lake Victoria, East Africa, Limnol. Oceanogr., 59, 1805–1838, https://doi.org/10.4319/lo.2014.59.6.1805, 2014.

Matthews, M. W.: A current review of empirical procedures of remote sensing in inland and near-coastal transitional waters, Int. J. Remote Sens., 32, 6855–6899, https://doi.org/10.1080/01431161.2010.512947, 2011.

Matthews, M. W. and Odermatt, D.: Improved algorithm for routine monitoring of cyanobacteria and eutrophication in inland and near-coastal waters, Remote Sens. Environ., 156, 374–382, https://doi.org/10.1016/j.rse.2014.10.010, 2015.

Matthews, M. W., Bernard, S., and Robertson, L.: An algorithm for detecting trophic status (chlorophyll-a), cyanobacterial-dominance, surface scums and floating vegetation in inland and coastal waters, Remote Sens. Environ., 124, 637–652, https://doi.org/10.1016/j.rse.2012.05.032, 2012.

McArdle, B. H.: The structural relationship: regression in biology, Can. J. Zool., 66, 2329–2339, https://doi.org/10.1139/z88-348, 1988.

Méndez-Tejeda, R., Rosado, G., Rivas, D. V., Montilla, T., Hernández, S., Ortiz, A., and Santos, F.: Climate Variability and Its Effects on the Increased Level of Lake Enriquillo in the Dominican Republic, Science and Education, 4, 26–36, 2016.

Meyer, A. and Huete-Pérez, J. A.: Nicaragua Canal could wreak environmental ruin, Nature, 506, 287–289, 2014.

Mironov, D.: Parameterization of lakes in numerical weather prediction. Description of a lake model, COSMO Technical Report, No. 11, Deutscher Wetterdienst, Offenbach am Main, Germany, available at: http://www.cosmo-model.org/content/model/documentation/techReports/docs/techReport11.pdf (last access: 25 July 2018), 2008.

Mobley, C. D., Stramski, D., Bisset, W. P., and Boss, E.: Optical modeling of ocean water: Is the case 1–case 2 classification still useful?, Oceanography, 17, 60–67, 2004.

Moore, T. S., Dowell, M. D., Bradt, S., and Ruiz Verdu, A.: An optical water type framework for selecting and blending retrievals from bio-optical algorithms in lakes and coastal waters, Remote Sens. Environ., 143, 97–111, https://doi.org/10.1016/j.rse.2013.11.021, 2014.

Morel, A. and Prieur, L.: Analysis of Variations in Ocean Color, Limnol. Oceanogr., 22, 709–722, 1977.

Nechad, B., Ruddick, K., Schroeder, T., Oubelkheir, K., Blondeau-Patissier, D., Cherukuru, N., Brando, V., Dekker, A., Clementson, L., Banks, A. C., Maritorena, S., Werdell, P. J., Sá, C., Brotas, V., Caballero de Frutos, I., Ahn, Y.-H., Salama, S., Tilstone, G., Martinez-Vicente, V., Foley, D., McKibben, M., Nahorniak, J., Peterson, T., Siliò-Calzada, A., Röttgers, R., Lee, Z., Peters, M., and Brockmann, C.: CoastColour Round Robin data sets: a database to evaluate the performance of algorithms for the retrieval of water quality parameters in coastal waters, Earth Syst. Sci. Data, 7, 319–348, https://doi.org/10.5194/essd-7-319-2015, 2015.

Neukermans, G., Ruddick, K., Loisel, H., and Roose, P.: Optimization and quality control of suspended particulate matter concentration measurement using turbidity measurements, Limnol. Oceanogr.-Meth., 10, 1011–1023, https://doi.org/10.4319/lom.2012.10.1011, 2012.

Odermatt, D., Gitelson, A., Brando, V. E., and Schaepman, M. E.: Review of constituent retrieval in optically deep and complex waters from satellite imagery, Remote Sens. Environ., 118, 116–126, https://doi.org/10.1016/j.rse.2011.11.013, 2012.

Odermatt, D., Gangkofner, U., Ratzmann, G., Ruescas, A. B., Stelzer, K., Philipson, P., and Brockmann, C.: Algorithm Theoretic Baseline Document v2.4, ESA DUE Project Diversity II, available at: www.diversity2.info/products/documents/ (last access: 25 July 2018), 2015a.

Odermatt, D., Brito, J., Philipson, P., Hahn, N., and Brockmann, C.: Products User Handbook Inland Waters v2.1, ESA DUE Project Diversity II, available at: http://www.diversity2.info/products/documents/ (last access: 25 July 2018), 2015b.

Ohte, N., Tayasu, I., Kohzu, A., Yoshimizu, C., Osaka, K., Makabe, A., Koba, K., Yoshida, N., and Nagata, T.: Spatial distribution of nitrate sources of rivers in the Lake Biwa watershed, Japan: Controlling factors revealed by nitrogen and oxygen isotope values, Water Resour. Res., 46, W07505, https://doi.org/10.1029/2009WR007871, 2010.

Palmer, S. C. J., Odermatt, D., Hunter, P. D., Brockmann, C., Présing, M., Balzter, H., and Tóth, V. R.: Satellite remote sensing of phytoplankton phenology in Lake Balaton using 10 years of MERIS observations, Remote Sens. Environ., 158, 441–452, https://doi.org/10.1016/j.rse.2014.11.021, 2015.

Payne, A. I.: The Ecology of Tropical Lakes and Rivers, John Wiley & Sons Ltd., New York, USA, 1986.

Peters, M.: beam-owt-classification Operator, Brockmann Consult, Geesthacht, Germany, available at: https://github.com/bcdev/coastcolour/tree/master/beam-owt-classification (last access: 25 July 2018), 2016.

Pfannkuche, J. and Schmidt, A.: Determination of suspended particulate matter concentration from turbidity measurements: particle size effects and calibration procedures, Hydrol. Process., 17, 1951–1963, https://doi.org/10.1002/hyp.1220, 2003.

Politi, E., MacCallum, S., Cutler, M. E. J., Merchant, C. J., Rowan, J. S., and Dawson, T. P.: Selection of a network of large lakes and reservoirs suitable for global environmental change analysis using Earth Observation, Int. J. Remote Sens., 37, 3042–3060, https://doi.org/10.1080/01431161.2016.1192702, 2016.

Rast, M., Bezy, J. L., and Bruzzi, S.: The ESA Medium Resolution Imaging Spectrometer MERIS a review of the instrument and its mission, Int. J. Remote Sens., 20, 1681–1702, https://doi.org/10.1080/014311699212416, 1999.

Ruescas, A. B., Brockmann, C., Stelzer, K., Tilstone, G. H., and Beltran, J.: Validation Report v1, ESA DUE Project CoastColour, Geesthacht, Germany, available at: http://www.coastcolour.org/documents/DEL-27ValidationReport_v1.pdf (last access: 25 July 2018), 2014.

Ruiz-Verdú, A., Simis, S. G. H., de Hoyos, C., Gons, H. J., and Peña-Martínez, R.: An evaluation of algorithms for the remote sensing of cyanobacterial biomass, Remote Sens. Environ., 112, 3996–4008, https://doi.org/10.1016/j.rse.2007.11.019, 2008.

Sandström, A., Philipson, P., Asp, A., Axenrot, T., Kinnerbäck, A., Ragnarsson-Stabo, H., and Holmgren, K.: Assessing the potential of remote sensing-derived water quality data to explain variations in fish assemblages and to support fish status assessments in large lakes, Hydrobiologia, 780, 71–84, https://doi.org/10.1007/s10750-016-2784-9, 2016.

Santer, R., Carrere, V., Dubuisson, P., and Roger, J. C.: Atmospheric correction over land for MERIS, Int. J. Remote Sens., 20, 1819–1840, https://doi.org/10.1080/014311699212506, 1999.

Santoro, M. and Wegmüller, U.: Multi-temporal Synthetic Aperture Radar Metrics Applied to Map Open Water Bodies, IEEE J. Sel. Top. Appl., 7, 3225–3238, https://doi.org/10.1109/JSTARS.2013.2289301, 2014.

Sayers, M. J., Grimm, A. G., Shuchman, R. A., Deines, A. M., Bunnell, D. B., Raymer, Z. B., Rogers, M. W., Woelmer, W., Bennion, D. H., Brooks, C. N., Whitley, M. A., Warner, D. M., and Mychek-Londer, J.: A new method to generate a high-resolution global distribution map of lake chlorophyll, Int. J. Remote Sens., 36, 1942–1964, https://doi.org/10.1080/01431161.2015.1029099, 2015.

Schroeder, T., Schaale, M., and Fischer, J.: Retrieval of atmospheric and oceanic properties from MERIS measurements: A new Case-2 water processor for BEAM, Int. J. Remote Sens., 28, 5627–5632, 2007.

Secretariat of the Convention on Biological Diversity: Global Biodiversity Outlook 3, Montréal, Canada, available at: www.cbd.int/gbo3/ (last access: 25 July 2018), 2010.

Seegers, B. N., Stumpf, R. P., Schaeffer, B. A., Loftin, K. A., and Werdell, P. J.: Performance metrics for the assessment of satellite data products: an ocean color case study, Opt. Express, 26, 7404–7422, 2018.

Simis, S., Ruizverdu, A., Dominguezgomez, J., Penamartinez, R., Peters, S., and Gons, H.: Influence of phytoplankton pigment composition on remote sensing of cyanobacterial biomass, Remote Sens. Environ., 106, 414–427, https://doi.org/10.1016/j.rse.2006.09.008, 2007.

Slater, J. A., Garvey, G., Johnston, C., Haase, J., Heady, B., Kroenung, G., and Little, J.: The SRTM Data "Finishing" Process and Products, Photogramm. Eng. Rem. S., 72, 237–247, https://doi.org/10.14358/PERS.72.3.237, 2006.

Spyrakos, E., O'Donnell, R., Hunter, P. D., Miller, C., Scott, M., Simis, S. G. H., Neil, C., Barbosa, C. C. F., Binding, C. E., Bradt, S., Bresciani, M., Dall'Olmo, G., Giardino, C., Gitelson, A. A., Kutser, T., Li, L., Matsushita, B., Martinez-Vicente, V., Matthews, M. W., Ogashawara, I., Ruiz-Verdú, A., Schalles, J. F., Tebbs, E., Zhang, Y., and Tyler, A. N.: Optical types of inland and coastal waters, Limnol. Oceanogr., 63, 846–870, https://doi.org/10.1002/lno.10674, 2018.

Swenson, S. and Wahr, J.: Monitoring the water balance of Lake Victoria, East Africa, from space, J. Hydrol., 370, 163–176, https://doi.org/10.1016/j.jhydrol.2009.03.008, 2009.

Tamatamah, R. A., Hecky, R. E., and Duthie, H.: The atmospheric deposition of phosphorus in Lake Victoria (East Africa), Biogeochemistry, 73, 325–344, https://doi.org/10.1007/s10533-004-0196-9, 2005.

Verpoorter, C., Kutser, T., Seekell, D. A., and Tranvik, L. J.: A global inventory of lakes based on high-resolution satellite imagery, Geophys. Res. Lett., 41, 6396–6402, https://doi.org/10.1002/2014GL060641, 2014.

Zaneveld, J. R. V., Barnard, A. H., and Boss, E.: Theoretical derivation of the depth average of remotely sensed optical parameters, Opt. Express, 13, 9052–9061, 2005.

3

Global and regional phosphorus budgets in agricultural systems and their implications for phosphorus-use efficiency

Fei Lun[1], Junguo Liu[2], Philippe Ciais[3], Thomas Nesme[4], Jinfeng Chang[3], Rong Wang[3], Daniel Goll[3], Jordi Sardans[5], Josep Peñuelas[6], and Michael Obersteiner[7]

[1]College of Resources and Environmental Sciences, China Agricultural University, Beijing 100193, China

[2]School of Environmental Science and Engineering, South University of Science and Technology of China, Shenzhen 518055, China

[3]IPSL – LSCE, CEA CNRS UVSQ, Centre d'Etudes Orme des Merisiers, 91191 Gif-sur-Yvette, France

[4]Bordeaux Sciences Agro, Université de Bordeaux, UMR 1391 ISPA, CS 40201, 33175 Gradignan CEDEX, France

[5]CREAF, Cerdanyola del Vallès 08193, Catalonia, Spain

[6]CSIC, Global Ecology Unit CREAF-CSIC-UAB, Cerdanyola del Vallès 08193, Catalonia, Spain

[7]International Institute for Applied Systems Analysis, 2361 Laxenburg, Austria

Correspondence: Junguo Liu (liujg@sustc.edu.cn, junguo.liu@gmail.com)

Abstract. The application of phosphorus (P) fertilizer to agricultural soils increased by 3.2 % annually from 2002 to 2010. We quantified in detail the P inputs and outputs of cropland and pasture and the P fluxes through human and livestock consumers of agricultural products on global, regional, and national scales from 2002 to 2010. Globally, half of the total P inputs into agricultural systems accumulated in agricultural soils during this period, with the rest lost to bodies of water through complex flows. Global P accumulation in agricultural soil increased from 2002 to 2010 despite decreases in 2008 and 2009, and the P accumulation occurred primarily in cropland. Despite the global increase in soil P, 32 % of the world's cropland and 43 % of the pasture had soil P deficits. Increasing soil P deficits were found for African cropland vs. increasing P accumulation in eastern Asia. European and North American pasture had a soil P deficit because the continuous removal of biomass P by grazing exceeded P inputs. International trade played a significant role in P redistribution among countries through the flows of P in fertilizer and food among countries. Based on country-scale budgets and trends we propose policy options to potentially mitigate regional P imbalances in agricultural soils, particularly by optimizing the use of phosphate fertilizer and the recycling of waste P. The trend of the increasing consumption of livestock products will require more P inputs to the agricultural system, implying a low P-use efficiency and aggravating P-stock scarcity in the future. The global and regional phosphorus budgets and their PUEs in agricultural systems are publicly available at https://doi.pangaea.de/10.1594/PANGAEA.875296.

1 Introduction

Population increases and dietary changes require higher food production, which increases global demand for fertilizers (Grote et al., 2005; Foley et al., 2011). Phosphorus (P) is an essential element for all organisms, and a lack of P limits growth. Fertilizer P enhances agricultural production, but P is also fixed in soils and can accumulate. In countries with high fertilizer use, much P is lost to leaching and run-off, leading to the eutrophication of both inland and coastal waters (Carpenter et al., 1998; MacDonald et al., 2011).

To supply the growing need for P in fertilizer, the mining of phosphate rock has quadrupled in the past half century, increasing from 46 Mt in 1961 to 198 Mt in 2011 (Scholz et al., 2013). Despite some short-term fluctuations in the price of phosphate rock, the global production of fertilizer P has been steadily increasing at a rate of 3 to 4 % annually during the half century before 2011 and is projected to increase by 50 to 100 % by 2050 (Cordell et al., 2009, 2012). Extractable phosphate rock is a non-renewable resource, and significant depletion of the resource is projected by the end of this century if the current intensive use continues, possibly leading to resource shortages (Cordell et al., 2009; van Vuuren et al., 2010; Peñuelas et al., 2013).

The mining of P and its application as fertilizer in cultivated land is a major anthropogenic perturbation of the natural biogeochemical P cycle (Carpenter and Bennett, 2011; Elser and Bennett, 2011; Steffen et al., 2015). The negative impacts of this perturbation on the natural environment depend on how much P is lost from regions with intensive fertilizer use (Smil, 2000; Bennett et al., 2001).

P application differs significantly between countries and crop types (Grote et al., 2005), and previous researchers have attempted to estimate the P flows in agricultural systems in Europe (Ott & Rechberger, 2012), the United States (Suh & Yee, 2011), China (Ma et al., 2011), France (Senthilkumar et al., 2012), Australia (Cordell et al., 2013), and the world (Smil, 2000; Liu et al., 2008; MacDonald et al., 2011; Schipanski & Bennett, 2012). International trade and regional agricultural policies affect P budgets by increasing or decreasing the gap between P inputs and P outputs in agricultural land (Grote et al., 2005). Previous research mainly focused on cropland, and P fluxes in pasture and livestock production systems have also received more and more attention recently, especially due to diet change. The differences in methodologies, system boundaries, and data sources have made it difficult to assess the differences in the phosphorus-use efficiencies (PUEs) among agricultural sectors and to extrapolate regional findings to the global scale.

To mitigate these problems, we (1) compiled a detailed and harmonized dataset of P fluxes in agriculture for countries around the world, including detailed analysis of input and output fluxes for cropland, managed grassland (hereafter, pasture), livestock, and human consumers of agricultural products; (2) characterized P budgets and P-use efficiencies in those different subsystems; and (3) examined how the international trade of phosphate fertilizer and agricultural commodities influences regional P fluxes. We performed this analysis on the scale of countries, regions, and the world; wherever possible, we distinguished different crop types. The study period was from 2002 to 2010, allowing us to study temporal trends.

2 Materials and methods

In this study, we obtained data for 224 countries (Table S1 in the Supplement). We defined the agriculture system as cropland and pasture ecosystems plus human and livestock consumers of agricultural production and of other products containing P (Fig. 1). External P inputs to the agriculture system came from mined phosphate rock and atmospheric deposition. Several processes cause P losses from the system into the external environment (here, defined as non-agricultural land and bodies of water). Figure 1 presents the fluxes of P into and out of the agriculture system on a global scale, including internal fluxes between ecosystems and consumers. We quantified these fluxes in the present study based on a mass-balance approach (Cordell et al., 2012). We defined the phosphorus-use efficiency (PUE) of the agricultural system and of its subsystems as the ratio of the total P harvested in economic outputs (e.g. crops, meat, milk, and eggs) to the total P input. International trade in fertilizer and food is discussed separately in Sect. 2.3. The data sources and an overview of the mass-balance equations are presented in the rest of this section; details and equations are presented in the Supplement.

2.1 P flows into and out of the agricultural system

Inputs into the agricultural system, which are within the grey box in Fig. 1, are from mined phosphate rocks and atmospheric deposition. We did not include P from the in situ weathering of soil particles because the rate of this process is insignificant compared with the magnitude of other inputs (Liu et al., 2008). Outputs included P emission into the atmosphere from fires and P loss to uncultivated land or bodies of water.

2.1.1 P inputs

Data on agricultural inputs of phosphate P in fertilizers were collected from the International Fertilizer Industry Association (http://www.fertilizer.org) and divided between cropland and pasture uses based on information from FAO (2002) and the FAOSTAT database (http://www.fao.org/faostat/en/#data). A small fraction (8 %) of P from mined phosphate rock is used to produce animal feed additives. Apart from fertilizer and animal feed additives, the rest of the mined P is used to produce detergents and other products directly consumed by humans (Ringeval et al., 2014). Atmospheric

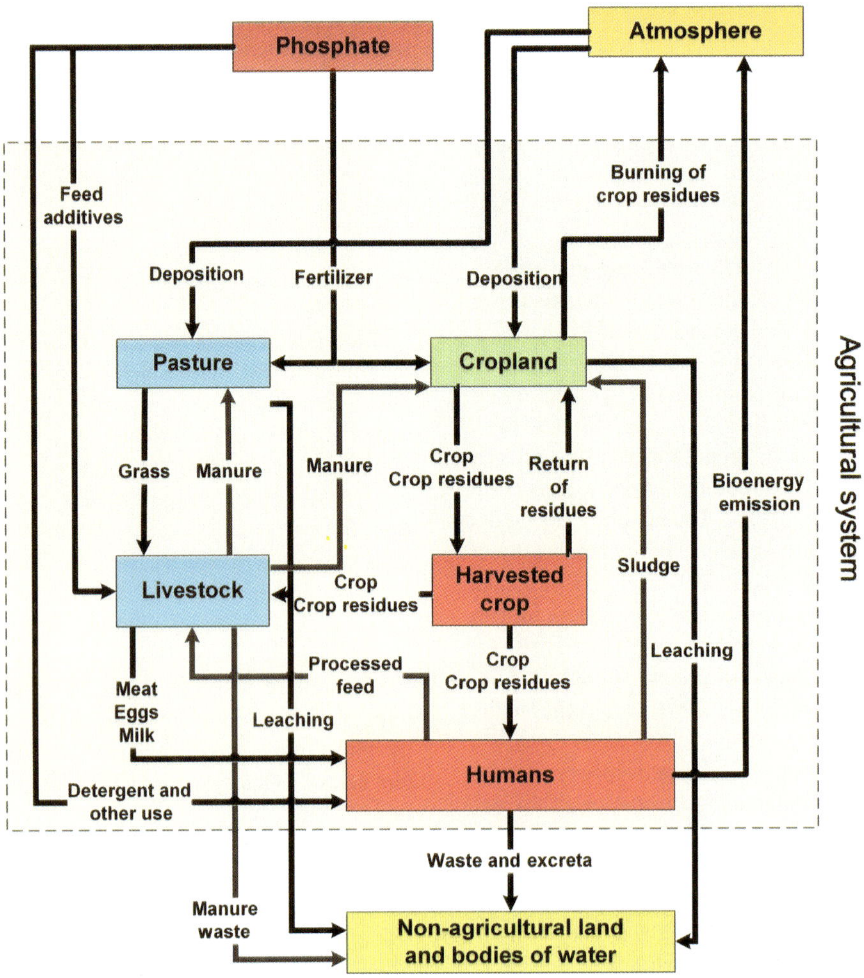

Figure 1. Scheme of the P pools and fluxes used to diagnose global P budgets for the agricultural sector. The agricultural sector (or system) in the grey box includes cropland and pasture soils, livestock, human consumers of livestock and crop products, and users of phosphate-derived products. National and regional P budgets are calculated using the same scheme, but including in addition exports and imports of P embedded in traded crop and livestock products and fertilizers.

P deposition in cropland and pasture areas was calculated separately in each country using gridded global P-deposition maps obtained using the LMDz-INCA aerosol chemistry transport model of Wang et al. (2014, 2015) and agricultural land-use maps. Details are provided in the Supplement (Table S2).

2.1.2 P outputs

P emissions from agricultural fires were obtained from the gridded dataset of Wang et al. (2015) and cover the burning of crop residues in the field, by households, and for the production of bioenergy from crop biomass. Leaching from cropland and pasture soils was assumed to be a constant fraction (12.5 %) of P inputs for each agricultural land-use type (Bouwman et al., 2013). P outputs from non-recycled livestock and human manure were calculated based on the mass balance. Note that erosion-induced losses of P are important in many agricultural regions (Quinton et al., 2010), but were not considered in this study because we lack data on the re-deposition of P in eroded soil material from agricultural soils. In future studies, it will be important to quantify this source of P, particularly in agricultural areas that receive large annual inputs of sediment (e.g. in river floodplains and sites on steep terrain that experience significant erosion farther up the slope followed by deposition).

2.2 P flows within the agricultural system

2.2.1 P in harvested crop biomass and crop residues

The flux of P in harvested crop biomass was estimated from yield data (FAOSTAT) using crop-specific P concentrations after grouping 178 different crops into 13 crop types (COMIFER, 2007; USDA-NRCS, 2009; Waller, 2010;

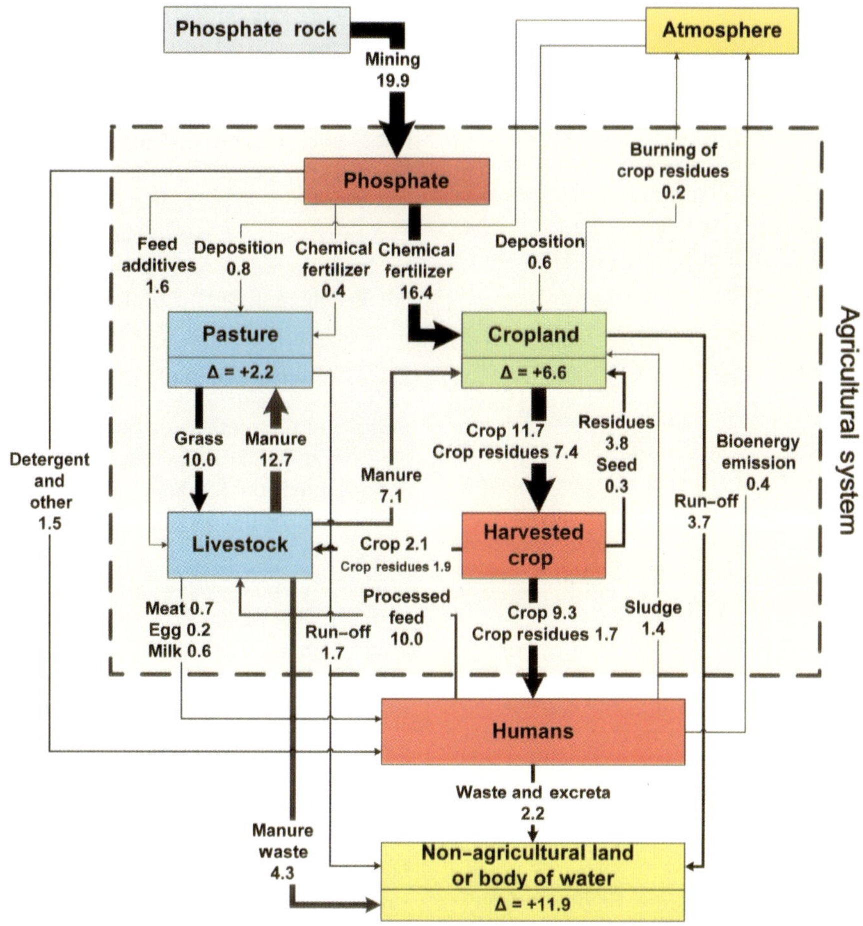

Figure 2. Annual P flows in the global agriculture system from 2002 to 2010. Values are Tg P yr^{-1}. The notation Δ denotes the average change of P in pasture and cropland soils, respectively. By convention, a positive value means accumulation. Note that livestock and human changes of P are assumed to be zero.

Table S2). P in harvested crop biomass was partitioned into crops (for human and livestock consumption) and crop residues (Fig. 1). We estimated the P fluxes of crop residues from FAOSTAT data and from Liu et al. (2008) to account for residue that is recycled in the field (50 %), transformed into livestock feed (25 %), and burned or used by other human activities (25 %).

2.2.2 P in grazed biomass

The P removed from pasture by livestock grazing was estimated by combining forage grass consumption data with the P concentrations in grass biomass (Antikainen et al., 2005; COMIFER, 2007; USDA-NRCS, 2009; Waller, 2010). Gridded data on grass biomass consumption by livestock were obtained by combining the global livestock production systems dataset of Herrero et al. (2013) with pasture net primary productivity simulated by the ORCHIDEE-GM global pasture model (Chang et al., 2013, 2015). We chose the ORCHIDEE-GM model for this analysis because it is able to separate the intake of grazed vs. cut forage grass.

2.2.3 P in animal feed products

Animal feed products used as complementary diet ("feed additives") represent direct inputs to the livestock subsystem (Fig. 1). This flux was deduced from the mass balance of the known input and output fluxes for the livestock P pool, but did not account for long-term changes in P storage in that pool. See the Supplement for more details.

2.2.4 P embedded in livestock products

This flux of P leaving the livestock subsystem and entering the human subsystem (Fig. 1) through the harvesting of products was calculated by multiplying the FAOSTAT production data for meat, eggs, and milk by the product-specific P concentrations reported by Grote et al. (2005).

2.2.5 P in livestock manure

We calculated the manure P production based on FAOSTAT data about N in livestock manure and P : N values for each types of livestock manure (MWPS-18, 1985; OECD Secretariat, 1991; Levington Agriculture, 1997; Sheldrick et al., 2003; ASAE, 2005; see Table S3). Once produced, manure P is either applied to cropland, left in the pasture, or lost to the environment as waste (Fig. 1), following the same partitioning as that for N in the manure from FAOSTAT.

2.2.6 P in human sewage sludge

We assumed that the P output from humans equaled the inputs from non-fertilizer P-ore products and the consumed crop and livestock products (Fig. 1) and used this to calculate the total P production in human excreta. P in human sewage sludge was estimated using population data and values of per capita production of P in excreta (Smil, 2000; Cordell et al., 2009). Following the method of Liu et al. (2008), we assumed that 30 % of the excreta P from urban populations and 70 % of P from rural populations were returned to cropland, either directly or after the treatment of sewage sludge, with the remaining P assumed to be lost to the environment (e.g. in landfills or bodies of water).

2.3 P flows from international trade

We compiled the flows of P in international trade both from the P embodied in crops and livestock products and in P embodied in fertilizers exchanged between countries. For agricultural commodities, we used FAOSTAT data that provided a matrix of commodities exchanged between countries and converted this data into P fluxes using commodity-specific P content data. For P fertilizers, we used the International Fertilizer Industry Association trade statistics. By convention, a positive trade balance for a country means that it is a net P importer. In addition, P fluxes associated with the international trade of fertilizers, food, feed, and fibre commodities can be associated with local cropland PUE and pasture PUE. We defined the dependency on chemical fertilizer imports (F_{fer}) as the ratio of the P in imported chemical fertilizers ($P_{\text{fer-imp}}$) to the P in all chemical fertilizers consumed by a country ($P_{\text{fer-con}}$). Similarly, we defined the dependency on food imports (F_{food}) as the ratio of P in food imports ($P_{\text{food-imp}}$) to the P in all food consumed by a country. Furthermore, we defined F_{total} as the ratio of the total P imported (food and fertilizers) to the total P consumed as fertilizers and food in a country. The equations for these calculations are presented in Sects. 2 to 6 of the Supplement.

2.4 Annual P budgets of cropland and pasture soils

Annual changes in P stocks in cropland and pasture soils (ΔP) were estimated as the difference between inputs and outputs (i.e. the budget); $\Delta P > 0$ indicates net P accumulation in the soil, $\Delta P < 0$ indicates a net deficit, and $\Delta P = 0$ represents no net change. ΔP calculated in this manner does not reflect the legacy effects from previous management and fertilization practices (Ringeval et al., 2014), but it is a useful metric to identify regions with a P surplus or deficit at any point in time and to compare countries.

Annual soil ΔP values were calculated as the differences between annual inputs and outputs. Details and the equations are presented in Sect. 2 of the Supplement.

2.5 Cumulative P budgets of cropland and pasture soils

Following the method of Sattari et al. (2012), we separated the P inputs to soils (except inputs in seeds) into two pools: (1) a stable P pool, which represents P that is unavailable to plants on an annual basis, such as the P absorbed onto iron and aluminum oxides (20 % of total P inputs, including fertilizers, manure, sludge, and deposition); and (2) a labile P pool that is assumed to be available for plant uptake (80 % of total P inputs). P can be exchanged between the two pools. If inputs of labile P are larger than P removal in crop biomass, we assumed that the surplus labile P gets transferred into the stable P pool at the end of the year. In the opposite case, in which inputs of labile P are lower than P removal, plants can take up P from the stable pool (Sattari et al., 2012). This approach assumes that the P loss by run-off and leaching into bodies of water is from the labile P pool only and that P stored in seeds does not belong to either the stable pool or the labile pool. This approach is simplistic, as more research will be required to allow for more realistic modelling of these two pools and the flows they are involved in.

2.6 Phosphorus-use efficiency

We defined PUE as the ratio of P in the harvested economic outputs to P in the inputs for the entire agricultural system (the grey area in Fig. 1) or for a given subsystem. PUE indicates how much of the input P is transferred into value-added products. If PUE > 1, the input of P is insufficient to sustain the output (harvested P), suggesting a net reduction of the system's P reservoir. For cropland PUE, we defined P in harvested crops as the economic P output of the crops and the sum of phosphate fertilizer, livestock manure, human sewage sludge, and P from atmospheric deposition as the P input. For pasture PUE, harvested P refers to the P consumed by grazing animals and the sum of phosphate fertilizer, livestock manure going to the pasture, and P from atmospheric deposition as the total inputs. For the livestock subsystem, the harvested P output represents the P in livestock products (meat, eggs, and milk), whereas the inputs represent the input into livestock. We also defined the PUE of human food ($\varepsilon_{\text{food}}$) as the ratio of the P content in human excreta to the total P input in human food; this represents an inconsistency with our previous definitions, since human excreta currently have no economic value. The equations for all the PUE terms are provided in Sect. 5 of the Supplement.

2.7 Uncertainty estimates

Uncertainties in each flux originate both from the material flux data and from data on the P concentration in each material considered by our analysis, including crop products, crop residues, livestock, meat, eggs, milk, livestock, and human excreta. Many of the global statistical datasets used in our analysis are not replicated, and no alternative dataset is available for establishing a range of uncertainty values for the different P fluxes. National datasets have usually not been formally analysed to determine their uncertainty, and many of the sources of uncertainty are difficult to trace (e.g. clerical errors, differences between countries in product definitions). Thus, we have only addressed the effect of uncertainties in the P concentration by means of Monte Carlo simulations (3000 iterations) using the range of P concentrations reported in the literature (Table S5).

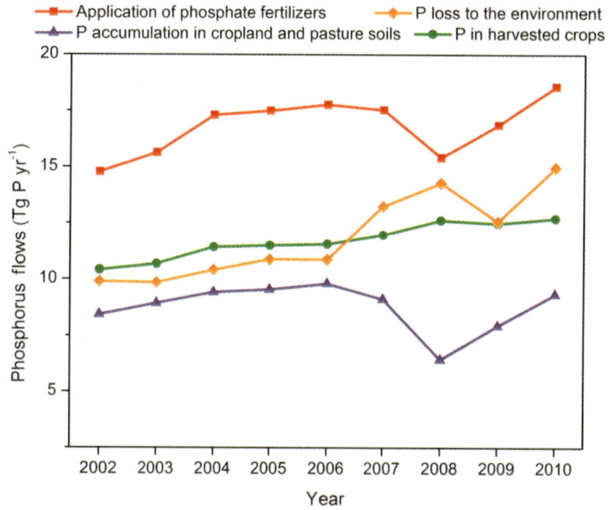

Figure 3. Time series in the four largest global annual P flows within, in, and out of the agriculture system from 2002 to 2010.

3 Results

3.1 Global agricultural P flows and their trends

3.1.1 Global P fluxes in and out of the agricultural system

Figure 2 summarizes the annual average of global P flows for the period from 2002 to 2010. P from phosphate fertilizers was the largest single input flux, representing 93 % of the 21.3 Tg P yr^{-1} of global input, and most of it (82.4 %) goes to cropland and pasture. Outputs from the agriculture system amounted to 12.5 Tg P yr^{-1}, which combines outputs from leaching and run-off into bodies of water (5.4), non-recycled manure waste (4.3), and sewage (2.2), bioenergy (0.4), and burned crop residues (0.2). The global annual P balance of agricultural systems was therefore positive during the entire study period, with 8.8 Tg P yr^{-1} accumulating in soil, of which 6.6 Tg P yr^{-1} accumulated in cropland and 2.2 Tg P yr^{-1} in pasture. On average, 41 % of the P input accumulated in soils from 2002 to 2010.

3.1.2 Temporal trends

Figure 3 shows the trends for the four largest P fluxes in the agriculture system, illustrating that chemical fertilizer inputs, P loss to the environment, and P harvested in crops presented significantly increasing trends during this period. The application of phosphate fertilizer increased at an average annual rate of 3.2 % from 2002 to 2010 despite a decrease in 2008 that reflected reduced fertilizer application at a time when the price of phosphate fertilizers increased (Cordell et al., 2009, 2012). The trend for P in harvested crop biomass was also a steady increase, but at a lower annual rate (2.4 %) and with no decrease in 2008, probably because of the availability of P that accumulated in the soil from previous years (as described in Sect. 2.5). Overall, P in agricultural soils increased by 1.3 % annually, whereas P losses to the environment increased faster (6.4 % yr^{-1}) than fertilizer inputs.

3.1.3 Global P fluxes in cropland

Cropland received the largest fraction (82 %) of phosphate fertilizer, 29 % of the manure produced by livestock, and all of the recycled human sewage sludge (Fig. 2). Atmospheric deposition contributed an additional 0.6 Tg P yr^{-1} of inputs to croplands. Harvesting of cropland removed 11.7 Tg P yr^{-1}, which can be divided into crop products used for human nutrition (9.3 Tg P yr^{-1}, including 5.3 for food, 2.7 for processing, 0.4 for waste, and 0.9 for other use) and for livestock feed (2.1 Tg P yr^{-1}), with a small pool in seeds returned to the cropland (0.3 Tg P yr^{-1}). On average, 50 % of the P contained in crop residues was recycled to cropland during the study period, with 0.2 Tg P yr^{-1} lost to the atmosphere from the burning of crop residues. The remaining 3.6 Tg P yr^{-1} contained in harvested crop residues is removed from cropland and redistributed to livestock and humans. Globally, 3.7 Tg P yr^{-1} was lost from cropland soils through leaching and run-off. The sum of all these fluxes results in an annual soil P accumulation of 6.6 Tg P yr^{-1} (Fig. 2).

The global cropland PUE averaged 0.46, with a maximum of 0.51 in 2008 and a minimum of 0.44 in 2006. The annual cropland P accumulation ratio (cropland soil P accumulation / total P input to cropland) was 23 %, which is lower than the accumulation ratio of 48 % found for the overall agriculture system. In countries where labile P inputs were lower than P removal in crops, the soil's labile P pool was depleted by 1.9 Tg P yr^{-1} through the harvesting of crop biomass. In countries where labile P inputs are higher than P removal by

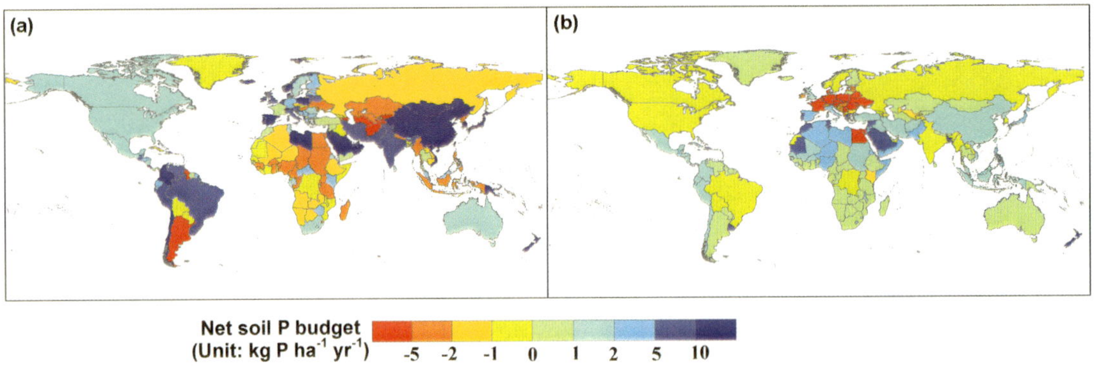

Figure 4. Map of global net soil P budgets (positive values, increase; negative values, decrease) for **(a)** cropland and **(b)** pasture.

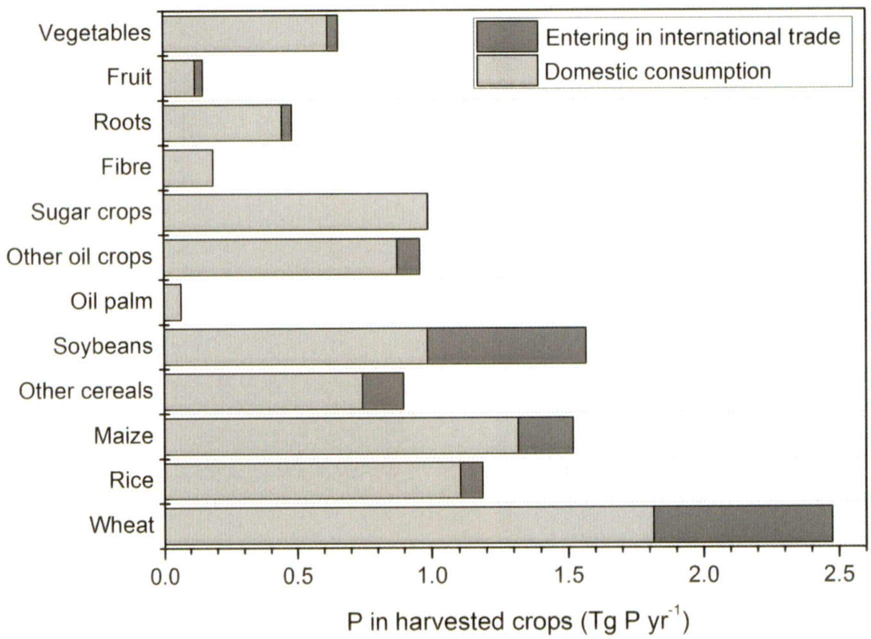

Figure 5. P flows embedded in different crop products, including the fraction of these flows entering into international trade circuits vs. being used for domestic consumption, for the year 2010.

crops, the accumulation of soil labile P was $6.0\,\mathrm{Tg\,P\,yr^{-1}}$. Thus, there is an asymmetry between these two groups of countries, with accumulation being larger than depletion on a global scale. In addition, the global stable P pool in cropland increased by an average of $5.6\,\mathrm{Tg\,P\,yr^{-1}}$ from 2002 to 2010.

3.1.4 Global P fluxes in pasture

Different from croplands, most P inputs to pasture were from livestock manure ($12.7\,\mathrm{Tg\,P\,yr^{-1}}$), with small additional contributions from atmospheric deposition ($0.8\,\mathrm{Tg\,P\,yr^{-1}}$) and phosphate fertilizers ($0.4\,\mathrm{Tg\,P\,yr^{-1}}$). The primary production of pasture incorporates $10.0\,\mathrm{Tg\,P\,yr^{-1}}$ of P into grass biomass that is digested by animals, and the leaching and run-off loss averages $1.7\,\mathrm{Tg\,P\,yr^{-1}}$. From all these fluxes, we estimated a global pasture PUE of 0.72 and a net accumulation of $2.2\,\mathrm{Tg\,P\,yr^{-1}}$ in the soil. In the countries where grass P removal exceeded the labile P inputs, the labile soil P pool was depleted by $1.4\,\mathrm{Tg\,P\,yr^{-1}}$. In the countries where the labile P input exceeded grass P removal, an average of $5.3\,\mathrm{Tg\,P\,yr^{-1}}$ was transferred from the labile to the stable soil P pool from 2002 to 2010.

3.1.5 Global P fluxes in livestock

The annual P input to livestock was $25.6\,\mathrm{Tg\,P\,yr^{-1}}$, with most of the contributions from grazed grass ($10.0\,\mathrm{Tg\,P\,yr^{-1}}$) and processed feed ($10.0\,\mathrm{Tg\,P\,yr^{-1}}$). The economic P output in the form of livestock products averaged $1.5\,\mathrm{Tg\,P\,yr^{-1}}$, which gives a PUE of 0.06. Averages of 29 and 56 % of the P produced in livestock manure were recycled into cropland and pasture, respectively; the rest of this manure (4.3 $TgPyr^{-1}$) was lost to the environment.

3.1.6 Global P fluxes in human use

Humans receive an annual input of 14.0 $TgPyr^{-1}$ from harvested crop products, livestock products, and the use of detergents and other products manufactured from phosphate rock. Although P inputs as food (crop food and livestock products) amounted to 6.8 $TgPyr^{-1}$, humans only absorbed 3.0 $TgPyr^{-1}$ (44 %); the remainder was either wasted before consumption (e.g. in food processing) or transferred back to livestock as processed feed. Thus, only 14.3 % of the total P inputs into the agriculture system end up as food being actually consumed by humans. P lost to the environment by human use amounts to 2.6 $TgPyr^{-1}$, which is divided among 2.2 $TgPyr^{-1}$ lost through inefficient processing and excreta and 0.4 $TgPyr^{-1}$ through bioenergy-related emissions. The fate of non-recycled P in human waste was not separated between bodies of water (untreated sewage) and landfills.

3.2 Regional P budgets

Globally, both cropland and pasture presented soil P accumulation from 2002 to 2010, with an accumulation of 59.6 and 19.4 Tg P, respectively. For croplands, the net P accumulation in the stable P pools amounted to 52.7 Tg P, and the remaining 6.9 Tg P accumulated in soil labile pools. For pasture, the accumulation in the stable P pool was 25.0 Tg P, but 5.6 Tg P was transferred from the stable P pool to be incorporated by grass in regions where P inputs are lower than grass P uptake.

Those global numbers mask large regional differences (Table 1 and Fig. 4) and there were also differences between cropland and grassland. About 32 % of the global cropland area (in 75 countries) had annual soil P deficits from 2002 to 2007, with a net cropland soil P accumulation of 6.20–7.66 $TgPyr^{-1}$. This fraction increased to 50 % in 2008 and 2009 but the net cropland soil P accumulation decreased to 4.38 $TgPyr^{-1}$ in 2008 and 5.39 $TgPyr^{-1}$ at the time of the global financial crisis as a result of high P prices and the resulting reduction in fertilizer application (Cordell et al., 2009, 2012). However, the fraction of cropland soil P deficits returned close to the decadal mean value in 2010, with a net soil P accumulation of 7.30 $TgPyr^{-1}$. On average, 48 % of cropland P uptake was supplied by stable P that accumulated in previous years according to the equations in Sect. 3 of the Supplement. Including the United States, France, Russia, Argentina, and Paraguay, 89 countries had labile P inputs into cropland that were lower than crop P removal from 2002 to 2010. However, if we consider stable P inputs, cropland soil still presented a net soil P surplus in the United States during the same period. Compared with cropland, a slightly larger proportion of the total global pasture area had a net annual soil P deficit from 2002 to 2010, mostly in Europe and North America. The deficit proportion of grassland was only about 38 % in 2002 and 2003, with an annual net soil P accumulation of 2.26 $TgPyr^{-1}$; however, it increased to 43 % during the period of 2004–2010 and the annual pasture soil P accumulation was about 2.10 $TgPyr^{-1}$, with the smallest of 2.00 $TgPyr^{-1}$ in 2009. However, only 48 countries had labile P inputs into pasture that were lower than the P removal in grass.

3.2.1 Regional cropland budgets

Examining Fig. 4a reveals that cropland in all African countries experienced an annual soil P deficit, especially in western and central Africa, with soil P loss rates per unit area ranging from 2.5 $kgPha^{-1}yr^{-1}$ in 2002 to 2.7 $kgPha^{-1}yr^{-1}$ in 2010. In contrast, cropland in eastern Asia accumulated 23.4 $kgPha^{-1}yr^{-1}$ during the period from 2002 to 2010, a cumulative storage equivalent to more than 4 years of P fertilizer application. Cropland in Oceania, Europe, and the Caribbean and Central America also annually accumulated P in their soils. Cropland soils in North America and South America accumulated P from 2002 to 2007, but experienced temporary P deficits from 2008 to 2010. Yet despite this, crop yields did not decrease from 2008 to 2010 in those two regions, probably because of the re-mobilization of P that accumulated in stable pools. Cropland soils in western and central Asia were nearly balanced, with a mean areal flux of 0.2 $kgPha^{-1}yr^{-1}$.

Considering the different countries (Fig. 4a), the largest cumulative soil P increase was found in China (34.6 Tg P) for the 9 years from 2002 to 2010, followed by India (11.4 Tg P) and Brazil (3.6 Tg P). Pakistan (1.8 Tg P), the United States (1.8 Tg P), and New Zealand (1.8 Tg P) also had net soil P accumulation, yet of a smaller magnitude. These six countries accounted for 77 % of the global accumulation of P in countries where cropland had a positive soil P balance. Furthermore, a large amount of P accumulated in the soil labile P pools of cropland in China and India, at about 20.0 and 4.5 Tg P, respectively; however, in the United States, about 6.0 Tg P accumulated in the cropland stable P pool from 2002 to 2010; thus, 4.2 Tg P was absorbed from the previous cropland soil P. In contrast, most African countries experienced persistent cropland soil P deficits from 2002 to 2010. This was especially true in Nigeria, which had a cumulative deficit of 1.7 Tg P (Fig. 4a). We also found cumulative soil P deficits in Russia, the Ukraine, and Kazakhstan, but with a smaller magnitude (1.1, 0.9, and 0.7 Tg P, respectively) for the 9 years. Comparing the rates of change of crop soil P per unit area, New Zealand had the fastest rate of increase (> 100 $kgPha^{-1}yr^{-1}$), whereas Argentina had the fastest rate of decrease (–7.9 $kgPha^{-1}yr^{-1}$). In terms of the difference between inputs and outputs, loss rates in Argentina were about 5 times input rates.

Table 1. Regional annual agricultural P budgets and P-use efficiency (PUE).

Subsystem	World	Eastern and southern Africa	Northern Africa	Western and central Africa	Eastern Asia	Southern and south-eastern Asia	Western and central Asia	Oceania	Europe	North America	Caribbean and Central America	South America
Agricultural land P budget ($Tg\,P\,yr^{-1}$)[a]												
Cropland	6.6	−0.1	−0.1	−0.2	4.1	1.4	0.0	0.3	0.7	0.3	0.0	0.3
Pasture	2.2	0.1	0.5	0.3	0.5	0.2	0.2	0.4	−0.3	−0.1	0.1	0.3
Agricultural land P budget per unit area ($kg\,P\,ha^{-1}\,yr^{-1}$)[a]												
Cropland	4.7	−1.0	−1.5	−2.7	23.4	4.1	0.2	5.2	2.8	1.5	3.8	2.3
Pasture	0.4	0.2	1.6	0.9	1.0	0.4	0.6	0.8	−0.4	−0.1	3.4	0.4
Food consumption ($Tg\,P\,yr^{-1}$)												
Crops	5.3	0.2	0.3	0.2	1.4	1.3	0.3	0.0	0.8	0.4	0.0	0.3
Meat	0.71	0.01	0.01	0.01	0.29	0.05	0.01	0.01	0.17	0.10	0.00	0.05
Eggs	0.16	0.00	0.00	0.00	0.07	0.02	0.01	0.00	0.03	0.02	0.00	0.01
Milk	0.60	0.01	0.03	0.00	0.04	0.13	0.04	0.01	0.18	0.09	0.01	0.05
PUE												
Cropland	0.46	0.80	0.84	1.51	0.27	0.43	0.64	0.31	0.54	0.57	0.53	0.63
Pasture	0.72	0.77	0.61	0.46	0.58	0.80	0.61	0.42	1.25	0.98	0.37	0.75
Livestock	0.06	0.02	0.02	0.01	0.08	0.05	0.04	0.04	0.09	0.08	0.03	0.03
Food	0.45	0.60	0.50	0.64	0.40	0.64	0.44	0.26	0.28	0.32	0.68	0.42
International trade of P in commodities ($Tg\,P\,yr^{-1}$)[b]												
Crops	–	0.02	0.12	0.03	0.35	0.03	0.12	−0.07	−0.02	−0.41	0.03	−0.24
Meat	–	0.000	0.001	0.000	0.013	0.000	0.002	−0.004	−0.002	−0.005	0.001	−0.010
Eggs	–	0.0001	0.0000	0.0000	0.0001	−0.0004	0.0001	0.0000	0.0000	−0.0002	0.0000	−0.0001
Milk	–	0.000	0.003	0.001	0.004	0.005	0.004	−0.015	−0.012	0.003	0.001	−0.001
Fertilizer	–	0.05	−0.67	0.03	0.00	1.45	−0.05	0.21	−0.37	−1.00	0.06	1.10

[a] The positive values represent a soil P surplus, whereas negative values represent a soil P deficit. [b] The positive values represent net P importers, whereas negative values represent net P exporters.

3.2.2 Regional pasture budgets

We found mainly net losses of P in pasture soils (Fig. 4b), most likely because of the net removal of P through animal grazing, followed by the export of manure P to enrich cropland soils. Pasture soil P loss rates per unit area in Europe averaged $0.4\,kg\,P\,ha^{-1}\,yr^{-1}$ and reached high values in countries (Denmark, Luxembourg, Germany, and Belgium) with intensive livestock production systems (Chang et al., 2015) and large grass consumption by livestock, with loss rates $> 10\,kg\,P\,ha^{-1}\,yr^{-1}$. North American pastures had a smaller average loss rate of about $0.1\,kg\,P\,ha^{-1}\,yr^{-1}$. The United States, India, and Russia had the largest cumulative P deficits at 2.1, 1.5, and 0.7 Tg P, respectively, from 2002 to 2010. In contrast, pasture in the Caribbean and Central America had greater P inputs than P removals. Consequently, these regions had the largest soil P accumulation rates. Pasture in northern and eastern Africa also had net soil P accumulation. For instance, Mauritania, Tunisia, and Morocco had net soil P accumulation rates of 9.8, 9.4, and $5.5\,kg\,P\,ha^{-1}\,yr^{-1}$, respectively. The reason for this excess is not clear, but one possibility is that these countries apply P fertilizer to some of their pasture.

3.3 Phosphorus-use efficiencies in different regions

Table 1 gives the values of PUE for cropland, pasture, livestock, and food (human use) in the world's different regions. Globally, 116 countries have cropland PUE values above the global mean value of 0.46, mostly in Africa, and these countries account for 64 % of the global cropland area. In addition, 16 % of the countries had a PUE of around 0.6 (0.55 to 0.65). In particular, African countries had the highest overall cropland PUE (≥ 0.80) because of their low P input. On the other hand, eastern Asia and Oceania have cropland PUE below the global average. Conversely, pasture had high PUE in Europe (1.25) and North America (0.98) but low values in Africa (≤ 0.77) and particularly low values in the Caribbean and Central America (0.37). P removal from pasture exceeded P inputs in Europe, resulting in pasture PUE > 1, largely because of P inputs from feed given to animals.

The livestock subsystem generally had a low PUE (< 0.1), with the highest values in Europe, North America, and eastern Asia (Table 1). Regarding human food PUE, our data indicate that only 25 to 40 % of the P in food products in eastern Asia, Oceania, Europe, and North America is actually consumed by humans (Table 1). The resulting low PUE of human use in these regions results from both large P inputs and high food waste. Eastern and southern Africa, western and central Africa, southern and southeastern Asia, and the Caribbean and Central America had the highest PUE for human use, with more than 60 % of P in food being consumed by humans. Globally, most of the P consumed by humans (78 %) originates from crops, and the fraction of P from livestock differs among regions; it ranges from 35 % of the total human food P consumption in Oceania, Europe, and North America to 10 % in less developed regions (Africa and the Caribbean and Central America) and to 4 % in western and central Africa.

3.4 P flows through international trade

Approximately 2.1 $Tg\,P\,yr^{-1}$ entered into international trade in 2010, amounting to about 17 % of the total harvested crop P (Fig. 5). The remainder (10.6 $Tg\,P\,yr^{-1}$) is consumed domestically. Differences in crop types as a result of their specific P content (Table S1) strongly determine the magnitude of the traded P fluxes. For example, 37 % of the P in soybeans and 27 % of the P in wheat produced each year were traded internationally in 2010. Also significant fractions of the P in maize, other cereals, and fruit were traded internationally, but almost all of the P in sugar crops and fibre were consumed or processed in the countries where they were grown.

Considering the P fluxes in phosphate fertilizers and food products, we examined how international trade influences regional P budgets and redistributes P between regions. We found that southern and southeastern Asia have the largest net P imports (Table 1), with imports of phosphate fertilizer amounting to 1.4 $Tg\,P\,yr^{-1}$ and P exports as food products being much smaller, mainly to China and South Korea. South America is the second-largest exporter of P in food, but imports 56 % of its P fertilizer. North America is a large exporter of P in both crop products and fertilizer, yet it also imports P-rich milk products. Most European countries imported nearly all their phosphate fertilizers, but Europe as a whole is a net exporter because of large exports (0.9 $Tg\,P\,yr^{-1}$) from Russia (Fig. 6). Western European countries were the main exporters of P-rich livestock products. Some northern African countries (especially Morocco and Tunisia, which have the largest mines of P-rich ores) exported a total of 0.7 $Tg\,P\,yr^{-1}$ in fertilizer. The remaining regions (eastern and southern Africa, northern Africa, and the Caribbean and Central America) imported P in both food and fertilizer, although much less than other regions (Table 1).

Figure 6 illustrates the disparities among countries with respect to the role of international trade in crops, livestock, and fertilizer for the main exporters and importers. Based on data for all 224 countries, a country can be categorized into one of the following four groups (Fig. 7).

Food and fertilizer P exporters. P storage in these countries has been decreasing due to their international exports of both fertilizer and food. Examples include the United States and Russia.

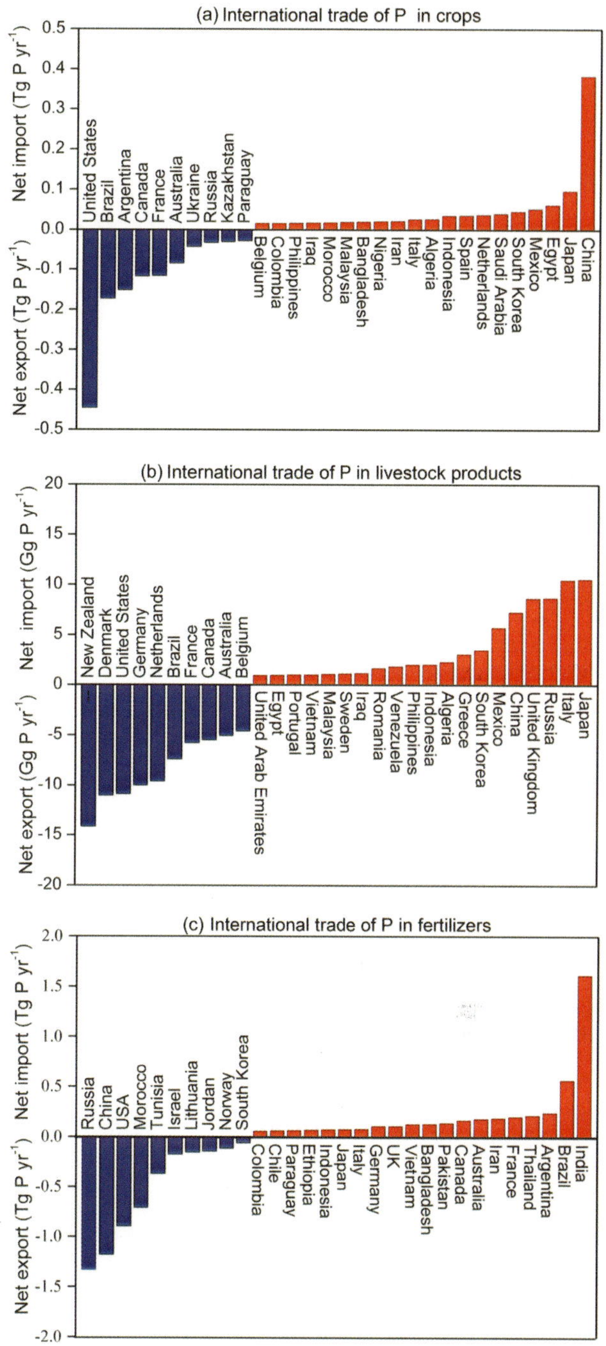

Figure 6. Annual P flows embedded in traded crop products **(a)**, livestock products **(b)**, and fertilizers **(c)** in 2010. By convention, a positive flow is P received (imported) by a country.

Food P importers and fertilizer P exporters. This group mainly comprises countries that export phosphate fertilizers and import food to meet domestic consumption. Examples includes Tunisia, Morocco, and China.

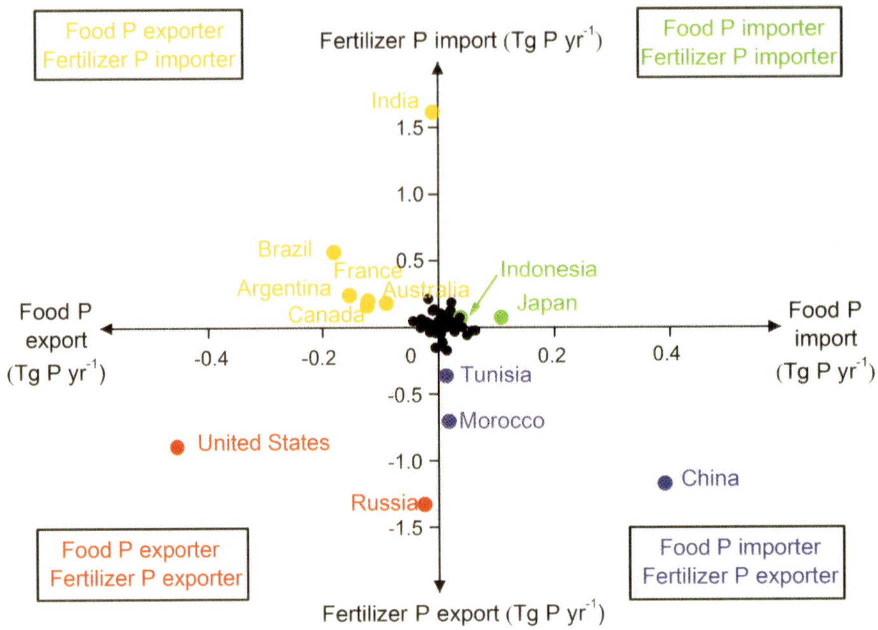

Figure 7. Groupings of the countries based on whether they import or export P through their international trade in food and fertilizer.

Food P exporters and fertilizer P importers. These countries have high food and livestock production, but this depends strongly on phosphate fertilizer imported from other countries. Examples include Brazil, Argentina, Canada, France, Australia, and India.

Food and fertilizer P importers. These countries depend on imports for both food and fertilizers; they are thus vulnerable to economic shocks that result from changing food prices. Examples include Japan and Indonesia.

International trade affects the global P cycle by physically moving the P contained in traded crops, livestock products, and phosphate fertilizers (Grote et al., 2005). Imports of P fertilizers accounted for 55 and 79 %, respectively, of the total application of P fertilizer for countries that are food P exporters and fertilizer P importers or food and fertilizer P importers. The P trade in food followed a similar trend. Countries that are food P importers and fertilizer P exporters or food and fertilizer P importers depended more on food imports than countries that are food and fertilizer P exporters or food P exporters and fertilizer P importers. International trade also increased the connections among countries (Table 2). For example, although the United States and China are clearly major P fertilizer exporters, they also import fertilizer from each other; 2.6 % of the P fertilizer applied in the United States originated in China, and 3.6 % of the phosphate fertilizer applied in China originated in the United States. In addition, 11.4 % of the phosphate fertilizer consumption in the United States originated from Russia, Morocco, Tunisia, and other countries. About 1.5 % of Chinese domestic P consumption originates from the United States, which is higher than the fraction of domestic P consumption in the United States from China. Countries with small or no reserves of P-containing minerals imported large amounts of phosphate fertilizer; for example, imports accounted for 61 and 46 % of total P consumed in France and Brazil (food P exporters and fertilizer P importers) and 76 % of total P consumed in Japan.

3.5 Uncertainties in soil P changes result from uncertain P concentrations

We estimated the net cropland soil P balance in 2000 by means of Monte Carlo simulations, as described in Sect. 2.7. We found a net accumulation of $5.8 \pm 0.6\,\mathrm{Tg\,P\,yr^{-1}}$. More detailed calculations suggest that uncertainty in the crop P concentrations contributed $\pm 0.2\,\mathrm{Tg\,P\,yr^{-1}}$ of the uncertainty in the net cropland soil P balance; this is because of the dominance of calculations using cereals, which have low uncertainty due to the narrow range of reported P concentrations (Antikainen et al., 2005; COMIFER, 2007; USDA-NRCS, 2009; Waller, 2010). Uncertainty in P concentrations in crop residues contributed an additional $\pm 0.2\,\mathrm{Tg\,P\,yr^{-1}}$ to the total uncertainty, and uncertainty in P concentrations in the livestock manure applied to cropland added $\pm 0.4\,\mathrm{Tg\,P\,yr^{-1}}$. In addition, the uncertainty in the pasture soil P balance attributed to uncertainty in the P concentrations in grass biomass and manure was $\pm 1.3\,\mathrm{Tg\,P\,yr^{-1}}$. This relative uncertainty is higher than that for the cropland soil P balance and the results from the large range of grass P concentrations found in our review of the available data. See Table S5 for more details.

Table 2. Proportions of total consumption and total international trade accounted for by P in fertilizer and food imports and exports.

Group	Proportion (%)			
	P fertilizer imports as a proportion of total consumption	P fertilizer exports as a proportion of the total international P fertilizer trade	P in food imports as a proportion of total consumption	P in food exports as a proportion of the total international P in the food trade
Group level				
Food and fertilizer exporter	22	43	7	31
Food importer and fertilizer exporter	5	48	22	5
Food exporter and fertilizer importer	55	5	5	48
Food and fertilizer importer	79	4	28	15
Country level				
United States (food and fertilizer exporter)	13	18	6	26
China (food importer and fertilizer exporter)	2	20	14	2
France (food exporter and fertilizer importer)	52	0	19	8
Brazil (food exporter and fertilizer importer)	44	1	4	10
Japan (food and fertilizer importer)	40	0	60	0

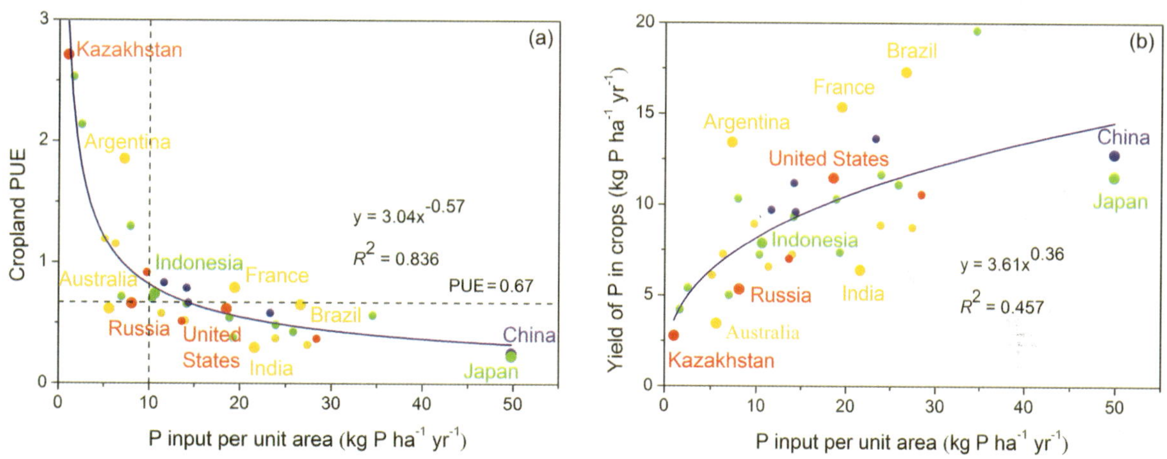

Figure 8. The relationships between P input per unit area of cropland and **(a)** phosphorus-use efficiency (PUE) The horizontal line at PUE = 0.67 represents the global average. **(b)** P in harvested crops for the 35 largest crop producers representing 90 % of global crops. The equations give the fit to the data represented by black curves.

4 Discussion

4.1 Cropland PUE and P in harvested crops as a function of cropland P inputs

Figure 8a shows the relationship between the cropland PUE and cropland P inputs for 35 countries that are large crop producers. PUE decreased exponentially with increasing input; that is, P was used most efficiently at low application rates. PUE decreased rapidly as P inputs increased to 10 kg P ha^{-1} yr^{-1} and then decreased more slowly. High PUE values were associated with countries that had a low P input and a soil P deficit. This suggests that there is a trade-off between the efficient use of P in cropland and the avoidance of soil P deficits that limit crop yields (Obersteiner et al., 2013). Figure 8a also indicates that cropland soils have a net soil P deficit if their inputs are lower than 10 kg P ha^{-1} yr^{-1}, which is a threshold value that corresponds to PUE = 0.67. Argentina, South Africa, Indonesia, Mexico, and Paraguay are below this threshold (Fig. 9).

P in harvested crops increased exponentially with increasing P inputs, but the response slowed at high P inputs (Fig. 8b). The P in harvested crops in countries with cropland PUE > 0.67 (except Argentina) is only half of that in countries with high P in the harvested crops, such as the United States and China. P in the harvested crops was very low in Australia due to low cropland P input, which was less

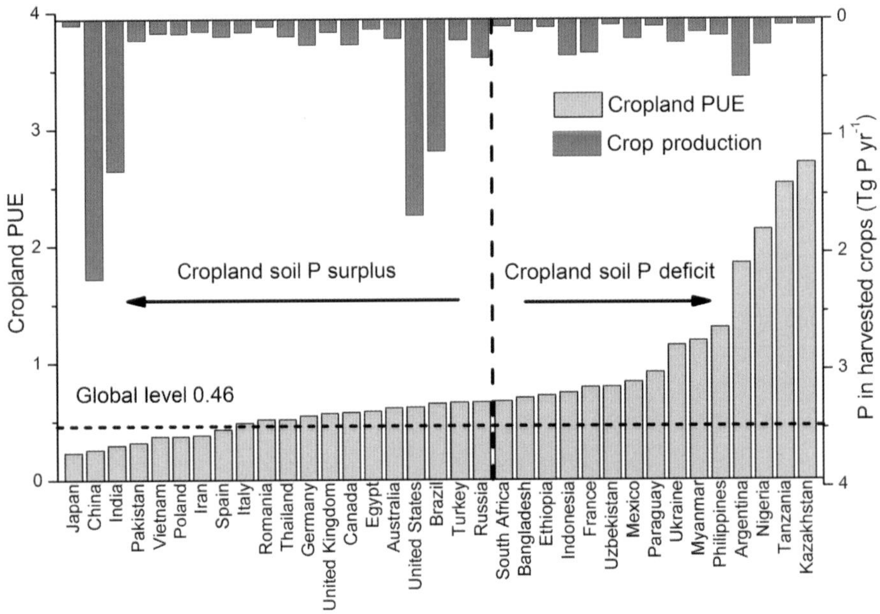

Figure 9. Phosphorus-use efficiency (PUE) and P in harvested crops for the 35 large countries shown in Fig. 8. Cropland soil P surplus or deficit is separated by the vertical dashed line.

than 25 % of the inputs in the United States and China. P already present in the soil may be sufficient to sustain high crop yields for some time without additional inputs in some countries (e.g. France) that formerly had large P fertilization rates, despite currently having a negative annual P balance. Comparing Fig. 8a and b suggests that total cropland P inputs of 20 to 25 $\mathrm{kg\,P\,ha^{-1}\,yr^{-1}}$ may be a good compromise that will achieve high yields while creating a near-equilibrium soil P balance. Both excessive P inputs (e.g. China and Japan) and low PUE (e.g. India) can lead to high P accumulation in cropland soil, leading to high losses into the environment.

The data in Fig. 8 indicate that different countries face different challenges for P resource management, implying a need for country-specific policy options and solutions. Countries like Kazakhstan and Argentina may have to increase P inputs to their cropland in order to prevent long-term depletion of soil P, which could be realized by increasing the application of phosphate fertilizer or reducing losses to leaching and erosion. Countries like France that are currently experiencing a net negative soil P balance (Fig. 9) following a period of sustained accumulation (Senthilkumar et al., 2012; van Dijk et al., 2016) may need to progressively adjust fertilizer inputs in coming years to balance inputs with removals and avoid the risk of a long-term soil fertility decline due to inadequate levels of P. In contrast, countries such as Japan and China are rapidly accumulating P in cropland soils due to high and sustained P inputs and will urgently need to consider how to improve their cropland PUE. This could be initiated by identifying crop types that are being over-fertilized and regions with excessive application of phosphate fertilizer; they can then consider a range of options such as precision agriculture (i.e. applying only as much P as the crop requires). We estimate that if Chinese cropland PUE could be increased to the global average of 0.46 (Fig. 9), China would save 3.8 $\mathrm{Tg\,P\,yr^{-1}}$ of phosphate fertilizer, which is equivalent to 60 % of its phosphate fertilizer consumption in 2010. Last, in countries like India where crop P harvests are lower than average despite high average P inputs and positive soil ΔP, improvements in agricultural management (such as the use of precision fertilization) appear necessary. We did not have access to subnational data for this study, but it is likely that in a country as large as India, excessive or insufficient P may occur in different regions, for different crop types, or for different region–crop-type combinations.

4.2 Pasture P budget, livestock consumption, and international trade

Figure 10 shows that the soil P balance is negatively related to the flux of P in livestock products per unit area of pasture. Several western European countries (Germany, the Netherlands, Denmark, and Belgium) achieve high P yields in livestock products (defined by the amount of P in livestock products per unit area of pasture), and all of these countries export livestock products. In these countries, only a small fraction of livestock manure is recycled to pasture, so there is currently a soil P deficit; in the long term, this may result in a loss of soil fertility. Therefore, these countries should increase P fertilization in pasture or import forage or feed to supply the P required to sustain high livestock production. New Zealand, Australia, and Canada are also large exporters of P in live-

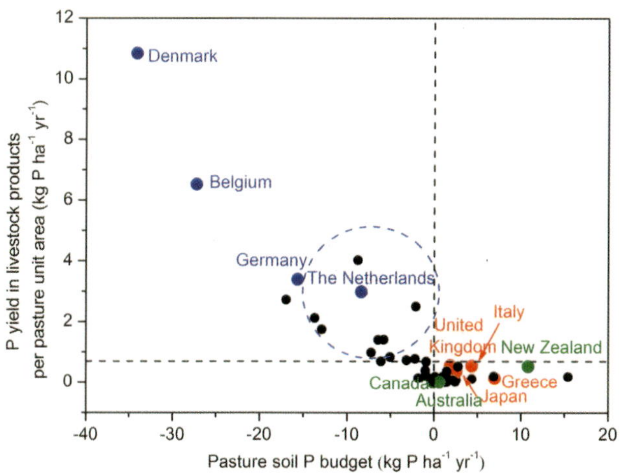

Figure 10. The relationship between the P yield of livestock products, defined by the amount of P in livestock products per unit area of pasture and the P balance of pasture soils.

stock products. However, given their low-input production systems and large areas of pasture (Fig. 4b), P removals per unit area through grazing are much lower than in western Europe, and the soil P balance of pasture ranges from slightly negative to slightly positive.

4.3 Livestock and human food PUE and trends in P consumption

The increasing consumption of livestock products by humans is an essential factor that is responsible for increasing P mining and increasing P inputs to agricultural systems (Metson et al., 2012; van Dijk et al., 2016). Where socioeconomic development is improving the income of residents, especially in Africa and the Caribbean and Central American region, residents are consuming more P from livestock products (Fig. S3). Unfortunately, the livestock PUE in countries in these two regions is much smaller (0.01 to 0.03) than the global average of 0.06 (Table 1), indicating that only a small proportion of livestock P inputs is used by humans. This may be because countries in these regions are primarily importers of livestock products. Therefore, animal husbandry has important implications for global P security and special attention will be required to improve livestock PUE (Wu et al., 2014). If livestock PUE reaches the global level of 0.06 in these two regions, both regions could more than double their livestock production by about $0.16\,\mathrm{Tg\,P\,yr^{-1}}$.

In addition, the management of manure differs greatly among regions due to different livestock production systems. The yield of livestock products is very low in African countries, resulting in low livestock PUE. Almost all livestock manure is applied to cropland for which this resource is an important P input. In contrast, with the application of phosphate fertilizer to pasture in Europe and eastern Asia, only a small fraction of livestock manure is recycled for pasture (36 and 17 %, respectively); a larger fraction of the manure is applied to cropland in eastern Asia (40 %) and Europe (60 %). Consequently, improving the manure utilization efficiency and applying more livestock manure to pasture will be important strategies in eastern Asia and Europe (Wu et al., 2014).

As shown in Sect. 3.3, only 45 % of the P that enters the food production subsystem was absorbed by humans; thus, large amounts of food (and the P it contains) are wasted, although some parts of the waste were consumed by livestock. Despite this recycling, $2.2\,\mathrm{Tg\,P\,yr^{-1}}$ flowed into the environment as waste either before or after food consumption, and only 14.3 % of the total P inputs to the agriculture system ended up in food consumed by humans. In eastern Asia, Oceania, Europe, and North America, the PUE of human food was very low, reflecting the high proportion of livestock products in the diet and a high degree of waste. Therefore, decreasing food waste before consumption, recycling P in food waste, and better treatment of organic waste could significantly decrease the amount of P required to support humans (Metson et al., 2012; van Dijk et al., 2016). In eastern Asia, Oceania, Europe, and North America, fully absorbing the 45 % of the P that enters food produced for humans could reduce agricultural inputs of P by $0.7\,\mathrm{Tg\,P\,yr^{-1}}$ globally. Thus, decreasing food waste and improving the PUE of human food represent key challenges that must be solved to achieve sustainable P management.

Population increases and dietary changes are requiring higher P inputs in cultivated land and the increased mining of P ores (Grote et al., 2005; Foley et al., 2011). From 2002 to 2010, this mining increased by 33 % in our estimate, during a period when the global population and per capita food P consumption increased by 10 and 5 %, respectively. In 2010, humans consumed 8.0 and 3.8 % more P in livestock products and crops, respectively. Since livestock PUE was much lower than cropland PUE, the consumption of more livestock products resulted in lower external P inputs in food that flowed into the human subsystem; this proportion decreased from 36 % in 2002 to 31 % in 2010. Therefore, consuming more livestock products will require increasing P inputs. Thus, human dietary shifts may have been responsible for half of the increase of P-ore mining.

4.4 International trade and global P flows

International trade also increased the connections among countries. Whether international trade is good or bad for humans and the environment in terms of its impact on the management of P resources is a complex question. International trade can increase cropland P deficits if countries that export large amounts of P in crop and livestock products do not counteract these exports by increasing inputs of phosphate fertilizer to soils. For example, Argentina exported lots of food to other countries (about $0.15\,\mathrm{Tg\,P\,yr^{-1}}$) and has developed a serious cropland soil P deficit of $0.38\,\mathrm{Tg\,P\,yr^{-1}}$

($10.3\,kg\,P\,ha^{-1}\,yr^{-1}$). Massive P imports through trade can result in an excess supply of P to cropland soils as manure (Schipanski and Bennett, 2012), with potentially significant negative environmental effects. On the one hand, trade can hamper the proper recycling of P resources from waste and manure to agricultural soils through local food webs (Schipanski and Bennett, 2012). On the other hand, trade may contribute to more efficient use of P resources if traded products flow from countries with lower PUE to countries with higher PUE, as is generally observed for water resources (Dalin et al., 2014). This confirms that more integrated studies are required to fully assess the effects of trade on P resource recycling, efficiency, and conservation. Our study identified world regions and countries with lower PUE and others with high PUE and regions and countries with a net loss of P in soils and others with a net gain. This provides valuable information to policymakers on how to improve the trade relationships for a global optimization of PUE and therefore global food security.

4.5 Comparison with previous studies

Previous studies have estimated P flows in agriculture on a global scale (Smil, 2000; Sheldrick et al., 2003; Liu et al., 2008; Cordell et al., 2009; Bouwman et al., 2009, 2013; Potter et al., 2010; MacDonald et al., 2011). However, to the best of our knowledge, the present analysis provides the first consistent multi-year overview of the P flows in agriculture. In addition, it provides national and regional P budgets, calculates agricultural PUE, and quantifies P fluxes in international trade based on a combination of datasets for cropland and pasture inputs (fertilizers, manure, atmospheric deposition, and recycling of crop residues) and outputs (crop harvests, residue removal, and P loss by burning and leaching or surface run-off into bodies of water). For data from 2000, our results are consistent with the abovementioned studies for most P flows (Table 3). For data from 2000, our results are generally consistent with those in the previous studies for cropland soil P inputs, harvested crop P, cropland soil P lost by erosion or surface run-off into bodies of water, pasture soil P inputs, and harvested grass P (Table 3). However, methods, data sources, and system boundaries differed among the studies, making an accurate comparison difficult. Our estimate of a net accumulation of $5.8 \pm 0.6\,Tg\,P\,yr^{-1}$ is in line with the reported net accumulation in soils, which ranged between 0 and $11.5\,Tg\,P\,yr^{-1}$ (Smil, 2000; Bennett et al., 2001; Bouwman et al., 2009; MacDonald et al., 2011), but disagrees with the estimate of Liu et al. (2008), who calculated a net loss of $9.6\,Tg\,P\,yr^{-1}$. The difference from the present results can be explained by accounting for large P losses ($19.3\,Tg\,P\,yr^{-1}$) due to soil erosion caused by land-use change and overgrazing. The quantification of erosional losses of P from arable land is prone to high uncertainties due to the unknown amount of redeposited soil material, and other studies have reported much lower losses (e.g. $2.5\,Tg\,P\,yr^{-1}$; Quinton et al., 2010).

The main cropland P fluxes estimated in our study agreed with previous results, except for the production and recycling of crop residues (Table 3). Smil (2000) and Liu et al. (2008) used harvest index data (defined as the ratio of total aboveground biomass to crop residues) for estimating the P in crop residues, whereas we estimated P in crop residues by combining data from Liu et al. (2008) and FAO. MacDonald et al. (2011) estimated that 29 % of the global cropland area was subject to soil P deficits in 2000, which is similar to our estimate (32 %) based on data from 2002 to 2010. In addition, our estimate of $22.3\,Tg\,P\,yr^{-1}$ in animal manure for the livestock subsystem in 2000 is within the reported range of 17.1 to $24.3\,Tg\,P\,yr^{-1}$ from Potter et al. (2010). We defined global cropland PUE as the ratio of P in harvested crops to total P inputs without accounting for the recycling of crop residues. Under this definition, global PUE was estimated to be 0.43 by Liu et al. (2008) and 0.40 by Smil (2000), both of which are comparable to our estimate of 0.46 from 2002 to 2010. Since we applied the same methods across the globe to calculate agricultural P fluxes, we were able to compare the P fluxes and budgets for different regions and countries on a consistent basis. This information is of critical importance for the development of more appropriate agricultural policy and to support the development of technological and other solutions for different types of countries, which better integrate cultivated ecosystems, livestock production, and the human food supply.

4.6 Limitations and novelty of our study

Due to limited data sources for some parameters, our study and most previous studies focused on P in livestock products and manure as the outputs of the livestock system and did not consider the fate of P in non-edible livestock products (e.g. bones, blood, leather products). Xu et al. (2005) pointed out that from 12 to 23 and 72 % of P were contained in livestock meat and bones, respectively. If these percentages are applied to our data, this gives an annual flux of $2.5\,Tg\,P\,yr^{-1}$ in the bones of slaughtered animals. Although most livestock bones are currently wasted or landfilled, some countries have begun to use them as fertilizers, protein sources, and condiments (Wu and Ma, 2005; Li, 2008). In addition, as we focused on the annual P budgets for livestock and human beings, we did not account for P accumulation in humans. From 2002 to 2010, the global population increased by 635×10^6 persons. If we assume that a typical adult body contains 600 g of P, then about 0.38 Tg more P would have accumulated in humans. Therefore, the annual human P accumulation would be $0.04\,Tg\,P\,yr^{-1}$, accounting for only 0.3 % of the P inputs into humans.

Despite the abovementioned limitations in our study, we were able to achieve some interesting and novel results. First, we have provided a detailed and harmonized summary of

Table 3. Comparison of the present results for P flows and budgets in 2000 with the results of other studies at a global level ($Tg\,P\,yr^{-1}$).

Global P flux		Previous studies	Our study	Reasons for differences
Cropland	Fertilizer input	14–15[1–3]	13.7	—
	Animal manure to cropland	6–8[2,3]	6.7 ± 0.4	Method
	Human sewage sludge to cropland	1.5[1,3]	1.3	Method
	Crop production	8.2–12.3[1–5]	10.2 ± 0.4	Boundary/data
	Crops (human food)	3.5[3]	4.8 ± 0.2	Method/data
	Crops (animal feed)	2.6[3]	1.9 ± 0.1	Data
	Crop residues	3.75–4.5[1–2]	6.7 ± 0.2	Method/data
	Recycling of residues	1–2.2[1–3]	3.5 ± 0.1	Method/data
	Leaching and run-off from cropland	4[6]	3.2	Method
Pasture	Livestock manure	17.1–24.3[5,7,8]	22.3 ± 1.3	Method/data
	Manure wasted (released into the environment)	2–8[1–3]	4.1 ± 0.2	Method/data
	Grass	6–12.1[3,4]	8.9 ± 1.3	Method/data
	Animal feed additives	0.9[3]	1.4	Data
	Leaching and run-off from pasture	1.0[5]	1.6	Method
Humans	Excreta	3–3.3[1,3]	2.8	Method

Sources: [1] Liu et al. (2008), [2] Smil (2000), [3] Cordell et al. (2009), [4] MacDonald et al. (2011), [5] Bouwman et al. (2009), [6] Bouwman et al. (2013), [7] Sheldrick et al. (2003), [8] Potter et al. (2010).

the P fluxes as inputs and outputs for the agricultural system and the internal P flows within the agricultural system on national, regional, and global scales. In addition, we have characterized the P budgets and P-use efficiencies in the subsystems of the overall agricultural system and discussed their influences and impacts. Finally, we have discussed how changes in population, diets, and food consumption have influenced the global mining of P ore and how international trade has influenced P fluxes. These insights will support the development of policies to use P more sustainably at national, regional, and global levels.

5 Data availability

The global and regional phosphorus budgets and their PUEs in agricultural systems are publicly available at https://doi.pangaea.de/10.1594/PANGAEA.875296.

6 Conclusions

The estimation of global and regional phosphorus budgets in agricultural systems and their PUE is a major effort by the anthropogenic nutrient cycle research community that requires lots of work. We quantified in detail the P inputs and outputs of cropland and pasture and the P fluxes through human and livestock consumers of agricultural products on global, regional, and national scales from 2002 to 2010. The results from this analysis confirmed that P from phosphate fertilizers is the largest single input flux into the agricultural system, while one-half of this input was lost to freshwaters and one-third accumulated in soils. Chemical fertilizer inputs, P losses to the environment, and P harvested in crop biomass all present increasing trends. The positive global P balance in cropland and pasture soils is the sum of regional excess and deficits, with differences in the sign of the balance between cropland and pastures as well. Compared with croplands, a slightly larger proportion of the global pasture area had a net annual soil P deficit. The hot spots of cropland P budgets are on the one hand eastern Asian countries with excessive inputs and sub-Saharan African countries with deficits. Hot spots of pasture P budgets are European and North American grazing lands showing significant soil P deficits. There are great differences in PUE for croplands, pastures, livestock products, and food products on global, regional, and national scales. Livestock products generally had the lowest PUE. We showed that cropland PUE decreased exponentially with increasing input implying that P is used more efficiently at low application rates. In parallel, P in harvested crop biomass increased with P inputs following a saturating exponential relationship. International trade plays a significant role in the P redistribution among countries when considering P embedded in imported livestock and food products consumed in each country and the trade of fertilizers. Nearly one-fifth of total harvested crop P entered into international trade during the period 2002–2010. Population increases and dietary changes are requiring higher P inputs in cultivated land and the increased production of P fertilizers from minerals. Human dietary shifts may have been responsible for half of the increase in P mining during the period 2002–2010. Regional P imbalances in agricultural soils can be mitigated both by optimizing phosphate fertilizer application and recycling P.

Competing interests. The authors declare that they have no conflict of interest.

Acknowledgements. This study was supported by the National Natural Science Foundation of China (41625001, 41571022), the Beijing Natural Science Foundation (grant 8151002), the Southern University of Science and Technology (grant no. G01296001), and a Synergy Grant (ERC-2013-SyG-610028 IMBALANCE-P) from the European Research Council.

Edited by: Attila Demény
Reviewed by: two anonymous referees

References

Antikainen, R., Lemola, R., Nousiainen, J. I., Sokka, L., Esala, M., Huhtanen, P., and Rekolainen, S.: Stocks and flows of nitrogen and phosphorus in the Finnish food production and consumption system, Agr. Ecosyst. Environ., 107, 287–305, 2005.

ASAE: Manure Production and Characteristics. D384.2, American Society of Agricultural Engineers, St. Joseph, MI, USA, 2005.

Bennett, E. M., Carpenter, S. R., and Caraco, N. F.: Human impact on erodable phosphorus and eutrophication: a global perspective: increasing accumulation of phosphorus in soil threatens rivers, lakes, and coastal oceans with eutrophication, BioScience, 51, 227–234, 2001.

Bouwman, A. F., Beusen, A. H., and Billen, G.: Human alteration of the global nitrogen and phosphorus soil balances for the period 1970–2050, Global Biogeochem. Cy., 23, https://doi.org/10.1029/2009GB003576, 2009.

Bouwman, L., Goldewijk, K. K., Van Der Hoek, K. W., Beusen, A. H., Van Vuuren, D. P., Willems, J., Rufino, M. C., and Stehfest, E.: Exploring global changes in nitrogen and phosphorus cycles in agriculture induced by livestock production over the 1900–2050 period, P. Natl. Acad. Sci. USA, 110, 20882–20887, 2013.

Carpenter, S. R. and Bennett, E. M.: Reconsideration of the planetary boundary for phosphorus, Environ. Res. Lett., 6, 014009, https://doi.org/10.1088/1748-9326/6/1/014009, 2011.

Carpenter, S. R., Caraco, N. F., Correll, D. L., Howarth, R. W., Sharpley, A. N., and Smith, V. H.: Nonpoint pollution of surface waters with phosphorus and nitrogen, Ecol. Appl., 8, 559–568, 1998.

Chang, J., Viovy, N., Vuichard, N., Ciais, P., Campioli, M., Klumpp, K., Matin, R., Leip, A., and Soussana, J. F.: Modeled changes in potential grassland productivity and in ruminant livestock density in Europe over 1961–2010, PLOS ONE, 10, e0127554, https://doi.org/10.1371/journal.pone.0127554, 2015.

Chang, J. F., Viovy, N., Vuichard, N., Ciais, P., Wang, T., Cozic, A., Lardy, R., Graux, A.-I., Klumpp, K., Martin, R., and Soussana, J.-F.: Incorporating grassland management in ORCHIDEE: model description and evaluation at 11 eddy-covariance sites in Europe, Geosci. Model Dev., 6, 2165–2181, https://doi.org/10.5194/gmd-6-2165-2013, 2013.

COMIFER: Teneur en P, K et Mg des organes végétaux récoltés pour les cultures de plein champ et les principaux fourrages, Comité Français d'Étude et de Développement de la Fertilisation Raisonneé, Paris, 2007 (in French).

Cordell, D., Drangert, J., and White, S.: The story of phosphorus: global food security and food for thought, Global Environ. Chang., 19, 292–305, 2009.

Cordell, D., Neset, T. S., and Prior, T.: The phosphorus mass balance: identifying "hotspots" in the food system as a roadmap to phosphorus security, Curr. Opin. Biotech., 23, 839–845, 2012.

Cordell, D., Jackson, M., and White, S.: Phosphorus flows through the Australian food system: identifying intervention points as a roadmap to phosphorus security, Environ. Sci. Policy, 29, 87–102, 2013.

Dalin, C., Hanasaki, N., Qiu, H., Mauzerall, D. L., and Rodriguez-Iturbe, I.: Water resources transfers through Chinese interprovincial and foreign food trade, P. Natl. Acad. Sci. USA, 111, 9774–9779, 2014.

Elser, J. and Bennett, E.: Phosphorus cycle: a broken biogeochemical cycle, Nature, 478, 29–31, 2011.

FAO: Fertilizer Use by Crop (5th Edition), Food and Agriculture Organization of the United Nations, Rome, 2012.

Foley, J. A., Ramankutty, N., Brauman, K. A., Cassidy, E. S., Gerber, J. S., Johnston, M., Muller, N. D., O'Connell, C., Ray, D. K., West, P. C., Balzer, C., Bennett, E. M., Carpenter, S. R., Hill, J., Monfreda, C., Polasky, S., Rockström, J., Sheehan, J., Siebert, S., Tilman, D., and Balzer, C.: Solutions for a cultivated planet, Nature, 478, 337–342, 2011.

Grote, U., Craswell, E., and Vlek, P.: Nutrient flows in international trade: ecology and policy issues, Environ. Sci. Policy, 8, 439–451, 2005.

Herrero, M., Havlík, P., Valin, H., Notenbaert, A., Rufino, M. C., Thornton, P. K., Blümmelb, M., Weissc, F., Grace, D., and Obersteiner, M.: Biomass use, production, feed efficiencies, and greenhouse gas emissions from global livestock systems, P. Natl. Acad. Sci. USA, 110, 20888–20893, 2013.

Levington Agriculture: A Report for the European Fertiliser Manufacturers' Association. Levington Agriculture Ltd., Ipswich, UK, 1997.

Li, J.: The development and prospective of nutrition in livestock bones, Meat Ind., 322, 41–44, 2008 (in Chinese).

Liu, Y., Villalba, G., Ayres, R. U., and Schroder, H.: Global phosphorus flows and environmental impacts from a consumption perspective, J. Ind. Ecol., 12, 229–247, 2008.

Ma, W., Ma, L., Li, J., Wang, F., Sisák, I., and Zhang, F.: Phosphorus flows and use efficiencies in production and consumption of wheat, rice, and maize in China, Chemosphere, 84, 814–821, 2011.

MacDonald, G. K., Bennett, E. M., Potter, P. A., and Ramankutty, N.: Agronomic phosphorus imbalances across the world's croplands, P. Natl. Acad. Sci. USA, 108, 3086–3091, 2011.

Metson, G. S., Bennett, E. M., and Elser, J. J.: The role of diet in phosphorus demand, Environ. Res. Lett., 7, 044043, https://doi.org/10.1088/1748-9326/7/4/044043, 2012.

MWPS-18: Livestock Waste Facilities Handbook, Midwest Plan Service, University of Missouri, Ames, IA, USA, 1985.

Obersteiner, M., Peñuelas, J., Ciais, P., Van Der Velde, M., and Janssens, I. A.: The phosphorus trilemma, Nat. Geosci., 6, 897–898, 2013.

OECD: Secretariat National Soil Surface Nutrient Balances, 1985 to 1995. Explanatory Notes. Coefficients to Convert Livestock Numbers Into Manure Nitrogen Quantities From National Sources, Organisation for Economic Cooperation and Development, Paris, 1991.

Ott, C. and Rechberger, H.: The European phosphorus balance, Resour. Conserv. Recy., 60, 159–172, 2012.

Penuelas, J., Poulter, B., Sardans, J., Ciais, P., Van Der Velde, M., Bopp, L., Boucher, O., Godderis, Y., Hinsinger, P., Llusia, J., Nardin, E., Vicca, S., Obersteiner, M., and Nardin, E.: Human-induced nitrogen–phosphorus imbalances alter natural and managed ecosystems across the globe, Nat. Commun., 4, 2934, https://doi.org/10.1088/1748-9326/7/4/044043, 2013.

Potter, P., Ramankutty, N., Bennett, E. M., and Donner, S. D.: Characterizing the spatial patterns of global fertilizer application and manure production, Earth Interact., 14, 1–22, 2010.

Quinton, J. N., Govers, G., Van Oost, K., and Bardgett, R. D.: The impact of agricultural soil erosion on biogeochemical cycling, Nat. Geosci., 3, 311–314, 2010.

Ringeval, B., Nowak, B., Nesme, T., Delmas, M., and Pellerin, S.: Contribution of anthropogenic phosphorus to agricultural soil fertility and food production, Global Biogeochem. Cy., 28, 743–756, 2014.

Sattari, S. Z., Bouwman, A. F., Giller, K. E., and van Ittersum, M. K.: Residual soil phosphorus as the missing piece in the global phosphorus crisis puzzle, P. Natl. Acad. Sci. USA, 109, 6348–6353, 2012.

Schipanski, M. E. and Bennett, E. M.: The influence of agricultural trade and livestock production on the global phosphorus cycle, Ecosystems, 15, 256–268, 2012.

Scholz, R. W., Ulrich, A. E., Eilittä, M., and Roy, A.: Sustainable use of phosphorus: a finite resource, Sci. Total Environ., 461, 799–803, 2013.

Senthilkumar, K., Nesme, T., Mollier, A., and Pellerin, S.: Regional-scale phosphorus flows and budgets within France: the importance of agricultural production systems, Nutr. Cycl. Agroecosys., 92, 145–159, 2012.

Sheldrick, W., Syers, J. K., and Lingard, J.: Contribution of livestock excreta to nutrient balances, Nutr. Cycl. Agroecosys., 66, 119–131, 2003.

Smil, V.: Phosphorus in the environment: natural flows and human interferences, Annu. Rev. Mater. Sci., 25, 53–88, 2000.

Steffen, W., Richardson, K., Rockström, J., Cornell, S. E., Fetzer, I., Bennett, E. M., Biggs, R., Carpenter, S. R., de Vries, W., de Wit, C. A., Folke, C., Gerten, D., Heinke, J., Mace, G. M., Persson, L. M., Ramanathan, V., Reyers, B., and Sörlin, S.: Planetary boundaries: guiding human development on a changing planet, Science, 347, 1259855, https://doi.org/10.1126/science.1259855, 2015.

Suh, S. and Yee, S.: Phosphorus use-efficiency of agriculture and food system in the US, Chemosphere, 84, 806–813, 2011.

USDA-NRCS: Crop Nutrient Tool: Nutrient Content of Crops. United States Department of Agriculture, Natural Resource Conservation Service, Washington, 2009.

van Dijk, K. C., Lesschen, J. P., and Oenema, O.: Phosphorus flows and balances of the European Union Member States, Sci. Total Environ., 542, 1078–1093, 2016.

Van Vuuren, D. P., Bouwman, A. F., and Beusen, A. H. W.: Phosphorus demand for the 1970–2100 period: a scenario analysis of resource depletion, Global Environ. Chang., 20, 428–439, 2010.

Waller, J. C.: Byproducts and unusual feedstuffs, Feedstuffs, 9, 18–22, 2010.

Wang, R., Tao, S., Balkanski, Y., Ciais, P., Boucher, O., Liu, J., Piao, S., Shen, H., Vuoloc, M. R., Valarie, M., Chen, H., Chen, Y., Cozic, A., Huang, Y., Li, B., Li, W., Shen, G., Wang, B., and Zhang, Y.: Exposure to ambient black carbon derived from a unique inventory and high-resolution model, P. Natl. Acad. Sci. USA, 111, 2459–2463, 2014.

Wang, R., Balkanski, Y., Boucher, O., Ciais, P., Peñuelas, J., and Tao, S.: Significant contribution of combustion-related emissions to the atmospheric phosphorus budget, Nat. Geosci., 8, 48–54, 2015.

Wu, H., Yuan, Z., Zhang, Y., Gao, L., and Liu, S.: Life-cycle phosphorus use efficiency of the farming system in Anhui Province, Central China, Resour. Conserv. Recy., 83, 1–14, 2014.

Wu, L. and Ma, M.: The comprehensive utilization of animals' bone in China, Modern Food Sci. Tech., 83, 38–46, 2005 (in Chinese).

Xu, J., Liu, X., Wang, F., Zhang, F., Ma, W., and Ma, L.: Phosphorus balance and environmental effect of animal production in China, Acta Ecol. Sinica, 11, 2911–2918, 2005 (in Chinese).

4

Photosynthesis–irradiance parameters of marine phytoplankton: synthesis of a global data set

Heather A. Bouman[1], Trevor Platt[2], Martina Doblin[3], Francisco G. Figueiras[4], Kristinn Gudmundsson[5], Hafsteinn G. Gudfinnsson[5], Bangqin Huang[6], Anna Hickman[7], Michael Hiscock[8], Thomas Jackson[2], Vivian A. Lutz[9], Frédéric Mélin[10], Francisco Rey[11], Pierre Pepin[12], Valeria Segura[9], Gavin H. Tilstone[2], Virginie van Dongen-Vogels[3], and Shubha Sathyendranath[2,13]

[1]Department of Earth Sciences, University of Oxford, Oxford, OX1 3AN, UK
[2]Plymouth Marine Laboratory, Prospect Place, The Hoe, PL1 3DH, UK
[3]Plant Functional Biology and Climate Change Cluster, Faculty of Science, University of Technology Sydney, P.O. Box 123 Broadway, Sydney, NSW 2007, Australia
[4]Instituto de Investigaciones Marinas (CSIC), Eduardo Cabello 6, 36208 Vigo, Spain
[5]Marine Research Institute, P.O. Box 1390, 121 Reykjavík, Iceland
[6]State Key Laboratory of Marine Environmental Science/Key Laboratory of Coastal and Wetland Ecosystems, Ministry of Education, Xiamen University, Xiamen, Fujian 361005, China
[7]Ocean and Earth Science, University of Southampton, National Oceanography Centre Southampton, European Way, Southampton, SO14 3ZH, UK
[8]United States Environmental Protection Agency, Ariel Rios Building, 1200 Pennsylvania Avenue, Washington D.C., 20460, USA
[9]Instituto Nacional de Investigacion y Desarrollo Pesquero, Mar del Plata, Argentina
[10]European Commission, Joint Research Centre, Ispra 21027, Italy
[11]Institute of Marine Research, c/o Department of Biological Sciences, University of Oslo, P.O. Box 1066, 0316, Oslo, Norway
[12]Fisheries and Oceans Canada, Northwest Atlantic Fisheries Centre, P.O. Box 5667, St John's, Newfoundland, A1C 5X1, Canada
[13]National Centre for Earth Observation, Plymouth Marine Laboratory, Prospect Place, Plymouth PL1 3DH, UK

Correspondence: Heather A. Bouman (heather.bouman@earth.ox.ac.uk)

Abstract. The photosynthetic performance of marine phytoplankton varies in response to a variety of factors, environmental and taxonomic. One of the aims of the MArine primary Production: model Parameters from Space (MAPPS) project of the European Space Agency is to assemble a global database of photosynthesis–irradiance (*P-E*) parameters from a range of oceanographic regimes as an aid to examining the basin-scale variability in the photophysiological response of marine phytoplankton and to use this information to improve the assignment of *P-E* parameters in the estimation of global marine primary production using satellite data. The MAPPS *P-E* database, which consists of over 5000 *P-E* experiments, provides information on the spatio-temporal variability in the two *P-E* parameters (the assimilation number, P_m^B, and the initial slope, α^B, where the superscripts B indicate normalisation to concentration of chlorophyll) that are fundamental inputs for models (satellite-based and otherwise) of marine primary production that use chlorophyll as the state variable. Quality-control measures consisted of removing samples with abnormally high parameter values and flags were added to denote whether the spectral quality of the incubator lamp was used to calculate a broad-band value of α^B. The MAPPS database provides a photophysiological data set that is unprecedented in number of observations and in spatial coverage. The database will be useful to a variety of research communities, including marine ecologists, biogeochemical modellers, remote-sensing scientists and algal physiologists. The compiled data are available at https://doi.org/10.1594/PANGAEA.874087 (Bouman et al., 2017).

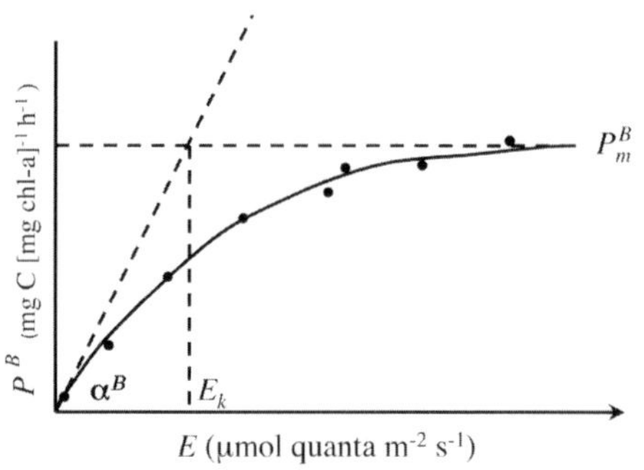

Figure 1. Photosynthesis–irradiance (*P-E*) curve showing the two biomass-normalised photophysiological parameters, the maximum photosynthetic rate P^B_m and the initial slope α^B.

1 Introduction

Although global estimates of marine primary production tend to converge on a number around 40–50 GT yr^{-1}, the accuracy and precision on regional scales of the estimation protocols remain relatively poor, partly as a result of an incomplete understanding of how the photosynthetic performance of marine phytoplankton varies in the global ocean (Carr et al., 2006; Lee et al., 2015). Photosynthesis–irradiance (*P-E*) parameters derived from carbon uptake experiments conducted over a controlled range of available-light levels provide a means of comparing the photosynthetic characteristics of marine phytoplankton across different natural populations and cultured isolates (Platt and Jassby, 1976; Prézelin et al., 1989; MacIntyre et al., 2002). The *P-E* experiment exposes algal cells to a range of light intensities from near-zero to those levels typically available at the sea surface (Lewis and Smith, 1983; Babin et al., 1994). The photosynthetic rates are then normalised to the concentration of chlorophyll *a* (a useful and practical index of phytoplankton biomass relevant for photosynthesis) found within the sample. This normalisation serves two purposes: first, dividing by pigment biomass reduces the variability of photosynthesis rates due to differences in biomass alone, facilitating the comparison of photosynthetic performance across trophic gradients, and second, chlorophyll-normalised photophysiological parameters may be applied in the estimation of primary production over large scales by using satellite-derived maps of chlorophyll concentration (Longhurst et al., 1995; Antoine and Morel, 1996). A schematic diagram showing the biomass-normalised data generated from these experiments plotted against the light intensity at which each bottle was incubated is shown in Fig. 1 to illustrate how the ensemble of data, when fitted to a suitable non-linear equation, forms a *P-E* curve. The curve may be represented by a variety of mathematical forms (Jassby and Platt, 1976; Platt et al., 1980). In cases where photoinhibition is negligible, all equations suitable for describing the *P-E* curve can be represented using just two parameters: the initial slope, α^B, which represents the photosynthetic efficiency under light levels close to zero, and the asymptote of the curve, P^B_m, which is the photosynthetic rate at light saturation (Jassby and Platt, 1976; Platt et al., 1980; Sakshaug et al., 1997).

2 Data

Chlorophyll *a* concentrations and photosynthesis–irradiance (*P-E*) parameters collected from four oceanic domains and 35 biogeochemical provinces (Longhurst, 2007, Table 1) were compiled from individual investigators and online data repositories (Table 2). *P-E* data were obtained by ^{14}C and ^{13}C (Argentine Sea) uptake experiments, with incubation times varying from 1.5 to 4 h. Chlorophyll concentrations used to normalise the carbon fixation rates were measured using either high-performance liquid chromatography (HPLC) or the standard fluorometric method (Mantoura et al., 1997). An intercomparison between HPLC and fluorometrically determined chlorophyll *a* concentrations revealed that pheopigment-correcting acidification methods such as Holm-Hansen et al. (1965) show a good overall correlation ($r^2 = 0.85$). However, the study noted that the presence of the accessory pigment chlorophyll *b* could lead to an underestimation of chlorophyll *a* concentration by 2–19 % (Mantoura et al., 1997). This potential source of bias in fluorometrically determined chlorophyll *a* concentration would result in an overestimation of the chlorophyll *a* normalised photosynthetic parameters of up to 19 % where relative chlorophyll *b* concentrations are high (e.g. the deep chlorophyll maxima of the subtropical gyres). Further details on the experimental methodology for individual field campaigns are provided in the original publications (see Table 2). The environmental variables and photosynthetic parameters included the MAPPS database and their corresponding units are listed in Table 3.

Table 2 includes information on which functional form was fitted to the *P-E* data for each of the data sets used in this study. In cases where photoinhibition was absent (photosynthetic rates stayed independent of irradiance in the light-saturated range), or where the fit was applied to data unaffected by photoinhibition, a two-parameter curve fit was

Table 1. Numbers corresponding to biogeochemical province and domain as described by Longhurst (2007) included in the MAPPS database.

Province number	Longhurst domain	Longhurst province
1	Polar	Boreal Polar Province
2	Polar	Atlantic Arctic Province
3	Polar	Atlantic Subarctic Province
4	Westerlies	North Atlantic Drift Province
5	Westerlies	Gulf Stream Province
6	Westerlies	North Atlantic Subtropical Gyre Province (West)
7	Trades	North Atlantic Tropical Gyre Province
8	Trades	Western Tropical Atlantic Province
9	Trades	Eastern Tropical Atlantic Province
10	Trades	South Atlantic Gyre Province
11	Coastal	North East Atlantic Shelves Province
12	Coastal	Canary Coastal Province
15	Coastal	North West Atlantic Shelves Province
17	Trades	Caribbean Province
18	Westerlies	North Atlantic Subtropical Gyre Province (East)
20	Coastal	Brazil Current Coastal Province
21	Coastal	South West Atlantic Shelves Province
22	Coastal	Benguela Current Coastal Province
30	Trades	Indian Monsoon Gyres Province
33	Coastal	Red Sea, Persian Gulf Province
34	Coastal	North West Arabian Upwelling Province
37	Coastal	Australia–Indonesia Coastal Province
50	Polar	North Pacific Epicontinental Province
51	Westerlies	Pacific Subarctic Gyres Province (East)
53	Westerlies	Kuroshio Current Province
58	Westerlies	Tasman Sea Province
60	Trades	N. Pacific Tropical Gyre Province
63	Trades	W. Pacific Warm Pool Province
64	Trades	Archipelagic Deep Basins Province
68	Coastal	Chile–Peru Current Coastal Province
69	Coastal	China Sea Coastal Province
80	Westerlies	S. Subtropical Convergence Province
81	Westerlies	Subantarctic Province
82	Polar	Antarctic Province
83	Polar	Austral Polar Province

used, of the form

$$P^B(E) = P_m^B \tanh\left(\frac{\alpha^B E}{P_m^B}\right), \tag{1}$$

where $P^B(E)$ is the chlorophyll-normalised photosynthetic rate ($\mathrm{mg\,C\,(mg\,Chl}\ a)^{-1}\,\mathrm{h}^{-1}$) and E is the available light, which in this study is expressed in $\mu\mathrm{mol\,quanta\,m^{-2}\,s^{-1}}$. The light saturation parameter, E_k, is defined by the following relationship,

$$E_k = \frac{P_m^B}{\alpha^B}, \tag{2}$$

and is illustrated in Fig. 1 by the drawing a line from the intersection of the initial slope with the plateau of the curve onto the abscissa and has dimensions of irradiance.

In most cases, however, data were fit to the three-parameter function of Platt et al. (1980), which also describes the decrease in photosynthetic rate with irradiances much higher than saturating light levels, as follows:

$$P^B(E) = P_s^B\left(1 - \exp\left(-\frac{\alpha^B E}{P_s^B}\right)\right)\exp\left(-\frac{\beta^B E}{P_s^B}\right), \tag{3}$$

where β^B is the photoinhibition parameter describing the decrease in photosynthetic rate at high irradiance and P_s^B is the hypothetical maximum photosynthetic rate in the absence of photoinhibition. Hence when $\beta^B = 0$, $P_s^B = P_m^B$. When photoinhibition was present, values of P_m^B were derived using the following equation:

$$P_m^B = P_s^B\left(\frac{\alpha^B}{\alpha^B + \beta^B}\right)\left(\frac{\beta^B}{\alpha^B + \beta^B}\right)^{\frac{\beta}{\alpha}}. \tag{4}$$

Table 2. Summary of contributions to the MAPPS database.

Dataset provider	Regions	Dates	N	Non-linear equation(s) fitted to experimental data	Database	Relevant publication(s)
Trevor Platt, Plymouth Marine Laboratory (tplatt@dal.ca)	Arctic, Arabian Sea, Azores, Caribbean Sea, Celtic Sea, Georges Bank, Grand Banks, Humboldt Current System, Hudson Bay, Labrador Sea, Mid-Atlantic Ridge, New England Seamounts, Sargasso Sea, Scotian Shelf, Vancouver Island	1977–2003	2146	Photoinhibition function (Platt et al., 1980)	BIOCHEM (www.meds-sdmm.dfo-mpo.gc.ca)	Bouman et al. (2005); Harrison and Platt (1986); Kyewalyanga et al. (1998); Platt et al. (1980); Platt et al. (1982); Platt et al. (1993); Sathyendranath et al. (1999)
Francisco Rey, Institute of Marine Research (pancho@IMR.no)	Barents Sea	1980–1989	223	Photoinhibition function (Platt et al., 1980)		Rey (1991)
Pierre Pepin, Fisheries and Oceans Canada, Northwest Atlantic Fisheries Centre (pierre.pepin@dfo-mpo.gc.ca)	Grand Banks	2004–2012	524	Photoinhibition function (Platt et al., 1980)		Unpublished
Heather Bouman, University of Oxford (heather.bouman@earth.ox.ac.uk)	Subtropical Atlantic, Greenland Sea, Norwegian Sea	1996, 2010, 2013	195	Photoinhibition function (Platt et al., 1980)		Bouman et al. (2000a); Jackson (2013); Bouman (unpublished)
Michael Hiscock, National Center for Environmental Research US Environmental Protection Agency (hiscock.michael@epa.gov)	Southern Ocean – Pacific sector	1997–1998	172	Photoinhibition function (Platt et al., 1980)		Hiscock (2004); Hiscock et al. (2003)
Vivian Lutz, Instituto Nacional de Investigación y Desarrollo Pesquero (vlutz@inidep.edu.ar)	Argentine Sea	2005–2006	69	Photoinhibition function (Platt et al., 1980)		Dogliotti et al. (2014); Lutz et al. (2010); Segura et al. (2013)
Gavin Tilstone, Plymouth Marine Laboratory (ghti@pml.ac.uk)	Benguela upwelling system, eastern tropical Atlantic, North Atlantic Subtropical Gyre, Canary coastal system, North Atlantic Drift Province	1998	129	Photoinhibition function (Platt et al., 1980)	BODC (www.bodc.ac.uk)	Tilstone et al. (2003)
Bangqin Huang, State Key Laboratory of Marine Environmental Science/Key Laboratory of Coastal and Wetland Ecosystems Ministry of Education, Xiamen University (bqhuang@xmu.edu.cn)	South China Sea	2010–2012	130	Photoinhibition function (Platt et al., 1980)		Xie et al. (2015)
Anna Hickman, National Oceanography Centre Southampton (a.hickman@noc.soton.ac.uk)	North Atlantic Subtropical Gyre, North Atlantic Drift Province, Canary coastal system, South Atlantic Subtropical Gyre, western tropical Atlantic	2004	31	Hyperbolic tangent function (Jassby and Platt, 1976).	BODC (www.bodc.ac.uk)	Hickman (2007); Lawrenz et al. (2013)
Kristinn Gudmundsson, Marine Research Institute, Iceland (kristinn@hafro.is)	Iceland and Faroes	1981–2007	559	Photoinhibition function (Platt et al., 1980) and hyperbolic tangent function (Jassby and Platt, 1976).		Gudmundsson (1998); Pálsson et al. (2012); Zhai et al. (2012)
Francisco G. Figueiras, Instituto de Investigaciones Marinas (CSIC) Eduardo Cabello 6, 36208 Vigo, Spain (paco@iim.csic.es)	Antarctic Peninsula	1995	51	Exponential without photoinhibition Webb et al. (1974)	JGOFS International Collection Volume 1: Discrete Datasets (1989–2000) DVD	Lorenzo et al. (2002)
Martina Doblin, University of Technology, Sydney (martina.doblin@uts.edu.au)	Southern Ocean, Antarctic Peninsula, Tasman Sea	1990–2013	1482	Photoinhibition function (Platt et al., 1980)	AADC (https://data.aad.gov.au/metadata) MARLIN (http://www.marine.csiro.au/marq/edd_search.search_choice?tFre=primary+production&ch1=freetext&cSub=%3E%3E) CSIRO Marine National Facility (http://www.marine.csiro.au/nationalfacility/voyages/datasets.htm) PANGAEA (http://doi.pangaea.de/10.1594/PANGAEA.103773, http://doi.pangaea.de/10.1594/PANGAEA.843554,	Mackey et al. (1995); Griffiths et al. (1999); Hanson et al. (2005a); Hanson et al. (2005b); Westwood et al. (2011)

Table 3. List of environmental variables and *P-E* parameters in the MAPPS database and their corresponding units.

Header	Description	Units
LAT	Latitude of sampling station	Decimal degrees
LON	Longitude of sampling station	Decimal degrees
DEPTH	Depth at which sample was collected	m
YEAR	Year of sample collection	
MONTH	Month of sample collection	
DAY	Day of sample collection	
TCHL	Chlorophyll concentration measured using either high-performance liquid chromatography or the fluorometric method.	$\text{mg Chl } a\,\text{m}^{-3}$
ALPHA	α^B, the initial slope of the photosynthesis–irradiance curve normalised to phytoplankton chlorophyll concentration.	$\text{mg C (mg Chl } a)^{-1}\,\text{h}^{-1}\,(\mu\text{mol quanta m}^{-2}\,\text{s}^{-1})^{-1}$
PMB	P_m^B, the rate of photosynthesis at saturating irradiance, normalised to phytoplankton chlorophyll concentration.	$\text{mg C (mg Chl } a)^{-1}\,\text{h}^{-1}$
EK	E_k, irradiance at which the onset of saturation occurs, calculated as the ratio of P_m^B to α^B.	$\mu\text{mol quanta m}^{-2}\,\text{s}^{-1}$
PROVNUM	The corresponding biogeochemical province defined by Longhurst (please refer to Table 2).	

3 Quality control for the MAPPS *P-E* database

3.1 Experimental conditions

The *P-E* experiments were performed in incubators that maintained samples under in situ temperature conditions using either temperature-controlled water baths or the ship's underway water supply. Samples where incubation temperatures differed from in situ temperatures by more than 2 °C were removed from the database. It is well known that the light spectrum has a significant effect on the magnitude of light-limited photosynthesis (α^B) and the derived light saturation parameter (E_k) (Kyewalyanga et al., 1997; Schofield et al., 1991). We have included in the database quality flags indicating whether a correction factor for the spectrum of the lamp was applied to obtain a readily intercomparable broad-band (white light) value (e.g. Kyewalyanga et al., 1997; Xie et al., 2015). This broad-band α^B combined with information on the shape of the phytoplankton absorption spectrum has been shown to provide an accurate estimate of the photosynthetic action spectrum $\alpha^B(\lambda)$. The correction factor X can be used to convert the measured α^B from the incubation experiment using a given artificial light source to an estimate of α^B if the sample were subject to a spectrally neutral light environment: it is the ratio of the unweighted mean absorption coefficient of phytoplankton ($\overline{a}_\text{p}$) to the mean absorption coefficient weighted by the shape of the emission spectrum of the lamp source ($\overline{a}_\text{L}$),

$$X = \frac{\overline{a}_\text{p}}{\overline{a}_T}, \tag{5}$$

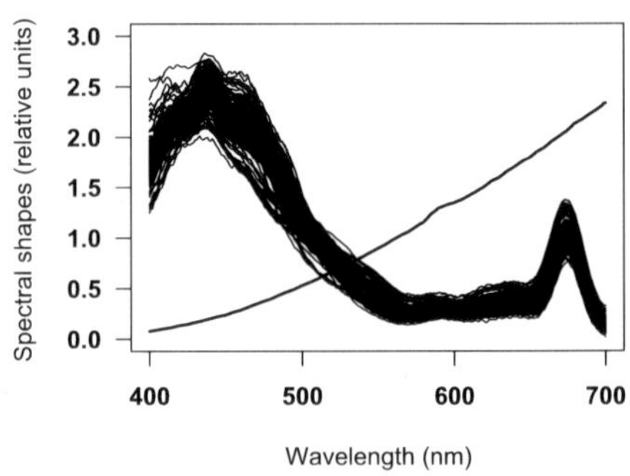

Figure 2. Spectral shapes of the tungsten-halogen lamp used in the incubations conducted by the Bedford Institute from 1984 to 2003 (red line) and in vivo absorption spectra of marine phytoplankton collected in the North Atlantic and subpolar waters (Labrador Sea). Phytoplankton and lamp spectra were normalised to their mean value to define the shape.

where $\overline{a}_\text{p}$ is determined by

$$\overline{a}_\text{p} = \frac{\int_{400}^{700} a_\text{p}(\lambda)\text{d}\lambda}{\int_{400}^{700}\text{d}\lambda}, \tag{6}$$

and $\overline{a}_T$ is computed as

$$\overline{a}_T = \frac{\int_{400}^{700} a_\text{p}(\lambda)E_T(\lambda)\text{d}\lambda}{\int_{400}^{700} E_T(\lambda)\text{d}\lambda}. \tag{7}$$

The incubators in this study used a range of light sources, including tungsten halogen, halogen, metal halide, and fluorescent lamps. Tungsten halogen lamps are the most commonly used light source in *P-E* experiments because they

Table 4. Information on the light sources and filters used for photosynthesis–irradiance experiments for each of the data set providers. Also noted is whether the spectral correction of Kyewalyanga et al. (1997) shown in equation was applied to values of α^B.

Dataset provider	Lamp source details	Spectral correction
Platt	GTE Sylvania PAR 150	N
	Combination of GTE Sylvania PAR 150 (irradiances < 200 W) and New Haline OHS tungsten halogen (irradiances 200–1000 W)	N
	2000 W tungsten-halogen lamp (New Haline OHS 2000) with a maximum intensity of 1000 W m^{-2} (PAR)	N
	Gilway Technical Lamp L 7391 tungsten halogen. Spectral correction applied to samples collected from 1994 onwards.	Y
Rey	Low-light incubator (LL, 0 to 390 mmol m^{-2} s^{-1}) was equipped with daylight-type fluorescent tubes (OSRAM 191 Daylight 5000 de Luxe). High-light incubator (HL, 0 to 1700 µmol m^{-2} s^{-1}) was equipped with a halogen-metal lamp (OSRAM Power Star, 400 W).	N
Pepin	ENH-type tungsten-halogen quartz projection lamps directed through a heat filter (solution of copper sulfate 20 g L^{-1}) to remove the infrared emission (no additional corrections made).	N
Bouman	Gilway Technical Lamp L 7391 tungsten halogen, Spectral correction applied.	Y
	Lee 1/4 CT blue filter was placed in front of incubator window to diminish the spectral dependency of the light source (2 × 50 W Sylvania 2315 tungsten halogen). Spectral correction applied.	Y
Hiscock	250 W tungsten-halogen slide projector lamp (Gray Co. #ENH), spectrally modified using a heat mirror, a broad-band cool mirror (Optical Coating Laboratory, Inc.), and blue stage-lighting screens (Cinemills Corp. #M144).	Y
Lutz	70 W Westinghouse halogen lamp. Spectral correction applied.	Y
Tilstone	AMT 6 and 20: tungsten-halogen lamps spectrally corrected. AMT 23: both tungsten-halogen and LED lamps used and spectrally corrected.	Y
Huang	150 W metal halide lamps with an ultraviolet filter. Spectral correction applied.	Y
Hickman	Tungsten halogen lamps behind blue light filters (no additional corrections made).	N
Figueiras	50 W (12 V) tungsten halogen lamp. Spectral correction applied.	Y
Gudmundsson	Fluorescent tubes (Philips TLF 20 W/33). No additional corrections made.	N
Doblin	Cool daylight fluorescent tubes (Philips TLD 36 W/54). A mix of grey and blue filters (Rosco 3402, 50 % neutral density filter, i.e. grey strips, and Rosco 3204 half blue) were used to attenuate the light intensity in the incubator.	N

provide intensities sufficiently high to resemble irradiances at the sea surface (~2000 µmol quanta m^{-2} s^{-1}). One limitation of using tungsten lamps is that they have a spectrum heavily weighted towards the red and infrared (see Fig. 2), unless the light first passes through a filter that removes the red emission. Table 4 describes the various lamps and filters used in the *P-E* incubators used in this study.

To estimate the impact of a tungsten halogen lamp compared with a white light source on the magnitude of the α^B and consequently E_k, which is derived from estimates of α^B (Eq. 2), we used a data set from the North Atlantic that spanned several decades. From 1994, *P-E* data have been corrected for the spectrum of the lamp source following the method of Kyewalyanga et al. (1997), whereas prior to 1994, no correction was made due to a lack of information on the excitation spectrum of the lamp and the absorptive properties of the phytoplankton communities. By comparing data from similar regions and seasons as lamp sources have changed, we are able to assess how the light source may cause variability in the photosynthetic parameter α^B. In the post-1994 data set, with corresponding lamp and absorption spectra, the correction factor X varied from 1.30 to 2.06 (mean = 1.70 with

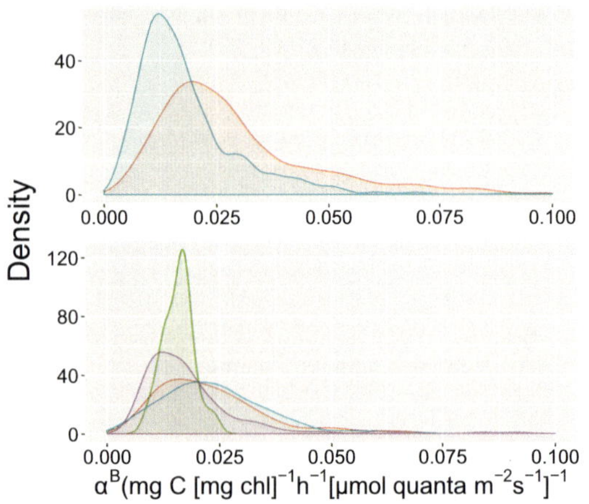

Figure 3. **(a)** Density plots of the initial slope (α^B) obtained using a tungsten halogen lamp (cyan) and corrected values using the shape of the phytoplankton absorption spectrum (red) as described in Kyewalyanga et al., 1997. **(b)** Density plots of data collected in the North Atlantic using different lamp types (red – GTE Sylvania PAR 150; green – combination of GTE Sylvania PAR 150 (irradiances < 200 W) and New Haline OHS tungsten halogen (irradiances 200–1000 W); cyan – New Haline OHS 2000 tungsten halogen; purple – Gilway Technical Lamp L 7391 tungsten halogen). Both plots are using data collected within the top 20 m of the water column.

a standard deviation of 0.15). This variation in X (Fig. 3a) is sufficient to account for difference in magnitudes of α^B obtained using incubators with different light sources across the North Atlantic cruise data set (Fig. 3b). Note that potential errors in the computation of primary production due to changes in α^B caused by spectral differences in light sources will be most acute deeper in the water column, where the influence of the magnitude of α^B on primary production is greatest (Ulloa et al., 1997; Bouman et al., 2000a) and thus errors for integrated water-column primary production will be modest since productivity rates are highest at the surface and decrease in an exponential manner once $E(z) \ll E_k$ (Ulloa et al., 1997; Bouman et al., 2000a).

3.1.1 Theoretical maxima

The photophysiological constraints of marine phytoplankton are well known and provide a useful check on the quality of the carbon-uptake experiments. The theoretical maximum quantum yield of carbon fixation (ϕ_m^T) is $0.125\,\mathrm{mol\,C\,(mol\,quanta)^{-1}}$ (Platt and Jassy, 1976; Sakshaug et al., 1997). The realised maximum quantum yield of photosynthesis (ϕ_m) is calculated by dividing α^B by $\overline{a}^*$, the chlorophyll-specific absorption coefficient of phytoplankton averaged over the visible spectrum (Platt and Jassby, 1976), and multiplying by a factor of 0.0231, which converts milligrams to moles of C, micromoles to moles of photons, and hours to seconds. Values of ϕ_m were calculated using either simultaneous measurements of $\overline{a}^*$, or estimates derived from a global relationship between chlorophyll concentrations and $\overline{a}^*$ (Bouman, unpublished data) and samples with ϕ_m well above the theoretical maximum ($> 0.15\,\mathrm{mol\,C\,(mol\,quanta)^{-1}}$) were discarded from the database. We also set a lower limit for the light saturation parameter P_m^B of $0.2\,\mathrm{mg\,C\,(mg\,Chl}\,a)^{-1}\,\mathrm{h}^{-1}$ and the initial slope α^B of $0.002\,\mathrm{mg\,C\,(mg\,Chl}\,a)^{-1}\,\mathrm{h}^{-1}\,(\mathrm{\mu mol\,quanta\,m^{-2}\,s^{-1}})^{-1}$. Data from experiments on sea-ice algae with Chl a concentrations exceeding $50\,\mathrm{mg\,Chl}\,a\,\mathrm{m}^{-3}$ were also removed. Using these criteria, 278 experiments were excluded from the global database.

4 Results

4.1 Spatio-temporal patterns of the MAPPS *P-E* database

In this study we adopt the Longhurst's (2007) geographical classification system of domains and provinces to partition the global data set according to the prevailing physical conditions that shape the structure and function of phytoplankton communities over large (basin) scales. The rationale behind using Longhurst's approach to estimate primary productivity is that physical forcing dictates the supply of nutrients and the average irradiance within the surface mixed layer and these factors directly impact the physiological capacity of algal cells. The four domains (also referred to as Longhurst biomes) are found in each ocean basin and are subject to distinct mechanisms of physical forcing: in the polar domain the density structure of the surface layer is strongly influenced by sea-ice melt; in the westerlies domain the mixed layer dynamics are governed by a local balance of heat-driven stratification and wind-driven turbulent mixing by winds; and in the trades domain, the depth of the mixed layer is governed by geostrophic responses to seasonal changes in the strength and location of the trade winds, while in the coastal domain, terrestrial influx of freshwater and interaction of local winds with topography play a critical role in governing ecosystem properties. The next level of partition is biogeochemical provinces, which embrace a wider set of environmental factors that govern regional ocean circulation and stratification that in turn influence ecological structure. Although it would be preferable that both domain and provincial boundaries were dynamic to accommodate seasonal, annual and decadal changes in ocean circulation (Devred et al., 2007), to exploit the entire MAPPS *P-E* parameter data set, which contains a large number samples that were collected prior to the launch of ocean-colour satellites, we used the fixed rectilinear boundaries of Longhurst (2007) with the understanding that some of the within-province variability may be the result of the estimated and actual provincial boundaries being spatially offset. Thus, for each data point, a Longhurst province was assigned based on geographic location and is

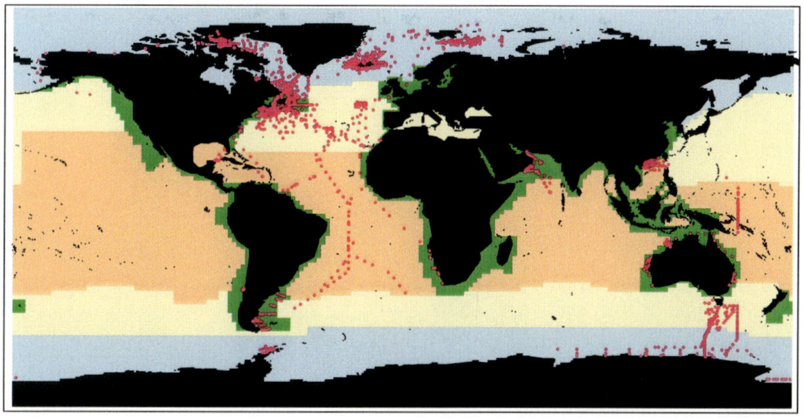

Figure 4. Global distribution of MAPPS *P-E* data set that passed quality control (5711 samples). The blocked colours represent the four primary biomes as described in Longhurst (2007): Polar (blue), Westerlies (yellow), Trades (orange) and Coastal (green).

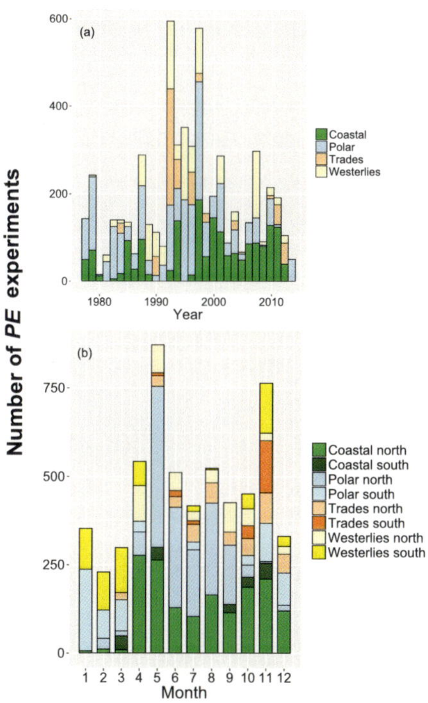

Figure 5. The number of *P-E* experiments in the MAPPS database for the four Longhurst oceanic domains by **(a)** year and **(b)** month and hemisphere (north and south).

denoted in the database by the province number as shown in Table 1.

Roughly half of the quality-controlled samples were collected from within the upper 20 m of the water column, accounting for approximately 54 % of the data set. Most of the data fall within the Atlantic Basin (Fig. 4), with a large region of the Pacific Basin being grossly undersampled in both space and time. The latitudinal coverage of the database is relatively sparse in the tropics and the mid-latitudes of the Southern Hemisphere. In this study seasons were divided into 3-month intervals in order for data to be used with monthly climatologies and satellite composite data. Thus, in the case of the Northern Hemisphere, "spring" covers the months of March to May, "summer" covers June to August, "autumn" covers September to November and "winter" covers December to February. The seasonal distribution of data shows the majority of samples were collected during the spring (40 %), summer (34 %) and autumn (22 %), and only 3 % of the samples collected during the winter period (Table 2, Fig. 5). Across all seasons, the data set covers a range of trophic conditions, with chlorophyll *a* concentrations representative of highly oligotrophic conditions ($0.02\,\mathrm{mg\,m^{-3}}$) to spring bloom conditions ($39.8\,\mathrm{mg\,m^{-3}}$) (Fig. 6). The dynamic range of the photosynthetic parameters was similar to that reported in other global studies, with values P_m^B ranging from 0.21 and $25.91\,\mathrm{mg\,C\,(mg\,Chl}\,a)^{-1}\,\mathrm{h}^{-1}$ with an average value of 3.11 and a standard deviation of 2.28, and values of α^B ranging from 0.002 to $0.373\,\mathrm{mg\,C\,(mg\,Chl}\,a)^{-1}\,\mathrm{h}^{-1}\,(\mathrm{\mu mol\,quanta\,m^{-2}\,s^{-1}})^{-1}$ with an average value of 0.043 and a standard deviation of 0.034.

The *P-E* parameters exhibited both spatial (between provinces) and temporal (between seasons) differences. In general, values of the assimilation number increased with decreasing latitude (Table 5) and tended to be higher during the summer months in temperate marine systems. However, the seasonal and latitudinal bias in data coverage has important implications for variability in parameter values in the data set because of the environmental conditions known to influence phytoplankton photophysiology. High-latitude samples will be associated with lower temperatures, which may limit their maximum photosynthetic rate for carbon fixation (Smith Jr. and Donaldson et al., 2015), and this is reflected in the generally low values of P_m^B in the boreal (BPLR) and austral (APLR) polar provinces. Geographical variation in surface irradiance may also explain the lower values of P_m^B in high latitudes compared with low latitudes. The combina-

Table 5. Seasonal mean values and standard deviation for the photosynthetic parameters P_m^B and α^B for each of the 35 Longhurst provinces. The blocked data represent the four primary biomes of the upper ocean: Polar (first quarter), Westerlies (second quarter), Trades (third quarter) and Coastal (fourth quarter).

PROV	Spring					Summer					Autumn					Winter				
		P_m^B		α^B			P_m^B		α^B			P_m^B		α^B			P_m^B		α^B	
	N	Mean	SD	Mean	SD	*N*	Mean	SD	Mean	SD	*N*	Mean	SD	Mean	SD	*N*	Mean	SD	Mean	SD
Polar																				
BPLR	194	1.86	0.91	0.037	0.032	385	1.93	1.29	0.030	0.031	50	2.10	1.12	0.029	0.022	13	1.56	0.76	0.019	0.007
ARCT	421	2.46	1.52	0.047	0.031	203	2.38	1.27	0.042	0.028	37	2.53	0.84	0.056	0.032	18	2.30	0.73	0.023	0.008
SARC	150	2.67	1.18	0.046	0.023	41	3.08	2.38	0.045	0.022	2	1.40	0.46	0.088	0.045	0				
ANTA	79	3.55	1.69	0.101	0.047	145	3.50	1.70	0.051	0.023	78	3.50	1.70	0.051	0.023	22	4.18	2.21	0.096	0.026
APLR	119	1.84	1.22	0.044	0.023	222	1.79	1.05	0.042	0.019	12	1.79	1.05	0.042	0.019	0				
BERS	0					0					3	3.43	2.03	0.059	0.043	0				
Westerlies																				
NADR	40	3.49	1.13	0.033	0.010	5	4.41	4.64	0.029	0.025	48	2.10	1.06	0.028	0.018	0				
GFST	55	4.41	2.40	0.031	0.013	12	3.57	2.30	0.016	0.007	32	3.41	1.85	0.056	0.039	0				
NASW	85	5.92	3.30	0.026	0.014	91	1.72	1.34	0.025	0.022	35	3.67	2.15	0.066	0.072	0				
NASE	15	2.63	1.42	0.030	0.006	6	4.02	1.88	0.031	0.013	37	3.17	2.65	0.057	0.042	0				
PSAE	0					0					2	3.41	1.05	0.034	0.010	0				
TASM	0					7	2.89	1.59	0.057	0.022	16	7.18	2.20	0.070	0.017	0				
SSTC	50	6.81	4.58	0.076	0.033	116	3.10	2.19	0.035	0.025	6	4.24	0.37	0.035	0.008	6	6.36	2.94	0.066	0.008
SANT	162	4.67	2.35	0.085	0.040	165	3.34	2.19	0.066	0.041	107	3.88	1.55	0.054	0.018	14	4.69	1.94	0.075	0.024
KURO	6	9.25	4.25	0.027	0.009	0					1	10.50		0.023		0				
Trades																				
NATR	6	3.33	2.29	0.037	0.017	66	2.54	1.50	0.057	0.048	20	2.97	2.69	0.041	0.035	0				
WTRA	0					8	5.08	4.53	0.064	0.015	17	1.34	0.63	0.022	0.020					
ETRA	11	3.00	0.96	0.027	0.011	0					0					0				
SATL	48	1.29	0.65	0.027	0.024	0					2	3.58	2.47	0.028	0.016	0				
CARB	20	3.71	2.16	0.011	0.005	0					15	2.15	1.75	0.026	0.020	0				
MONS	0					0					62	4.05	2.04	0.024	0.007	4	5.66	1.60	0.026	0.005
WARM	0					64	2.23	1.67	0.039	0.027	167	2.44	1.65	0.032	0.019	0				
ARCH	70	3.62	3.41	0.034	0.016	0					30	5.86	3.72	0.025	0.009	18	3.89	3.09	0.034	0.030
NPTG	0					0					4	2.62	1.99	0.054	0.074	0				
Coastal																				
CHIL	7	2.50	0.93	0.029	0.004	0					0					0				
NECS	29	3.20	1.07	0.023	0.010	0					1	2.22		0.062		0				
CNRY	0					0					23	4.62	1.40	0.065	0.017	0				
AUSW	44	3.04	2.02	0.038	0.015	15	2.61	1.93	0042	0.026	16	2.12	1.67	0.032	0.017	0				
BRAZ	9	2.37	1.86	0.041	0.035	5	3.63	2.05	0.014	0.005	0					1	0.87		0.019	
REDS	0					4	4.13	2.07	0.022	0.006	0					0				
ARAB	0					132	4.07	1.67	0.021	0.008	79	5.61	1.85	0.027	0.007	0				
FKLD	25	2.96	2.85	0.040	0.035	1	1.71		0.010		19	1.85	1.00	0.017	0.006	9	1.06	0.43	0.012	0.003
BENG	0					0					20	3.82	2.13	0.027	0.011	0				
CHIN	24	7.50	3.68	0.029	0.012	37	6.87	5.18	0.023	0.016	17	8.92	4.45	0.033	0.015	0				
NWCS	543	2.78	1.93	0.039	0.033	259	3.34	1.90	0.025	0.030	405	3.86	1.94	0.042	0.034	42	2.79	1.47	0.042	0.034

tion of lower surface irradiances and deep convective mixing in high latitudes results in markedly lower light levels within the mixed layer, which may result in photoacclimation to lower light levels, by modulating pigment content per cell and hence the carbon-to-chlorophyll ratio (Cullen et al., 1982; Sathyendranath et al., 2009). However, it is important to note that some of the polar samples were collected in regions highly influenced by sea-ice melt, which may lead to the formation of a fresh, shallow and highly stable mixed layer, and consequently higher average light level than would be the case for deeper mixing.

The paucity of winter data reduces the number of samples with cells acclimated to low growth irradiances. The number of observations is also low within the tropical and subtropical oceans, which are characterised by warm and hence strongly stratified mixed layers. Phytoplankton cells in these regions tend to have higher upper bounds of P_m^B, due to the combined effect of warmer sea-surface temperatures and the acclimation to high-light conditions. Such spatio-temporal patterns in the *P-E* parameters are likely driven by changes in oceanographic conditions (temperature, stratification, macro- and micronutrient availability) (Geider et al., 1996) as well as in community structure and other biotic processes that may consume cellular energy at the expense of carbon fixation (Puxty et al., 2016).

To examine whether spatial variation in the photosynthetic parameters for a given season was captured in the boundaries of the Longhurst provinces, we conducted the Bonferroni adjusted pairwise *t* tests to analyse differences between adjacent provinces for each season (Fig. 7). For both photosynthetic parameters differences across the boundaries were detected in polar regions and in seasons and provinces where

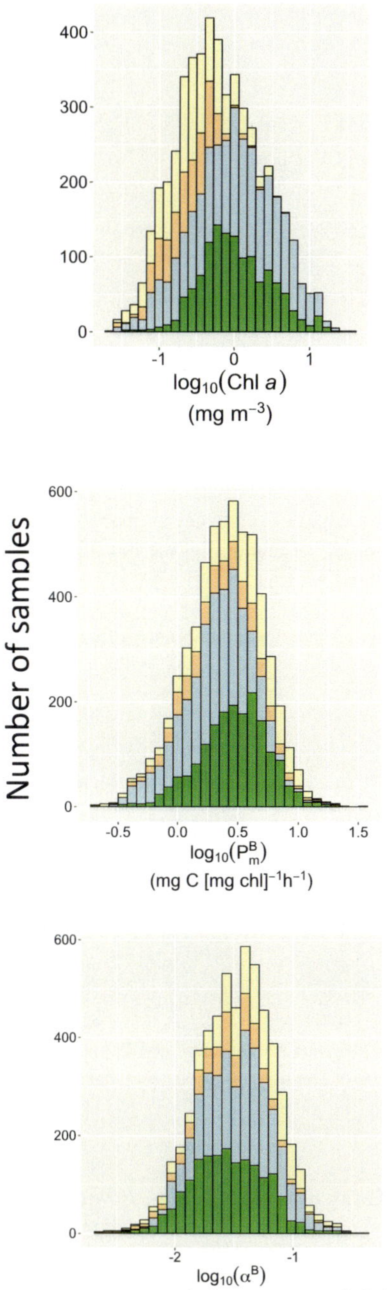

Figure 6. Histograms illustrating the sample distribution across concentrations of chlorophyll *a* and range of *P-E* parameter values for the four Longhurst oceanic domains.

the number of observations tended to be high. The static nature of the province boundaries and the uneven spatial distribution of the data may explain in part the small number of differences in the *P-E* parameters between adjacent provinces.

4.2 Relationship between the maximum photosynthetic rate and the initial slope

Strong correlations have been reported between the two *P-E* parameters P_m^B and α^B, which have been explained on both ecological and photophysiological grounds (Platt and Jassby, 1976; Côté and Platt, 1983; Behrenfeld et al., 2004). The MAPPS data set shows that the data fall largely within the bounds of E_k values between 20 and 300 μmol quanta m^{-2} s^{-1} (Fig. 8). In general, high-latitude samples (> 65°) tend to have lower P_m^B values for a given value of α^B (E_k values averaging 57.7 μmol quanta m^{-2} s^{-1}, with 10.0 % of the data falling above 100 μmol quanta m^{-2} s^{-1}) when compared with low-latitude samples between 40° N and 40° S (E_k values averaging 152.4 μmol quanta m^{-2} s^{-1}, with 57.1 % of the values falling above 100 μmol quanta m^{-2} s^{-1}).

When E_k is plotted as a function of latitude (Fig. 9) for open-ocean samples within the top 25 m of the water column, a clear pattern emerges, with higher latitude samples being characterised by lower values, whereas data from the mid- to low latitudes had, on average, higher values, although considerable scatter was observed over the entire range of temperatures. To illustrate the depth-dependent change in E_k due to vertical changes in irradiance, data from cruises that predominantly sampled stratified, oligotrophic regions (DCM and AMT cruises) are plotted against the sample depth (Fig. 10). The strong depth dependence of the photoacclimation parameter is consistent with other open-ocean studies (Babin et al., 1996). The latitudinal and depth dependence of E_k was also reported in a study which used a subset ($N = 1862$) of the MAPPS database from the North Atlantic spanning the tropics to the Arctic: 55 % of the variance in E_k could be explained using depth, latitude, temperature, nitrate and surface noon irradiance as predictive variables (Platt and Sathyendranath, 1995).

5 Discussion

Predicting the photosynthetic efficiency of phytoplankton cells remains one of the major challenges in determining marine primary production using remote sensing data (Carr et al., 2006). The MAPPS database of *P-E* parameters allows us to assess the global variability in phytoplankton photophysiological parameters and could be used to validate models that aim to provide a mechanistic understanding of changes in the photosynthetic parameters. Here, we attempt to explain the spatial patterns in the data set drawing on our current understanding of the key environmental factors governing variability in both P_m^B and α^B.

A positive correlation between P_m^B and α^B has been attributed to a variety of physiological and ecological factors, including changes in the allocation of ATP and NADPH to carbon fixation (Behrenfeld et al., 2004), as well as changes in phytoplankton community structure (Côté and

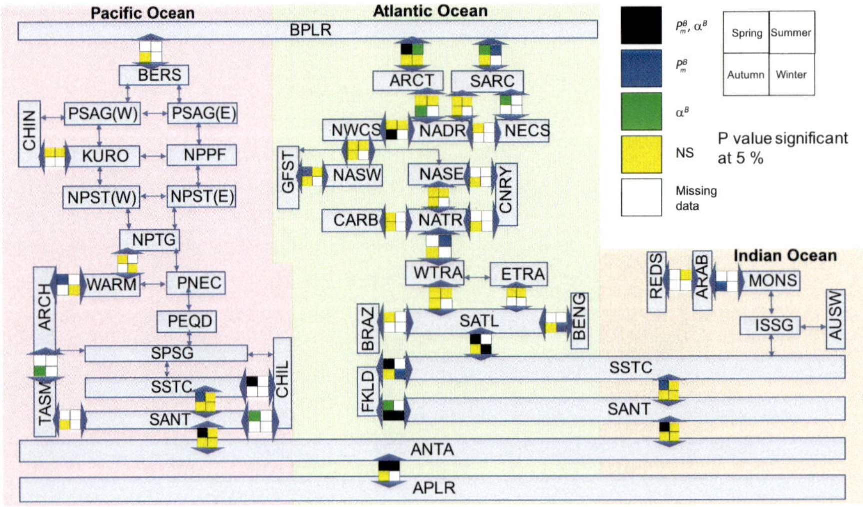

Figure 7. Diagram illustrating the seasonal differences of the *P-E* parameters between adjacent provinces by pairwise comparisons using Bonferroni adjusted *t* tests. Colours of blocks denote significance at the 5 % level. Note the orientation of the four blocks representing the seasonal difference remains as shown in the legend for both vertical and horizontal comparisons.

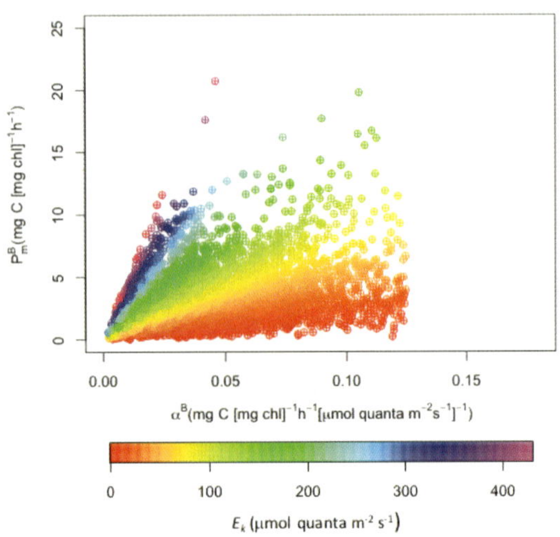

Figure 8. The assimilation number (P_m^B) plotted against the initial slope (α^B). Symbol colours represent the value of E_k in units of μmol quanta $m^{-2}\,s^{-1}$.

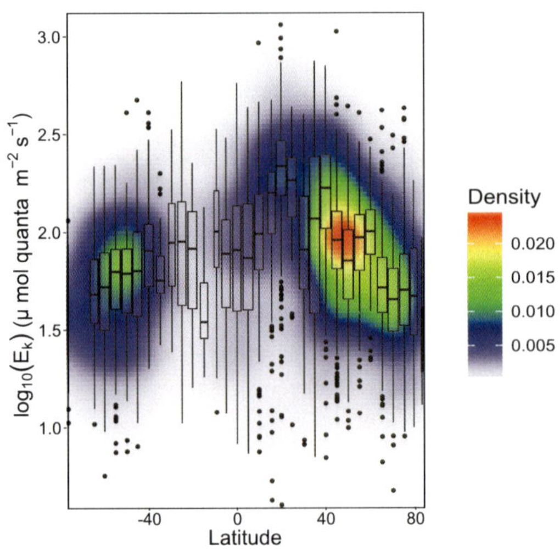

Figure 9. Density plot and box plots showing the variation in the photoadaptation parameter (E_k) with latitude. Middle horizontal line of boxplot represents the median value and lower and upper boundaries correspond to the first and third quartiles and the length of the whiskers is 1.5 times the inter-quantile range of the boundary. Heat map indicates probability density estimates. Filled circles denote outliers.

Platt, 1983). To disentangle the ecological from the physiological sources of variability is not straightforward, unless additional information on the taxonomic composition and photoacclimatory status of natural phytoplankton samples is available. Moreover, culture studies have invoked viral infection as another potential source of variability that is poorly understood in natural marine systems (Puxty et al., 2016).

Both taxon-specific and size-specific differences in P_m^B and α^B have been reported in both culture and field studies (Bouman et al., 2005; Côté and Platt, 1983, 1984; Huot et al., 2013; Xie et al., 2015). As new remote-sensing algorithms are now starting to yield information on the size and taxonomic structure of phytoplankton, it would be useful to derive additional information on the *P-E* response of key phytoplankton taxa and size classes, especially those implicated as playing key roles in ocean biogeochemical cycles (LeQuéré et al., 2005; Nair et al., 2008; Bracher et al., 2017). Although detailed information on the taxonomic and size structure of ship-based experiments was lacking for several of the samples included in this data set, more recent studies

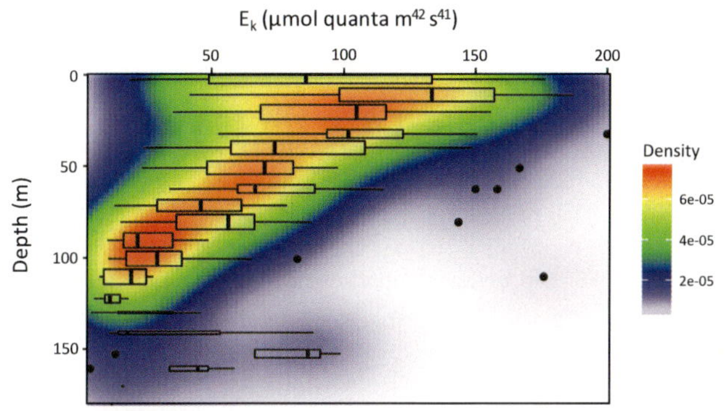

Figure 10. Density plot and box plots showing variation in the photoadaptation parameter (E_k) with depth for research cruises focussed on the oligotrophic gyres (DCM, AMT6, AMT15, AMT20 and AMT22). Middle horizontal line of boxplot represents the median value and lower and upper boundaries correspond to the first and third quartiles and the length of the whiskers is 1.5 times the inter-quantile range of the boundary. Heat map indicates probability density estimates. Filled circles denote outliers.

include some measure of phytoplankton community structure, whether it be from use of pigment markers, size fraction of pigment and/or productivity, or cell counts. As information on the global distribution of key phytoplankton groups is becoming available from global studies of phytoplankton pigment markers and flow cytometric counts (Buitenhuis et al., 2012; Peloquin et al., 2013; Swan et al., 2015), links between phytoplankton biogeography and large-scale pattern in photophysiology as revealed through the *P-E* parameters may be explored. Although there is a question as to what the standard indices of community structure should be that can help account for community-based variation in the photophysiological parameters across oceanographic data sets (Bracher et al., 2017), it is likely that information on gross community structure alone will not account for a large fraction of the variability in *P-E* parameters, especially across regions or seasons with different environmental forcing (Bouman et al., 2005; Smith Jr. and Donaldson, 2015) or resident ecotypes (Geider and Osborne, 1991). Establishing relationships between taxonomic composition and phytoplankton photophysiology will require simultaneous measurements of community structure alongside photosynthesis–irradiance experiments.

The high range of photosynthetic parameters recorded at lower latitudes is largely caused by depth-dependent changes due to photoacclimation and photoadaptation (Babin et al., 1996; Bouman et al., 2000b; Huot et al., 2007) in highly stratified waters. Strong vertical gradients in nutrient supply and growth irradiance lead to a vertical layering of ecological niches, resulting in strong vertical gradients in species composition, and in the case of marine picocyanobacteria, high-light and low-light ecotypes are observed (Johnson et al., 2006; Zwirglmaier et al., 2007). Although depth-dependent variability in the photosynthetic parameters can be examined in the MAPPS data set, in particular E_k (Fig. 10), it has been argued that optical depth may be a more useful predictor of changes in the *P-E* parameters resulting from vertical changes in the photoacclimatory status of phytoplankton cells (Babin et al., 1996; Bouman et al., 2000b). In highly stratified and stable seas such as the oligotrophic gyres this may be the case, yet in more dynamic ocean conditions such as the Beaufort Sea, optical depth has been shown to have no more predictive skill, and sometimes less, than using depth alone (Huot et al., 2013). It is important to note that diel changes in the *P-E* parameters were not taken into account in this meta-analysis due to a lack of information on the time of sample collection in a significant number of observations, which can be a significant source of variability (MacCaull and Platt, 1977; Prézelin and Sweeney, 1977; Prézelin et al., 1986; Harding et al., 1983; Cullen et al., 1992; Bruyant et al., 2005). However, as noted in the study of Babin and co-authors (1996) such diel and day-to-day variability in the *P-E* parameters is likely to be far smaller when compared with differences across biogeochemical provinces subject to markedly different environmental forcing.

Clear latitudinal differences in the range of E_k values are revealed in the MAPPS data set (Fig. 9), which suggests that E_k may be controlled by environmental factors that vary strongly with latitude, such as temperature and the availability of light. Figure 9 shows that samples collected from high-latitude environments, such as the Labrador Sea, the (sub)Arctic and the Southern Ocean have markedly lower E_k values, reflecting the physical constraints of low temperatures and, in some cases, low light levels (Harrison and Platt, 1986). The physical dynamics of the upper ocean and their impact on temperature and light conditions have been shown to play a dominant role in governing the photosynthetic performance of polar and temperate marine phytoplankton (Harrison and Platt, 1986; Bouman et al., 2005). Although large-scale shifts in the photophysiological status of the phytoplankton are observed in the data set, spatial differences between adjacent provinces were only observed at higher latitudes and seasons where the number of experimental observations tended to be higher (Fig. 7). It is clear from this analysis that more effort must be focused on obtaining information on photophysiology in oceanic regimes that are highly undersampled both in space and time.

6 Data availability

The compiled data set containing 5711 individual photosynthesis–irradiance experiments and corresponding metadata are available at PANGAEA (https://doi.pangaea.de): https://doi.org/10.1594/PANGAEA.874087 (Bouman et al., 2017).

7 Recommendations for use

The MAPPS database of photosynthesis–irradiance parameters may be used to compute marine primary production using remotely sensed data on ocean colour (e.g. Longhurst et al., 1995). The data present in Table 2 provide seasonal values of the photosynthetic parameters for Longhurst biogeochemical provinces, whose geographical boundaries can be found online at http://www.marineregions.org/. This data set will also serve as a useful resource for validation of other methods used to extrapolate *P-E* parameters to basin scales using ocean observables (Antoine and Morel, 1996; Platt et al., 2008; Saux-Picart et al., 2014). The MAPPS database will allow such extrapolation procedures to be further tested and refined over a much larger range of geographical and biogeochemical domains or for new approaches of parameter assignment to be developed.

Competing interests. The authors declare that they have no conflict of interest.

Acknowledgements. The authors wish to thank Alex Poulton, Lilian Krug and one anonymous reviewer, whose helpful suggestions improved the manuscript. The authors also thank the research scientists, technicians, students and crew who contributed to the collection of these data. In particular, we acknowledge the significant contributions made by Brian Irwin, Jeff Anning, Gary Maillet, Christine Hanson, Brian Griffiths, James McLaughlin, Richard Matear, and Andy Steven. This work was funded through the European Space Agency's MAPPS (MArine primary Production: model Parameters from Space) project as part of the Support to Science Element (STSE) Pathfinders Program. This is a contribution to the NERC-funded project Arctic PRIZE (NE/P006507/1) and the Simons Foundation CBIOMES Project. This work is also a contribution to the National Centre for Earth Observation, of NERC, UK.

Edited by: Francois Schmitt
Reviewed by: Alex Poulton and one anonymous referee

References

Antoine, D. and Morel, A.: Oceanic primary production 1. Adaptation of a spectral light-photosynthesis model in view of application to satellite chlorophyll observations, Global Biogeochem. Cy., 10, 43–55, 1996.

Babin, M., Morel, A., and Gagnon, R: An incubator designed for extensive and sensitive measurements of phytoplankton photosynthetic parameters, Limnol. Oceanogr., 39, 694–702, 1994.

Babin, M., Morel, A., Claustre, H., Bricaud, A., Kolber, Z., and Falkowski, P. J.: Nitrogen- and irradiance-dependent variations of maximum quantum yield of carbon fixation in eutrophic, mesotrophic and oligotrophic marine systems, Deep-Sea Res., 43, 1241–1272, 1996.

Behrenfeld, M. J., Prasil, O., Babin, M., and Bruyant, F.: In search of a physiological basis for covariations in light-limited and light-saturated photosynthesis, J. Phycol., 40, 4–25, 2004.

Bouman, H. A., Platt, T., Sathyendranath, S., Irwin, B. D., Wernand, M. R., and Kraay, G. W.: Bio-optical properties of the subtropical North Atlantic. II. Relevance to models of primary production, Mar. Ecol.-Prog. Ser., 200, 19–34, 2000a.

Bouman, H. A., Platt, T., Kraay, G. W., Sathyendranath, S., and Irwin, B. D.: Bio-optical properties of the subtropical North Atlantic. I. Vertical variability, Mar. Ecol.-Prog. Ser., 200, 3–18, 2000b.

Bouman, H. A., Platt, T., Sathyendranath, S., and Stuart, V.: Dependence of light-saturated photosynthesis on temperature and community structure, Deep-Sea Res. Pt .I, 52, 1284–1299, 2005.

Bouman, H. A., Platt, T., Doblin, M. A., Figueiras, F. G., Gudmundsson, K., Gudfinnsson, H. G., Huang, B., Hickman, A., Hiscock, M. R., Jackson, T., Lutz, V. A., Melin, F., Rey, F., Pepin, P., Segura, V., Tilstone, G., van Dongen-Vogels, V., and Sathyendranath, S.: A global dataset of photosynthesis-irradiance parameters for marine phytoplankton, https://doi.org/10.1594/PANGAEA.874087, 2017.

Bracher, A., Bouman, H. A., Brewin, R. J. W., Bricaud, A., Brotas, V., Ciotti, A. M., Clementson, L., Devred, E., Di Cicco, A., Dutkiewicz, S., Hardman-Mountford, N. J., Hickman, A. E., Hieronymi, M., Hirata, T., Losa, S. N., Mouw, C. B., Organelli, E., Raitsos, D. E., Uitz, J., Vogt, M., and Wolanin, A.: Obtaining phytoplankton diversity from ocean color: a scientific roadmap for future development, Front. Mar. Sci., 4, 55, https://doi.org/10.3389/fmars.2017.00055, 2017.

Bruyant, F., Babin, M. Genty, B., Prasil, O., Berhrenfeld, M. J., Claustre, H., Bricaud, A., Garczarek, L., Holtzendorff, J., Koblizek, M., Dousova, H., and Partensky, F.: Diel variations in the photosynthetic parameters of *Prochlorococcus* strain PCC 9511: combined effects of light and cell cycle, Limnol. Oceanogr., 50, 850–863, 2005.

Buitenhuis, E. T., Li, W. K. W., Vaulot, D., Lomas, M. W., Landry, M. R., Partensky, F., Karl, D. M., Ulloa, O., Campbell, L., Jacquet, S., Lantoine, F., Chavez, F., Macias, D., Gosselin, M., and McManus, G. B.: Picophytoplankton biomass distribution in the global ocean, Earth Syst. Sci. Data, 4, 37–46, https://doi.org/10.5194/essd-4-37-2012, 2012.

Carr, M. E., Friedrichs, M., Schmeltz, M., NoguchiAita, M., Antoine, D., Arrigo, K. R., Asanuma, I., Aumont, O., Barber, R., Behrenfeld, M. J., Bidigare, R., Buitenhuis, E., Campbell, J. W., Ciotti, A., Dierssen, H. M., Dowell, M., Dunne, J., Esaias, W., Gentili, B., Gregg, W. W., Groom, S. B., Hoepffner, N., Ishizaka, J., Kameda, T., Le Quere, C., Lohrenz, S., Marra, J., Mélin, F., Moore, K., Morel, A., Reddy, T., Ryan, J., Scardi, M., Smyth, T., Turpie, K., Tilstone, G. H., Waters, K., and Yamanaka, Y.: A comparison of global estimates of marine primary production from ocean color, Deep-Sea Res. Pt. II, 53, 741–770, 2006.

Cullen, J. J.: The deep chlorophyll maximum: comparing vertical profiles of chlorophyll a, Can. J. Fish. Aquat. Sci., 39, 791–803, 1982.

Cullen, J. J., Lewis, M. R., Davis, C. O., and Barber, R. T.: Photosynthetic characteristics and estimated growth rates indicate grazing is the proximate control of primary production in the equatorial Pacific, J. Geophys. Res., 97, 639–654, 1992.

Côté, B. and Platt, T.: Day-to-day variations in the spring-summer photosynthetic parameters of coastal marine phytoplankton, Limnol. Oceanogr., 28, 320–344, 1983.

Côté, B. and Platt, T.: Utility of the light-saturation curve as an operational model for quantifying the effects of environmental conditions on phytoplankton photosynthesis, Mar. Ecol.-Prog. Ser., 18, 57–66, 1984.

Devred, E., Sathyendranath, S., and Platt, T.: Delineation of ecological provinces using ocean colour radiometry, Mar. Ecol.-Prog. Ser., 346, 1–13, 2007.

Dogliotti, A. I., Lutz, V. A., and Segura, V.: Estimation of primary production in the southern Argentine continental shelf and shelf-break regions using field and remote sensing data, Remote Sens. Environ., 140, 497–508, 2014.

Geider, R. J. and Osborne, B. A.: Algal Photosynthesis: The Measurement of Gas Exchange and Related Processes, Chapman and Hall, New York, 256 pp., 1991.

Geider, R. J, MacIntyre, H. L., and Kana, T. M.: A dynamic model of photoadaptation in phytoplankton, Limnol. Oceanogr., 41, 1–15, 1996.

Griffiths, F. B., Bates, T. S., Quinn, P. K., Clementson, L. A., and Parslow, J. S.: Oceanographical context of the first Aerosol Characterisation Experiment (ACE-1): a physical, chemical and biological overview, J. Geophys. Res., 104, 21649–21671, 1999.

Gudmundsson, K.: Long-term variation in phytoplankton productivity during spring in Icelandic waters, ICES J. Mar. Sci., 55, 635–643, 1998.

Hanson, C. E., Pattiaratchi, C. B., Waite, A. M.: Sporadic upwelling on a downwelling coast: phytoplankton responses to spatially variable nutrient dynamics off the Gascoyne region of Western Australia, Cont. Shelf Res., 25, 1561–1582, 2005a.

Hanson, C. E., Pattiaratchi, C. B., and Waite, A. M.: Seasonal production regimes off south-western Australia: influence of the Capes and Leeuwin Currents on phytoplankton dynamics, Mar. Freshwater Res., 56, 1011–1026, 2005b.

Harding, L. W., Meeson, B. W., and Tyler, M. A.: Photoadaptation and diel periodicity of photosynthesis in the dinoflagellate *Prorocentrum mariaelebouriae*, Mar. Ecol.-Prog. Ser., 13, 73–85, 1983.

Harrison, W. G. and Platt, T.: Photosynthesis–irradiance relationships in polar and temperate phytoplankton populations, Polar Biol., 5, 153–164, 1986.

Hickman, A. E.: The photophysiology and primary productivity of phytoplankton within the deep chlorophyll maximum, doctoral thesis, University of Southampton, Southampton, UK, 237 pp., 2007.

Hiscock, M. R.: The regulation of primary productivity in the Southern Ocean, doctoral thesis, Duke University, Beaufort, USA, 120 pp., 2004.

Hiscock, M. R., Marra, J., Smith Jr., W. O., Goericke, R., Measures, C., Vink, S., Olson, R. J., Sosik, H. M., and Barber, R. T.: Primary productivity and its regulation in the Pacific Sector of the Southern Ocean, Deep-Sea Res. Pt. II, 50, 533–558, 2003.

Holm-Hansen, O., Lorenzen, C. J., Holmes, R. W., and Strickland, J. D. H.: Fluorometric determination of chlorophyll, J. Cons. Perm. Int. Explor. Mer., 30, 3–15, 1965.

Huot, Y., Babin, M., Bruyant, F., Grob, C., Twardowski, M. S., and Claustre, H.: Relationship between photosynthetic parameters and different proxies of phytoplankton biomass in the subtropical ocean, Biogeosciences, 4, 853–868, https://doi.org/10.5194/bg-4-853-2007, 2007.

Huot, Y., Babin, M., and Bruyant, F.: Photosynthetic parameters in the Beaufort Sea in relation to the phytoplankton community structure, Biogeosciences, 10, 3445–3454, https://doi.org/10.5194/bg-10-3445-2013, 2013.

Jackson, T.: Phytoplankton community structure, photophysiology and primary production in the Atlantic Arctic, doctoral thesis, University of Oxford, Oxford, UK, 222 pp., 2013.

Jassby, A. D. and Platt, T.: Mathematical formulation of the relationship between photosynthesis and light for phytoplankton, Limnol. Oceanogr., 21, 540–547, 1976.

Johnson, Z. I., Zinser, E. R., Coe, A., McNulty, N. P., Woodward, E. M. S., and Chisholm, S. W.: Niche partitioning among *Prochlorococcus* ecotypes along ocean-scale environmental gradients, Science, 311, 1737–40, 2006.

Kyewalyanga, M. N., Platt, T., and Sathyendranath, S.: Estimation of the photosynthetic action spectrum: implication for primary production models, Mar. Ecol.-Prog. Ser., 146, 207–223, 1997.

Kyewalyanga, M. N., Platt, T., Sathyendranath, S., Lutz, V. A., and Stuart, V.: Seasonal variation in physiological parameters of phytoplankton across the North Atlantic, J. Plankton Res., 20, 17–42, 1998.

Lawrenz, E., Silsbe, G., Capuzzo, E., Ylöstalo, P., Forster, R. M., Simis, S. G. H., Prášil, O., Kromkamp, J. C., Hickman, A. E., Moore, C. M., Forget, M.-H., Geider, R. J., and Suggett, D. J.: Predicting the electron requirement for carbon fixation in seas and oceans, PLoS ONE, 8, e58137, https://doi.org/10.1371/journal.pone.0058137, 2013.

Lee, Y. J., Matrai, P. A., Friedrichs, M. A. M., Saba, V. S., Antoine, D., Ardyna, M., Asanuma, I., Babin, M., Bélanger, S., Benoît-Gagné, M., Devred, E., Fernández-Méndez, M., Gentili, B., Hirawake, T., Kang, S.-H., Kameda, T., Katlein, C., Lee, S. H., Lee, Z., Mélin, F., Scardi, M., Smyth, T. J., Tang, S., Turpie, K. R., Waters, K. J., and Westberry, T. K.: An assessment of phytoplankton primary productivity in the Arctic Ocean from satellite ocean color/in situ chlorophyll *a* based models, J. Geophys. Res., 120, 6508–6541, 2015.

Le Quéré, C., Harrison, S. P., Prentice, I. C., Buitenhuis, E. T., Aumont, O., Bopp, L., Claustre, H., da Cunha, L. C., Geider, R., Giraud, X., Klaas, C., Kohfeld, K. E., Legendre, L., Manizza, M., Platt, T., Rivkin, R. B., Sathyendranath, S., Uitz, J., Watson, A. J., and Wolf-Gladrow, D.: Ecosystem dynamics based on plankton functional types for global ocean biogeochemistry models, Glob. Change Biol., 11, 2016–2040, 2005.

Lewis, M. R. and Smith, J. C.: A small volume, short-incubation time method for measurement of photosynthesis as a function of incident irradiance, Mar. Ecol.-Prog. Ser., 13, 99–102, 1983.

Longhurst, A. R.: Ecological Geography of the Sea, Academic Press, London, 2007.

Longhurst, A., Sathyendranath, S., Platt, T., and Caverhill, C.: An estimate of global primary production in the ocean from satellite radiometer data, J. Plankton Res., 17, 1245–1271, 1995.

Lorenzo, L. M., Arbones, B., Figueiras, F. G., Tilstone, G. H., and Figueroa, F. L.: Photosynthesis, primary production and phytoplankton growth rates in Gerlache and Bransfield Straits during Austral summer: cruise FRUELA 95, Deep-Sea Res. Pt. II, 49, 707–721, 2002.

Lutz, V. A., Segura, V., Dogliotti, A. I., Gagliardini, D. A., Bianchi, A. A., and Balestrini, C. F.: Primary production in the Argentine Sea during spring estimated by field and satellite models, J. Plankton Res., 32, 181–195, 2010.

MacCaull, W. A. and Platt. T.: Die l variations in the photosynthetic parameters of coastal marine phytoplankton, Limnol. Oceanogr., 22, 723–731, 1977.

MacIntyre, H. L., Kana, T. M., Anning, T., and Geider, R. J.: Photoacclimation of photosynthesis irradiance response curves and photosynthetic pigments in microalgae and cyanobacteria, J. Phycol., 38, 17–38, 2002.

Mackey, D. J., Parslow, J. S., Higgins, H. W., griffiths, F. B., and O'Sullivan, J. E.: Plankton productivity and biomass in the western equatorial Pacific: biological and physical controls, Deep-Sea Res. Pt. II, 42, 499–533, 1995.

Mantoura, R. F. C., Jeffrey, S. W., Llewellyn, C. A., Claustre, H., and Morales, C. E.: Comparison between spectrophotometric, fluorometric and HPLC methods for chlorophyll analysis, in: Phytoplankton Pigments in Oceanography: Guidelines to Modern Methods, edited by: Jeffrey, S. W., Mantoura, R. F. C., and Wright, S. W., UNESCO, Paris, p. 361, 1997.

Nair, A., Sathyendranath, S., Platt, T., Morales, J., Stuart, V., Forget, M.-H., Devred, E., and Bouman, H.: Remote sensing of phytoplankton functional types, Remote Sens. Environ., 112, 3366–3375, 2008.

Pálsson, Ó. K., Gislason, A., Guðfinnsson, H. G., Gunnarsson, B., Ólafsdóttir, S. R., Petursdottir, H., Sveinbjörnsson, S., Thorisson, K., and Valdimarsson, H. Ecosystem structure in the Iceland Sea and recent changes to the capelin (*Mallotus villosus*) population, ICES J. Mar. Sci., 69, 1242–1254, https://doi.org/10.1093/icesjms/fss071, 2012.

Peloquin, J., Swan, C., Gruber, N., Vogt, M., Claustre, H., Ras, J., Uitz, J., Barlow, R., Behrenfeld, M., Bidigare, R., Dierssen, H., Ditullio, G., Fernandez, E., Gallienne, C., Gibb, S., Goericke, R., Harding, L., Head, E., Holligan, P., Hooker, S., Karl, D., Landry, M., Letelier, R., Llewellyn, C. A., Lomas, M., Lucas, M., Mannino, A., Marty, J.-C., Mitchell, B. G., Muller-Karger, F., Nelson, N., O'Brien, C., Prezelin, B., Repeta, D., Jr. Smith, W. O., Smythe-Wright, D., Stumpf, R., Subramaniam, A., Suzuki, K., Trees, C., Vernet, M., Wasmund, N., and Wright, S.: The MAREDAT global database of high performance liquid chromatography marine pigment measurements, Earth Syst. Sci. Data, 5, 109–123, https://doi.org/10.5194/essd-5-109-2013, 2013.

Platt, T. and Jassby, A. D.: The relationship between photosynthesis and light for natural assemblages of coastal marine phytoplankton, J. Phycol., 12, 421–430, 1976.

Platt, T., Gallegos, C. L., and Harrison, W. G.: Photoinhibition of photosynthesis in natural assemblages of marine phytoplankton, J. Mar. Res., 38, 687–701, 1980.

Platt, T., Harrison, G., Irwin, B., and Horne, E. P.: Photosynthesis and photoadaptation in the arctic, Deep-Sea Res. Pt. I, 29, 1159–1170, 1982.

Platt, T., Sathyendranath, S., Joint, I., and Fasham, M. J. R.: Photosynthesis characteristics of phytoplankton in the Celtic Sea during late spring, Fish. Oceanogr., 2, 191–201, 1993.

Platt, T. and Sathyendranath, S.: Latitude as a factor in the calculation of primary production, in: Ecology of Fjords and Coastal Waters, edited by: Skjoldal, H. R., Hopkins, C., Erikstad, K. E., and Leinaas, H. P., Elsevier Science, Tromsø, Norway, 1995.

Platt, T., Sathyendranath, S., Forget, M.-H., White III, G. N., Caverhill, C., Bouman, H., Devred, E., and Son, S.: Operational estimation of primary production at large geographical scales, Remote Sens. Environ., 112, 3437–3448, 2008.

Prézelin, B. B., Meeson, B. W., and Sweeney, B. M.: Characterization of photosynthetic rhythms in marine dinoflagellates. 1. Pigmentation, photosynthetic capacity and respiration, Plant Physiol., 60, 384–387, 1977.

Prézelin, B. B., Putt, M., and Glover, H. E.: Diurnal patterns in photosynthetic capacity and depth-dependent photosynthesis–irradiance relationships in *Synechococcus* spp., and larger phytoplankton in three water masses in the northwest Atlantic Ocean, Mar. Biol., 92, 205–217, 1986.

Prézelin, B. B., Glover, H. R., Ver Hoven, B. M., Stenberg, D., Matlick, H. A., Schofield, O., Nelson, N. B., Wyman, M., and Campbell, L.: Blue-green light effects on light-limited rates of photosynthesis: relationship to pigmentation and productivity estimates for *Synechococcus* populations from the Sargasso Sea, Mar. Ecol.-Prog. Ser., 54, 121–136, 1989.

Puxty, R. J., Millard, A. D., Evans, D. J., and Scanlan, D. J.: Viruses inhibit CO_2 fixation in the most abundant phototrophs on earth, Curr. Biol., 26, 1585–1589, 2016.

Rey, F.: Photosynthesis–irradiance relationships in natural phytoplankton populations of the Barents Sea, Polar Res., 10, 105–116, 1991.

Sakshaug, E., Bricaud, A., Dandonneau, Y., Falkowski, P. G., Kiefer, D. A., Legendre, L., Morel, A., Parslow, J., and Takahashi, M.: Parameters of photosynthesis: definitions, theory and interpretation of results, J. Plankton Res., 19, 1637–1670, 1997.

Sathyendranath, S., Stuart, V., Irwin, B. D., Maass, H., Savidge, G., Gilpin, L., and Platt, T.: Seasonal variation in bio-optical properties of phytoplankton in the Arabian Sea (1999), Deep-Sea Res. Pt. II, 46, 633–353, 1999.

Sathyendranath, S., Stuart, V., Nair, A., Oka, K., Nakane, T., Bouman, H., Forget, M.-H., Maass, H., and Platt, T.: Carbon-to-chlorophyll ratio and growth rate of phytoplankton in the sea, Mar. Ecol.-Prog. Ser., 383, 73–84, 2009.

Saux-Picart, S., Sathyendranath, S., Dowell, M., Moore, T., and Platt, T.: Remote sensing of assimilation number for marine phytoplankton, Remote Sens. Env., 146, 87–96, 2014.

Schofield, O., Prézelin, B. B., Smith, R. C., Stegmann, P. M., Nelson, N. B., Lewis, M. R., and Baker, K. S.: Variability in spectral and nonspectral measurements of photosynthetic light utilization efficiencies, Mar. Ecol.-Prog. Ser., 78, 253–271, 1991.

Segura, V., Lutz, V. A., Dogliotti, A. I., Silva, R., Negri, R., Akselman, R., and Benavides, H.: Phytoplankton functional types and primary production in the Argentine Sea, Mar. Ecol.-Prog. Ser., 491, 15–31, 2013.

Smith Jr., W. O. and Donaldson, K.: Photosynthesis–irradiance responses in the Ross Sea, Antarctica: a meta-analysis, Biogeosciences, 12, 3567–3577, https://doi.org/10.5194/bg-12-3567-2015, 2015.

Swan, C., Vogt, M., Gruber, N., and Laugkoetter, C.: A global seasonal surface ocean climatology of phytoplankton types based on CHEMTAX analysis of HPLC pigments, Deep-Sea Res. Pt. I, 109, 137–156, https://doi.org/10.1016/j.dsr.2015.12.002, 2015.

Tilstone, G. H., Figueiras, F. G., Lorenzo, L. M., and Arbones, B.: Phytoplankton composition, photosynthesis and primary production during different hydrographic conditions at the Northwest Iberian upwelling system, Mar. Ecol.-Prog. Ser., 252, 89–104, 2003.

Ulloa, O., Hoepffner, N., and Larkin, D.: Depth and wavelength dependence of phytoplankton photosynthesis: implications for the remote sensing of marine primary production, Proc. SPIE 2963, Ocean Optics XIII, 1996, Halifax, Nova Scotia, Canada, https://doi.org/10.1117/12.266456, 1997.

Webb, W. L., Newton, M., and Starr, D.: Carbon dioxide exchange of Alnus rubra: a mathematical model, Oecologia, 17, 281–291, 1974.

Westwood, K. J., Griffiths, F. B., Webb, J. P., and Wright, S. W.: Primary production in the Sub-Antarctic and Polar Frontal zones south of Tasmania, Australia; SAZ-Sense survey, 2007, Deep-Sea Res. Pt. II, 58, 2162–2178, 2011.

Xie, Y., Bangqin, H., Lin, L., Laws, E. A., Wang, L., Shang, S., Zhang, T., and Dai, M.: Photosynthetic parameters in the northern South China Sea in relation to phytoplankton community structure, J. Geophys. Res., 120, 4187–4204, https://doi.org/10.1002/2014JC010415, 2015.

Zhai, L., Gudmundsson, K., Miller, P., Peng, W., Guðfinnsson, H., Debes, H., Hátún, H., White III, G. N., Walls, R. H., Sathyendranath, S., and Platt, T.: Phytoplankton phenology and production around Iceland and Faroes, Cont. Shelf Res., 37, 15–25, 2012.

Zwirglmaier, K., Heywood, J. L., Chamberlain, K., Woodward, E. M. S., Zubkov, M. V., and Scanlan, D. J.: Basin-scale distribution patterns of picocyanobacterial lineages in the Atlantic Ocean, Environ. Microbiol., 9, 1278–1290, 2007.

5

Copepod species abundance from the Southern Ocean and other regions (1980–2005) – a legacy

Astrid Cornils[1], **Rainer Sieger**[†], **Elke Mizdalski**[1], **Stefanie Schumacher**[1], **Hannes Grobe**[1], **and Sigrid B. Schnack-Schiel**[†]

[1]Alfred-Wegener-Institut Helmholtz-Zentrum für Polar- und Meeresforschung, Bremerhaven, Germany
[†]deceased

Correspondence: Astrid Cornils (astrid.cornils@awi.de)

Abstract. This data collection originates from the efforts of Sigrid Schnack-Schiel (1946–2016), a zooplankton ecologist with great expertise in life cycle strategies of Antarctic calanoid copepods, who also investigated zooplankton communities in tropical and subtropical marine environments. Here, we present 33 data sets with abundances of planktonic copepods from 20 expeditions to the Southern Ocean (Weddell Sea, Scotia Sea, Amundsen Sea, Bellingshausen Sea, Antarctic Peninsula), one expedition to the Magellan region, one latitudinal transect in the eastern Atlantic Ocean, one expedition to the Great Meteor Bank, and one expedition to the northern Red Sea and Gulf of Aqaba as part of her scientific legacy. A total of 349 stations from 1980 to 2005 were archived. During most expeditions depth-stratified samples were taken with a Hydrobios multinet with five or nine nets, thus allowing inter-comparability between the different expeditions. A Nansen or a Bongo net was deployed only during four cruises. Maximum sampling depth varied greatly among stations due to different bottom depths. However, during 11 cruises to the Southern Ocean the maximum sampling depth was restricted to 1000 m, even at locations with greater bottom depths. In the eastern Atlantic Ocean (PS63) sampling depth was restricted to the upper 300 m. All data are now freely available at PANGAEA via the persistent identifier https://doi.org/10.1594/PANGAEA.884619.

Abundance and distribution data for 284 calanoid copepod species and 28 taxa of other copepod orders are provided. For selected species the abundance distribution at all stations was explored, revealing for example that species within a genus may have contrasting distribution patterns (*Ctenocalanus*, *Stephos*). In combination with the corresponding metadata (sampling data and time, latitude, longitude, bottom depth, sampling depth interval) the analysis of the data sets may add to a better understanding how the environment (currents, temperature, depths, season) interacts with copepod abundance, distribution and diversity. For each calanoid copepod species, females, males and copepodites were counted separately, providing a unique resource for biodiversity and modelling studies. For selected species the five copepodite stages were also counted separately, thus also allowing the data to be used to study life cycle strategies of abundant or key species.

1 Introduction

Copepoda (Crustacea) are probably the most successful metazoan group known, being more abundant than insects, although far less diverse (Humes, 1994; Schminke, 2007). They occur in all aquatic ecosystems, from freshwater to marine and hypersaline environments, and from polar waters to hot springs (Huys and Boxshall, 1991). Although copepods are evolutionary of benthic origin (Bradford-Grieve, 2002), they have also successfully colonised the pelagic marine environment, where they can account for 80 %–90 % of the total zooplankton abundance (Longhurst, 1985). In the Southern Ocean, copepods are the most important zooplankton organisms next to Antarctic krill and salps, both in abundance and biomass (e.g. Pakhomov et al., 2000; Shreeve et al., 2005; Smetacek and Nicol, 2005; Ward et al., 2014; Tarling et al., 2017). In the Southern Ocean, copepods are also the most diverse zooplankton taxon, accounting for more than 300 species (Kouwenberg et al., 2014). However, only a few species dominate the Antarctic epipelagic assemblage: the large calanoids *Calanoides acutus*, *Calanus propinquus*, *Metridia gerlachei*, and *Paraeuchaeta antarctica*; the small calanoids *Microcalanus pygmaeus* and *Ctenocalanus citer*; and the cyclopoids *Oithona* spp. and species of the family Oncaeidae (e.g. Hopkins, 1985; Atkinson, 1998; Schnack-Schiel, 2001; Tarling et al., 2017). Together these taxa can comprise up to 95 % of the total abundance and up to 80 % of the total biomass of copepods (Schnack-Schiel et al., 1998). However, the smaller calanoid species alone can account for up to 80 % of the abundance of calanoid copepods (Schnack-Schiel, 2001). Stage-resolve counts for selected species will also allow future users to study life cycle strategies of abundant or key species.

Numerous studies on zooplankton have been conducted in the past in the Atlantic sector of the Southern Ocean (e.g. Boysen-Ennen and Piatkowski, 1988; Hopkins and Torres, 1988; Boysen-Ennen et al., 1991; Pakhomov et al., 2000; Dubischar et al., 2002; Ward et al., 2014; Tarling et al., 2017). A major zooplankton monitoring programme in the Southern Ocean is the Continuous Plankton Recorder survey (SO-CPR), providing a large-scale coverage of surface Antarctic zooplankton species distribution abundances over the last 25 years (Hosie et al., 2003; McLeod et al., 2010). A recent review summarizes the present knowledge on abundance and distribution of Southern Ocean zooplankton (Atkinson et al., 2012). Especially in the Weddell Sea occurrence data of copepods and other zooplankton species are scarce. One of our aims is to fill this gap with the here-presented data sets from the Southern Ocean, collected by Sigrid Schnack-Schiel (1946–2016) over the period of 1982 to 2005.

In recent years there has been ample evidence that marine ecosystems are greatly affected by climate change and ocean acidification (e.g. Beaugrand et al., 2002; Edwards and Richardson, 2004; Rivero-Calle et al., 2015; Smith et al., 2016). In the Southern Ocean, the pelagic ecosystem is likely to be severely affected by increasing water temperatures and the resulting reduction of sea ice coverage in the Southern Ocean (Zwally, 1994; Smetacek and Nicol, 2005). It has already been observed over decades that the biomass of Antarctic krill decreases (Atkinson et al., 2004), but little is known about the environmental effects on copepods. Within the pelagic ecosystem zooplankton communities and thus copepods are good indicators for ecosystem health and status due to their short life cycles und their rapid response to changing environments (Reid and Edwards, 2001; Chust et al., 2017). Furthermore, they are generally not commercially exploited and thus are likely to reflect impacts of environmental changes more objectively. To better understand the effects of environmental change on planktonic copepods, e.g. via biodiversity analyses and ecological niche modelling, data on species occurrence, abundance and distribution are essential. However, modelling studies are often limited by the scarcity of available plankton data (Chust et al., 2017). Thus, freely available data sets on abundance and presence or absence of copepod species are of great importance for future studies on environmental changes in the pelagic realm. The data sets presented here on copepod species and life stages (female, male, copepodites) occurrences and abundance from the Southern Ocean, the eastern Atlantic Ocean, the Magellan region and the Red Sea provide a unique resource for biodiversity and modelling studies. They may also help to enhance our understanding of how the environment (currents, temperature, depths, season) interacts with copepod abundance, distribution and diversity.

2 Methods

2.1 Sampling locations

The presented data sets were collected during 24 research cruises with several research vessels from 1980 to 2005 (Table 1; Cornils and Schnack-Schiel, 2018). Most of the data sets (28 data sets from 20 cruises) are based on samples from the Southern Ocean (Fig. 1), collected on-board R/V *Polarstern* (25 data sets from 16 cruises), R/V *Meteor* (1 data set), R/V *John Biscoe* (1 data set) and R/V *Polarsirkel* (1 data set). Southern Ocean sampling locations were restricted to the Weddell Sea, the Scotia Sea, the Antarctic Peninsula, the Bellingshausen Sea and the Amundsen Sea (Fig. 1).

Additionally, four data sets were collected in other regions (Table 1). In 1994 net samples were collected on-board R/V *Victor Hensen* in the Magellan region. Two data sets are based on research cruises with R/V *Meteor*, to the Great Meteor Bank in the North Atlantic (1998) and to the northern Red Sea and the Gulf of Aqaba (1999). In 2002, plankton net samples were taken during a research cruise with R/V *Polarstern* along a transect in the eastern tropical Atlantic Ocean (Table 1).

Maximum sampling depth varied greatly among stations due to different bottom depths (Table 1). However, during 11 cruises to the Southern Ocean the maximum depth was restricted to 1000 m, even at locations with greater bottom depths. In the eastern Atlantic Ocean (PS63) sampling depth was restricted to the upper 300 m.

2.2 Sampling gear

Three types of plankton nets were deployed: Bongo nets, single opening–closing Nansen nets and multiple opening–closing nets. During all expeditions vertical hauls were taken, thus allowing no movement of the vessel.

Table 1. Overview of the sampling periods and research cruises. Abbreviations of regions: Antarctic Peninsula (AP), Weddell Sea (WS), eastern Weddell Sea (EWS), Scotia Sea (SCO), Bellingshausen Sea (BS), Amundsen Sea (AS), western Weddell Sea (WWS), eastern Atlantic Ocean (EAO), Magellan region (MR), Great Meteor Bank (GMB), Red Sea (RS); MN: Multinet. * Data sets with abundance only for non-calanoid copepod species.

Cruise no.	Sampling period	Region	No. of stations	Net type	Max. sampling depth (m)	Mesh size (μm)	DOI of data sets	No. of taxa	Available CTD profiles/information on hydrography
PS04 ANT-II/2	1983-10-24–1983-11-10	AP	14	MN midi/ Nansen net	1900	200	https://doi.org/10.1594/PANGAEA.876508	110	
PS06 ANT-III/3	1985-01-07–1985-02-24	AP, EWS WS AP EWS	30 61 6 4	MN midi Bongo MN midi MN midi	2880 245 400 2500	100/200 300 200 100	https://doi.org/10.1594/PANGAEA.876726 https://doi.org/10.1594/PANGAEA.878275 https://doi.org/10.1594/PANGAEA.878276 https://doi.org/10.1594/PANGAEA.879771	258 28 21 14*	https://doi.org/10.1594/PANGAEA.860066
PS09 ANT-V/1	1986-05-21–1986-05-31	AP	8	MN midi	1850	200	https://doi.org/10.1594/PANGAEA.876734	162	https://doi.org/10.1594/PANGAEA.860066
PS10 ANT-V/2	1986-07-18–1986-09-05	EWS	18 3	MN midi	600	100	https://doi.org/10.1594/PANGAEA.878277 https://doi.org/10.1594/PANGAEA.878278	22 169	https://doi.org/10.1594/PANGAEA.860066
PS10 ANT-V/3	1986-10-05–1986-11-24	EWS	7 24	MN midi	1000	100	https://doi.org/10.1594/PANGAEA.879772 https://doi.org/10.1594/PANGAEA.878874	10* 211	https://doi.org/10.1594/PANGAEA.860066
PS14 ANT-VII/2	1988-11-09–1988-11-13	SCO	6	MN midi	1000	200	https://doi.org/10.1594/PANGAEA.879290	172	
PS14 ANT-VII/3	1988-11-26–1989-01-04	SCO	11 12	MN midi	1000	200	https://doi.org/10.1594/PANGAEA.879231 https://doi.org/10.1594/PANGAEA.879232	192 52	https://doi.org/10.1594/PANGAEA.860066
PS14 ANT-VII/4	1989-01-17–1989-01-19	SCO	5	MN midi	1000	200	https://doi.org/10.1594/PANGAEA.879230	166	https://doi.org/10.1594/PANGAEA.860066
PS16 ANT-VIII/2	1989-09-14–1989-10-06	WS	12	MN midi	1000	100	https://doi.org/10.1594/PANGAEA.879308	186	https://doi.org/10.1594/PANGAEA.860066
PS18 ANT-IX/2	1990-11-22–1990-12-15	WS	9	MN midi	1000	100	https://doi.org/10.1594/PANGAEA.879508	226	https://doi.org/10.1594/PANGAEA.860066
PS21 ANT-X/3	1992-04-11–1992-05-02	EWS	12	MN midi	1000	100	https://doi.org/10.1594/PANGAEA.879536	227	
PS23 ANT-X/7	1992-12-18–1993-01-16	WS	16	MN midi	1000	100	https://doi.org/10.1594/PANGAEA.879562	240	https://doi.org/10.1594/PANGAEA.860066
PS29 ANT-XI/3	1994-01-28–1994-03-03	BS, AS	20 6	MN midi	1000	55	https://doi.org/10.1594/PANGAEA.879712 https://doi.org/10.1594/PANGAEA.879718	220 42*	https://doi.org/10.1594/PANGAEA.860066
PS35 ANT-XII/4	1995-04-12–1995-04-17	AS	6 5	MN midi	1000	55	https://doi.org/10.1594/PANGAEA.879774 https://doi.org/10.1594/PANGAEA.879773	204 35*	https://doi.org/10.1594/PANGAEA.860066
PS58 ANT-XVIII/5b	2001-04-18–2001-04-30	BS	9	MN midi	650	55	https://doi.org/10.1594/PANGAEA.880375	143	https://doi.org/10.1594/PANGAEA.860066
PS63 ANT-XX/1	2002-11-02–2002-11-20	EAO	19	MN midi	300	100	https://doi.org/10.1594/PANGAEA.880927	353	Schnack-Schiel et al. (2010b)
PS65 ANT-XXI/2	2003-12-09–2004-01-01	EWS	10	MN midi	900	100	https://doi.org/10.1594/PANGAEA.880331	128	https://doi.org/10.1594/PANGAEA.860066
PS67 ANT-XXII/2	2004-12-01–2005-01-02	WWS	9	MN midi	1000	100	https://doi.org/10.1594/PANGAEA.880983	172	https://doi.org/10.1594/PANGAEA.742627
DAE1979/80	1980-01-01–1980-02-08	WS	50	Nansen net	700	250	https://doi.org/10.1594/PANGAEA.880239	61	
JB03	1982-02-02–1982-03-02	BS, AP, SCO	45	Nansen net	2850	200	https://doi.org/10.1594/PANGAEA.880568	182	https://www.bodc.ac.uk/resources/inventories/cruise_inventory/report/5916/ (last access: 10 February 2018)
VH1094	1994-11-12–1994-11-24	MR	17	MN midi	400	300	https://doi.org/10.1594/PANGAEA.880202	105	
M11/4	1989-12-27–1990-01-08	BS, AP	22	MN midi	2500	200	https://doi.org/10.1594/PANGAEA.880173	193	https://doi.org/10.1594/PANGAEA.742745
M42/3	1998-09-01–1998-09-16	GMB	17	MN midi	2500	100	https://doi.org/10.1594/PANGAEA.882283	349	Beckmann and Mohn (2002), Mohn and Beckmann (2002)
M44/2	1999-02-22–1999-03-07	RS	15 5	MN maxi	1300 500	150	https://doi.org/10.1594/PANGAEA.881899 https://doi.org/10.1594/PANGAEA.880901	186 52	Cornils et al. (2005), Plähn et al. (2002)

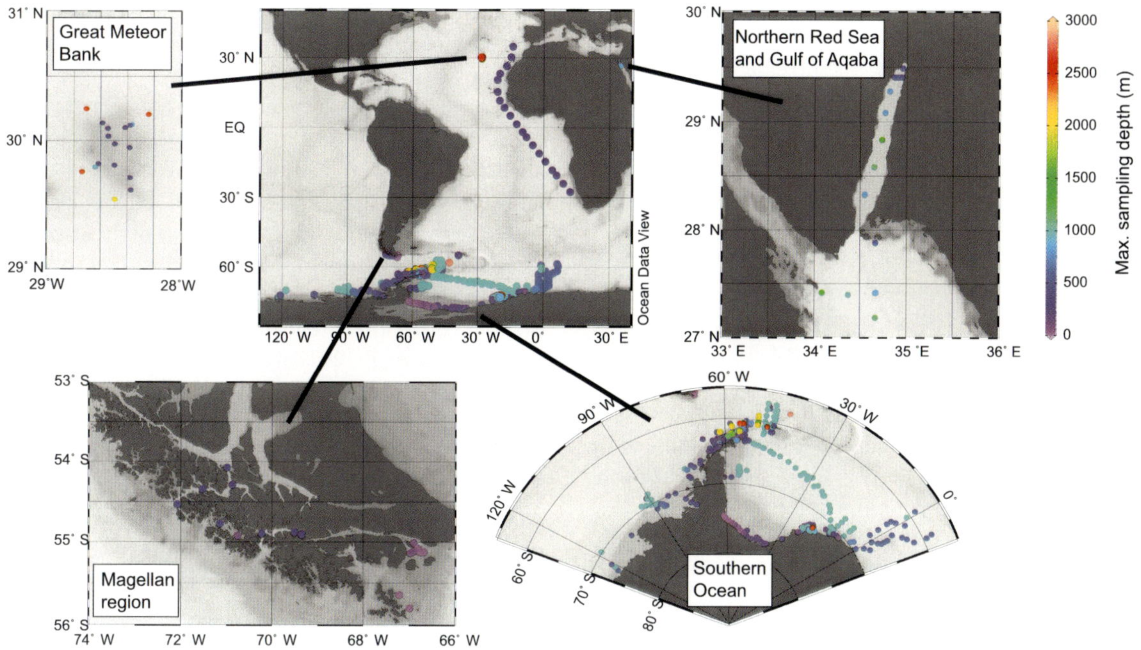

Figure 1. Overview of all stations and sampling regions, including the maximum sampling depths (colour scale bar) of the data set.

2.2.1 Nansen net

During the expeditions PS04, DAE1979/80 and JB03 net sampling was carried out with a Nansen net (Table 1). The Nansen net is an opening–closing plankton net for vertical tows (Nansen, 1915; Currie and Foxton, 1956). Thus, it is possible to sample discrete depth intervals to study the vertical distribution of zooplankton. The Nansen net has an opening of 70 cm diameter and is usually 3 m long. Two different mesh sizes were used: 200 µm for the cruises PS04 and JB03, and 250 µm for DAE1979/80. To conduct discrete depth intervals the net is lowered to maximum depth and then hauled to a certain depth and closed via a drop weight. Then the net is hauled to the surface and the sample is removed. This process of sampling depth intervals can be repeated until the surface layer is reached. The volume of filtered water was calculated using the mouth area and depth interval due to the lack of a flowmeter.

2.2.2 Multinet systems

Most presented data sets are based on plankton samples taken with a multinet system (MN) from Hydrobios (Table 1) a revised version (Weikert and John, 1981) of the net described by Bé et al. (1959). The multinet is equipped with five (midi) or nine (maxi) plankton nets, with a mouth area of 0.25 and 0.5 m^2, respectively. These nets can be opened and closed at depth on demand from the ship via a conductor cable. Thus, they allow sampling of discrete water layers. The net system was hauled at a general speed of 0.5 $m\,s^{-1}$. Mesh sizes varied between the data sets from 55 to 300 µm (Table 1). In the Southern Ocean the mesh sizes were consistent within regions: in the Weddell Sea 100 µm mesh size was used with a few exceptions during PS06. In the Scotia Sea and near the Antarctic Peninsula a mesh size of 200 µm was employed. In the Bellingshausen Sea and the Amundsen Sea multinet hauls with 55 µm mesh sizes were carried out. In other regions mesh sizes of 100 µm (PS63, M42/3), 150 µm (M44/2) and 300 µm (VH1094) were used. The MN maxi was only deployed during the research cruise M44/2 in the northern Red Sea.

Generally, the volume of filtered water was calculated from the surface area of the net opening (midi: 0.25 m^2, maxi: 0.5 m^2) and the sampling depth interval. For the data sets from PS63, PS65, PS67 and M44/2 a mechanical digital flowmeter was used to record the filtering efficiency and to calculate the abundances (see Skjoldal et al., 2013, p. 4). The flowmeter is situated in the mouth area of the net and measures the water flow, providing more accurate volume values of the filtering efficiency.

2.2.3 Bongo net

During one research cruise (PS06) 61 additional samples were taken with the Bongo net (McGowan and Brown, 1966) to study selected calanoid copepod species. The Bongo net contains two nets that are lowered simultaneously for vertical plankton tows. The opening diameter is 60 cm, and the length of the nets is 2.5 m with a mesh size of 300 µm. The volume of filtering water was recorded with a flowmeter and used for the calculation of abundance.

2.2.4 Effects of variable net types and mesh sizes

Quantitative sampling of copepods and zooplankton is challenging. Major sources of error are patchiness, avoidance of nets and escape through the mesh (Wiebe, 1971; Skjoldal et al., 2013). These errors are defined by mesh sizes and net types, in particular the mouth area. The effect of patchiness cannot be investigated here due to the lack of replicates.

To our knowledge the sampling efficiency of the Nansen net and the MN midi have not been compared directly (Wiebe and Benfield, 2003; Skjoldal et al., 2013). However, it has been stated that the catches with Nansen net are considerably lower than with the WP-2 net (Hernroth, 1987), although the WP-2 net is considered as a modified Nansen net with a cylindrical front section of 95 cm and a smaller mouth area (57 cm^2, Skjoldal et al., 2013). The WP-2 net with 200 µm mesh size however, is in its sampling efficiency, measured as total zooplankton biomass, comparable to the MN midi with 200 µm mesh size (Skjoldal et al., 2013). Thus, it has to be taken into account during future analysis that the abundance values from the Nansen net are not directly comparable to those from the MN midi.

The mesh size has a different effect on the zooplankton catch. It is well known that small-sized copepod species (< 1 mm), and thus in particular non-calanoid species (e.g. Oithonidae, Oncaeidae) and also juvenile stages from calanoid copepods (e.g. *Microcalanus*, *Calocalanus*, *Disco*), pass through coarse mesh sizes ($\geq$ 200 µm), while they are retained in finer mesh sizes (Hopcroft et al., 2001; Paffenhöfer and Mazzocchi, 2003). Thus, abundances of smaller specimens as well as the species and life stage composition may vary considerably, when comparing samples from the Bellingshausen and Amundsen Seas (55 µm mesh size), around the Antarctic Peninsula (200 µm) and the Weddell Sea (100 µm).

2.3 Sample processing and analysis

All samples were preserved immediately after sampling in a 4 % formaldehyde–seawater solution. Samples were stored at room temperature until they were sorted in the laboratory. The formaldehyde solution was removed, the samples were rinsed and copepods were identified and counted under a stereomicroscope, using a modified mini-Bogorov chamber with high transparency as described in the ICES Zooplankton Methodology Manual (Postel et al., 2000). Abundant species were sorted from one-quarter or less of the sample while the entire sample was screened for rare species. Samples were divided with a Motoda plankton splitter (Motoda, 1959; Van Guelpen et al., 1982). Abundance was calculated using the surface area of the net opening and the sampling depth interval or the recordings of the flowmeter. Samples for reanalysis are only available for the cruises M42/3 and M44/2.

Except for five data sets (Cornils and Schnack-Schiel, 2017; Cornils et al., 2017a, b, c, d) all data sets were sorted and identified by Elke Mizdalski. Thus, the taxonomic concept has been used consistently throughout the data sets. A wide variety of identification keys and species descriptions have been used to identify the copepods, which cannot all be named here. References for the species descriptions and drawings of all identified marine planktonic species can be found in Razouls et al. (2005–2018). Calanoid copepods were identified to the lowest taxa possible, in general genus or species. Furthermore, for each identified taxon, females, males and copepodite (juvenile) stages were separated. Cyclopoid copepods were identified to species level in four data sets (Cornils et al., 2017a, b, c, d).

Previously published data sets were revised to ensure consistency of species names throughout the data set collection (Michels et al., 2012; Schnack-Schiel et al., 2007, 2010; Schnack-Schiel, 2010a). In the present compilation we have used the currently acknowledged copepod taxonomy as published in WoRMS (World register of Marine Species, WoRMS Editorial Board, 2018) and in Razouls et al. (2005–2018). Species names have been linked to the WoRMS database, so future changes in taxonomy will be tracked. In the parameter comments the "old" names are archived that were used initially when the specimens were identified. All used species names can be found in the "Copepod species list" under "Further details" at https://doi.org/10.1594/PANGAEA.884619 or at http://hdl.handle.net/10013/epic.65463ec2-e309-4d57-8fe3-0cebdd7dce70 (last access: 10 February 2018). We also provided the unique identifier (Aphia ID) from WoRMS (WoRMS Editorial Board, 2018) and notes on the distribution of each species.

When specimens could not be identified due to the lack of identification material, uncertainties in the taxonomy or missing parts, they were summarized under the genus name (e.g. *Disco* spp., *Diaixis* spp., *Paracalanus* spp., *Microcalanus* spp.) or family name (e.g. Aetideidae, copepodites). In most data sets some individuals could not be assigned to any family or genus. These are summarized as Calanoida indeterminata, female; Calanoida indeterminata, male; and Calanoida indeterminata, copepodites.

3 Data sets

3.1 Metadata

Each data set has its own persistent identifier. The metadata are consistent among all data sets, thus ensuring the comparability of the data sets and document their quality.

The following metadata can be found in each data set:

- "Related to" includes the corresponding cruise report, related data sets, and scientific articles of Sigrid Schnack-Schiel and others that have used part of the data previously.

- "Other version" in a few cases we have revised a previously published version of the data to ensure consistent species names throughout all data sets (for more information see Sect. 2.3).

- "Projects" shows internal projects or those with external funding. In the present case all data sets are related to internal projects of the AWI (Alfred Wegener Institut Helmholtz Centre for Polar and Marine Research) research program.

- "Coverage" gives the minimum and maximum values of the georeferences (latitude–longitude) of all stations.

- "Event(s)" comprises a list of station labels, a combination of cruise abbreviation and station number. Latitude and longitude of the position (units are in decimals with six decimal places), date and time of start and end of station, and elevation giving the bottom depth. Latitude and longitude, date and time and elevation were all recorded by the systems of the respective scientific vessel. The campaign name contains the cruise label (including optional labels), the basis of which is the name of the research vessel. The name of the device contains the net type that was deployed, and the comment may show further details of the station operation.

- "Parameter(s)" is a list of parameters used in the data set with columns containing the full and short name, the unit, the PI (which in this data compilation is always Sigrid Schnack-Schiel, except for one data set (https://doi.org/10.1594/PANGAEA.880239), and the method with a comment. The parameter "Date/Time of event" is not always identical with "Date/Time" given in the event. This is the case when the "Device" in the event is set to "Multiple Investigations" and thus the starting time of all investigations at this event is given. "Date/Time of event", however, is the time when the plankton net haul started. "Date/Time" recorded on R/V *Polarstern* and during the cruises M42/3 and JB03 was UTC (Coordinated Universal Time) and during cruise M44/2 local time was recorded (UTC+2). No information on "Date/Time" was found for the cruises DAE1979/80, M11/4 and VH1094. "Elevation" provides information on the bottom depth of the plankton station, if available. Three parameters describe the sampling depths interval. "Depth, water" is the mean depth of the sampled depth interval. "Depth top" and "Depth bot" describe the upper and lower limit of the sampling depth interval, respectively. "Volume" is the amount of water that was filtered during each net tow, calculated either using the mouth area of the net and depth interval or with a flowmeter (Sect. 2.2.2). "Comment" gives the detailed information on the net type, the net number and mesh size.

In the following list of parameters are the copepod taxa for which abundance data were recorded. Calanoid taxa are separated into female, male and copepodites. Species names are consistent throughout all data sets, which ensures the comparability of the data sets. Clicking the link on the species names leads to their respective WoRMS ID (see Sect. 2.3). The "short names" of each taxon consist of the first letter of the generic name and the name of the species. In nine cases this results in identical short names (*Pleuromamma antarctica*, *Paraeuchaeta antarctica* = *P. antarctica*; *Temoropia minor*, *Temorites minor* = *T. minor*; *Chiridius gracilis*, *Centropages gracilis* = *C. gracilis*; *Clausocalanus minor*, *Calanopia minor* = *C. minor*; *Heterostylites longicornis*, *Haloptilus longicornis* = *H. longicornis*; *Scolecithricella abyssalis*, *Spinocalanus abyssalis* = *S. abyssalis*; *Scaphocalanus magnus*, *Spinocalanus magnus* = *S. magnus*). Thus, we advise to use the full scientific names of these species in further analyses.

3.2 Temporal station distribution

While samples of the Magellan region (November 1994), the Gulf of Aqaba and the northern Red Sea (February–March 1999), the Great Meteor Bank (September 1998), and the eastern Atlantic Ocean (November 2002) were restricted to 1 year and 1 season, the Southern Ocean was sampled multiple times (Table 1). Samples in the Southern Ocean were taken from 1980 to 2005 (Table 1, Fig. 2a, b). The highest number of zooplankton samples was taken in the 1980s (Fig. 2b). In the 1980s the sampling effort was concentrated to the Antarctic Peninsula, the Scotia Sea and the Weddell Sea (Fig. 2a). Samples were taken in multiple years. In the 1990s until 2005 most samples were taken in the Bellingshausen and Amundsen Sea, with fewer samples in the western and eastern Weddell Sea. Two transects were sampled across the Weddell Sea in the 1990s in austral summer and autumn (Fig. 2b). In general, most stations were sampled during summer (December to February), followed by autumn (March to May) and spring (September to November), while winter samples are only available from 1986 in the eastern Weddell Sea (Fig. 2b, c). Summer and autumn samples are widely distributed from the Amundsen Sea to the eastern Weddell Sea (Fig. 2b), while spring and autumn samples are mostly present from the Scotia Sea and eastern Weddell Sea. Most samples were taken in January and February (Fig. 2d). Samples are scattered throughout the entire day (Fig. 3).

It should be taken into account that several copepod species in regions with pronounced seasonality of primary production, e.g. in high latitudes or upwelling regions (Conover, 1988; Schnack-Schiel, 2001), undergo seasonal vertical migration (e.g. *Rhincalanus*, *Calanoides*). They reside in deep water layers during periods of food scarcity and rise to the surface layers when the phytoplankton blooms start. Furthermore, other species undergo pronounced diel vertical migrations (e.g. *Pleuromamma*) from

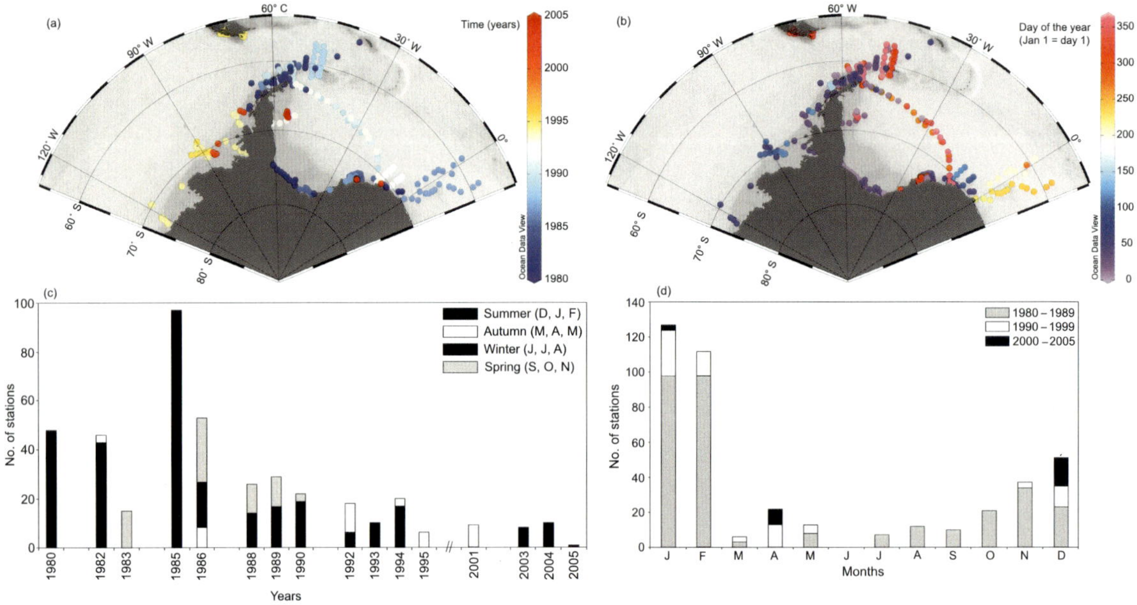

Figure 2. Sampling effort in the Southern Ocean: **(a)** station distribution in years, **(b)** station distribution in the annual cycle, **(c)** number of stations per year and season, **(d)** number of stations per month and year.

mesopelagic layers during day-time to avoid predators, moving to epipelagic waters at night to feed (Longhurst and Harrison, 1989). Thus, to avoid biases in the comparison of the vertical distribution of copepod species, season and day-time should be considered during further analysis of the data sets.

3.3 Copepoda

In total, specimens from six copepod orders were recorded in the compiled data sets.

However, in 29 data sets only calanoid copepods were identified on species level. Specimens of other copepod orders were comprised in families or orders.

3.3.1 Calanoida

In total 284 calanoid species could be separated into 29 data sets (see "Copepod species list" at https://doi.org/10.1594/PANGAEA.884619). These species are representatives of 28 families and 91 genera (Table 2). Abundance and distribution data for 96 calanoid species in the Southern Ocean were archived. In the eastern Atlantic Ocean 125 and around the Great Meteor Bank 135 calanoid copepod species could be identified (Table 2). These numbers already indicate the well-known fact that species richness in the tropical and subtropical open oceans is much higher than in the polar Southern Ocean (e.g. Rutherford et al., 1999; Tittensor et al., 2010). Compared to these the number of calanoid species (60) in the subtropical northern Red Sea is low, which is expected due to the shallow sills at the entrance of the Red Sea and the high salinity (see Cornils et al., 2005). The lowest number of calanoid species (35) was found in the Magellan Region. Calanoid copepod families with the highest number of species were Aetideidae (33), Augaptilidae (27) and Scolecitrichidae (40; Table 2).

For selected species from the Southern Ocean and the northern Red Sea and Gulf of Aqaba, the five copepodite stages were also counted individually (Table 3), providing valuable information on the seasonal and vertical distribution of the five copepodite stages. During four cruises, *Rhincalanus gigas* nauplii were also counted (PS09, PS21, PS23, PS29). In the 1990s Sigrid Schnack-Schiel used these data to publish a series of papers on life cycle strategies of Antarctic calanoid copepods such as *Calanoides acutus*, *Rhincalanus gigas*, *Microcalanus* cf. *pygmaeus* or *Stephos longipes* (e.g. Schnack-Schiel and Mizdalski, 1994; Schnack-Schiel et al., 1995; Ward et al., 1997; Schnack-Schiel, 2001). However, the stage-resolved copepod data of most species in Table 3 have not been analysed.

It is notable that none of the calanoid species were found in all five regions (see "Copepod species list" at https://doi.org/10.1594/PANGAEA.884619). In contrast, many species were only recorded in one region: 60 species were found only in the Southern Ocean, while 43 and 38 were found only in the data sets from the Great Meteor Bank and the transect in the eastern Atlantic Ocean, respectively. A total of 24 species were found only in the Red Sea and 6 were identified only from samples in the Magellan region. Of the 28 calanoid families, 11 were distributed in all five regions (Table 2).

As an example of the geographical and vertical distribution of the copepods, three abundant genera were chosen

Table 2. List of calanoid copepod families and genera, cyclopoid families, and other orders compiled in this data collection. The number of species for each genus is written in parentheses. The presence of the calanoid and cyclopoid families and other copepod orders in the five different regions is marked (X). For a complete overview of all species see the "Copepod species list" at https://doi.org/10.1594/PANGAEA.884619.

Order	Family	Genus	Distribution				
			Southern Ocean	Magellan region	Great Meteor Bank	Northern Red Sea	Eastern Atlantic Ocean
Calanoida	Acartiidae	*Acartia (3)*		X	X	X	X
	Aetideidae	*Aetideopsis (3), Aetideus (5), Chiridiella (1), Chiridius (3), Chirundina (1), Euchirella (7), Gaetanus (8), Pseudeuchaeta (1), Pseudochirella (2), Undeuchaeta (2)*	X	X	X		X
	Arctokonstantinidae	*Foxtonia (1)*	X		X		
	Arietellidae	*Arietellus (3), Paraugaptilus (1)*					X
	Augaptilidae	*Augaptilus (4), Euaugaptilus (7), Haloptilus (10), Pseudaugaptilus (1), Pseudhaloptilus (1)*	X	X	X	X	X
	Bathypontiidae	*Temorites (2)*	X		X		
	Calanidae	*Calanoides (3), Calanus (4), Mesocalanus (1), Nannocalanus (1), Neocalanus (3), Undinula (1)*	X	X	X	X	X
	Candaciidae	*Candacia (13)*	X	X	X	X	X
	Centropagidae	*Centropages (5)*		X	X	X	X
	Clausocalanidae	*Clausocalanus (12), Ctenocalanus (2), Drepanopus (1), Farrania (1), Microcalanus (1)*	X	X	X	X	X
	Diaixidae	*Diaixis (1)*				X	
	Discoidae	*Disco (1)*	X		X	X	X
	Eucalanidae	*Eucalanus (2), Pareucalanus (2), Subeucalanus (5)*	X	X	X	X	X
	Euchaetidae	*Euchaeta (7), Paraeuchaeta (6)*	X	X	X	X	X
	Fosshageniidae	*Temoropia (3)*	X		X	X	X
	Heterorhabdidae	*Disseta (1), Heterorhabdus (7), Heterostylites (2), Mesorhabdus (1), Microdisseta (1), Paraheterorhabdus (3)*	X	X	X		X
	Lucicutiidae	*Lucicutia (13)*	X	X	X	X	X
	Metridinidae	*Metridia (8), Pleuromamma (8)*	X	X	X	X	X
	Nullosetigeridae	*Nullosetigera (3)*			X		
	Paracalanidae	*Acrocalanus (5), Calocalanus (3), Delibus (1), Mecynocera (1), Paracalanus (4), Parvocalanus (1)*		X	X	X	X
	Phaennidae	*Cephalophanes (2), Cornucalanus (1), Onchocalanus (4), Phaenna (1), Xanthocalanus (2)*	X		X	X	X
	Pontellidae	*Calanopia (2), Labidocera (1), Pontellina (2), Pontellopsis (3)*			X	X	X
	Rhincalanidae	*Rhincalanus (4)*	X	X	X	X	X
	Scolecitrichidae	*Amallothrix (3), Archescolecithrix (1), Bradfordiella (1), Bradyidius (1), Cenognatha (1), Landrumius (1), Lophothrix (2), Macandrewella (1), Mixtocalanus (1), Pseudoamallothrix (5), Racovitzanus (2), Scaphocalanus (10), Scolecithricella (5), Scolecithrix (2), Scolecitrichopsis (1), Scottocalanus (1)*	X	X	X	X	X
	Spinocalanidae	*Mimocalanus (2), Mospicalanus (1), Spinocalanus (9), Teneriforma (2)*	X	X	X		X
	Stephidae	*Stephos (2)*	X			X	
	Temoridae	*Temora (1)*					X
	Tharybidae	*Tharybis (2), Undinella (1)*	X	X	X	X	X
Cyclopoida	Corycaeidae				X	X	X
	Hemicyclopinidae		X				
	Lubbockiidae		X		X	X	X
	Oithonidae		X	X	X	X	X
	Oncaeidae		X	X	X	X	X
	Sapphirinidae				X	X	X
	Incertae sedis				X	X	X
Harpacticoida			X	X	X	X	X
Monstrilloida			X		X		X
Mormonilloida			X		X		X
Siphonostomatoida			X				

Table 3. List of species and cruises to the Southern Ocean where copepodite stages 1–5 were separated and counted. Species with asterisks were separated from samples of cruise M44/2 to the northern Red Sea and the Gulf of Aqaba.

Species	ANT-II/2	ANT-III/3	ANT-V/1	ANT-V/2	ANT-V/3	ANT-VII/2	ANT-VII/3	ANT-VII/4	ANT-VIII/2	ANT-IX/2	ANT-X/3	ANT-X/7	ANT-XI/3	ANT-XII/4	ANT-XVIII/5b	ANT-XXI/2	ANT-XXII/2	M 11/4
Amallothrix dentipes		1		1	1					1	1	1						
Calanoides acutus	1	3	1	2	1	1	2	1		1	1	1	1	1	1	1	1	1
Calanus propinquus		3	1	2	1	1	2	1		1	1	1	1	1	1	1	1	1
Calanus simillimus						1	2	1						1				
Ctenocalanus citer		1		1	1				1	1	1	1	1	1	1	1	1	
Heterorhabdus austrinus	1	1	1	1	1	1	1	1	1	1	1	1	1	1				1
*Mesocalanus tenuicornis**																		
Metridia curticauda		1	1	1	1	1	2	1	1	1	1	1	1	1				1
Metridia spp. (*M. gerlachei, M. lucens*)	1	3	1	1	1	1	2	1	1	1	1	1	1	1	1	1	1	1
Microcalanus spp.		1		1	1				1	1	1	1	1	1	1	1	1	
*Nannocalanus minor**																		
Paraheterorhabdus farrani	1	1	1	1	1	1	1	1	1	1	1	1	1	1				1
Pleuromamma antarctica						1	2	1										
*Pleuromamma indica**																		
Pseudoamallothrix cenotelis		1		1	1					1	1	1						
Rhincalanus gigas	1	2	1	2	1	1	2	1		1	1	1	1	1	1	1	1	1
*Rhincalanus nasutus**																		
Scolecithricella minor		1		1	1					1	1	1						
Spinocalanus antarcticus		1			1					1	1	1						
Spinocalanus longicornis		1		1	1				1	1	1	1						
Spinocalanus terranovae		1		1	1				1	1	1	1						
Stephos longipes		1		1	1				1	1	1	1	1	1	1	1		
*Undinula vulgaris**																		

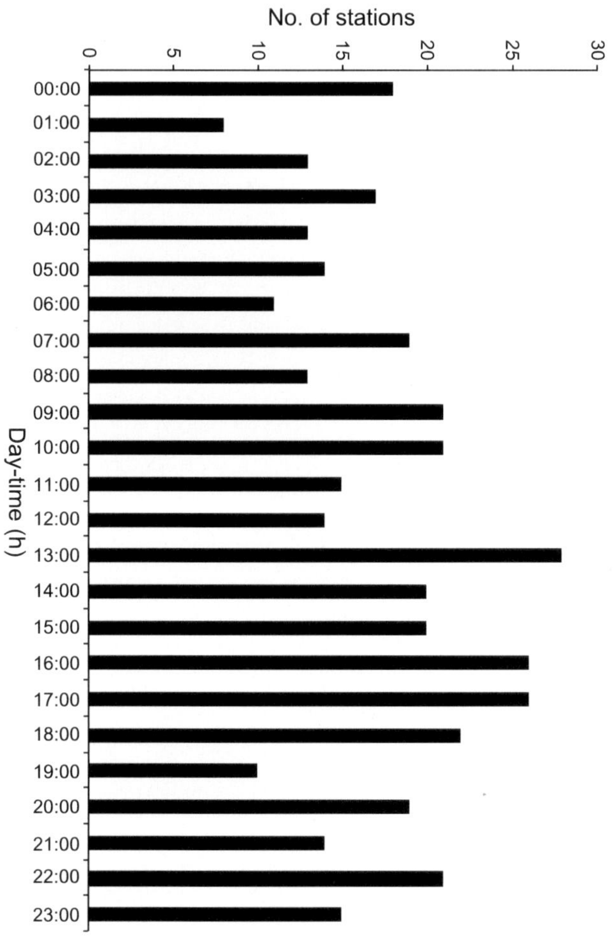

Figure 3. Sampling effort during day-time in the Southern Ocean. Day-time is important to understand the behaviour of diel vertical migrators. The number of stations is summarized for every hour of the day, e.g. the bar at 00:00 contains all stations taken between 00:00 and 00:59.

(Fig. 4). While *Microcalanus* spp. (not separated into species due to uncertainties in the taxonomy) and *Spinocalanus* spp. (nine species; Table 2) are abundant down to 1000 m, the two species of *Ctenocalanus* (two species, Fig. 4) and *Stephos* occur mainly in the epipelagic layer of the ocean. This is in accordance with their known vertical distribution (Schnack-Schiel and Mizdalski, 1994; Bode et al., 2018). Comparing the abundance of *Spinocalanus* and *Microcalanus* from all regions suggests that the abundance of these taxa is far greater in the Southern Ocean than in the warmer regions of the ocean. This picture, however, has to be treated with caution, since the tropical Atlantic was only sampled in the upper 300 m of the water column and was thus too shallow for the meso- and bathypelagic genera (Bode et al., 2018).

In the case of *Ctenocalanus* and *Stephos* our data sets reveal that closely related species within a genus may have contrasting distribution patterns. *Stephos longipes* and *Ctenocalanus citer* are restricted to colder and polar waters of the Southern Hemisphere, while *Ctenocalanus vanus* occurs in both the Red Sea and the warm Atlantic Ocean. *Stephos maculosus* occurs only in the Red Sea (see arrow in Fig. 4). Furthermore, the distribution patterns reveal that of the four genera only *C. citer* has a higher abundance in the samples from the Bellingshausen and Amundsen Seas, and around the Antarctic Peninsula, while *S. longipes*, *Microcalanus* spp. and *Spinocalanus* spp. all have higher abundances in the eastern Weddell Sea. This may be due to the lower water depth at the Peninsula since *Microcalanus* and *Spinocalanus* are considered as mesopelagic to bathypelagic. Thus, they are often not found at shallow stations (< 300 m depth). In case of the sea-ice-associated *S. longipes*, low sea-ice conditions and offshore stations may have caused the restricted distribution. *S. longipes* occurred mainly in the upper water layers, but it was also recorded with low abundances in deeper layers (Fig. 4). This pattern may be due to its life cycle, shifting seasonally from a sea-ice-associated to a bentho-pelagic life cycle (Schnack-Schiel et al., 1995).

3.3.2 Other Copepoda

In total, 28 non-calanoid taxa were recorded. Four data sets provide only abundance and distribution data for non-calanoid copepod orders (PS06, PS10, PS29, PS35; Table 1), in particular on species of the order Cyclopoida from the families Oithonidae (two species) and Oncaeidae (six species; Table 2). They were separated into female, male, and copepodite stages 1, 2, 3, 4 and 5. During VH1094 *Oithona* species were also identified (Table 2). In all other data sets species of these two families were not separated. In all regions representatives of the family Lubbockiidae were recorded. In the subtropical and tropical samples of PS63, M44/2 and M42/3 abundances of species of the families Corycaeidae and Sapphirinidae, and of the genus *Pachos* were also recorded. Except for PS65, species of the order Harpacticoida were not separated. In the latter five species were identified, mainly sea-ice-associated harpacticoids (Table 2; Schnack-Schiel et al., 1998). Also, specimens of the orders Monstrilloida, Mormonillida and Siphonostomatoida were counted.

In most data sets, copepod nauplii are also recorded as one parameter. However, due to the small size of nauplii they were not sampled quantitatively and should be discarded in further analysis.

3.4 Further remarks on the usage of the data compilation

Generally, the cruise reports have been linked to each data set. The cruise reports provide valuable information on the itinerary, zooplankton sampling procedures and on other scientific activities on-board that could be useful for the data analysis (e.g. CTD data). Abundance data of selected species

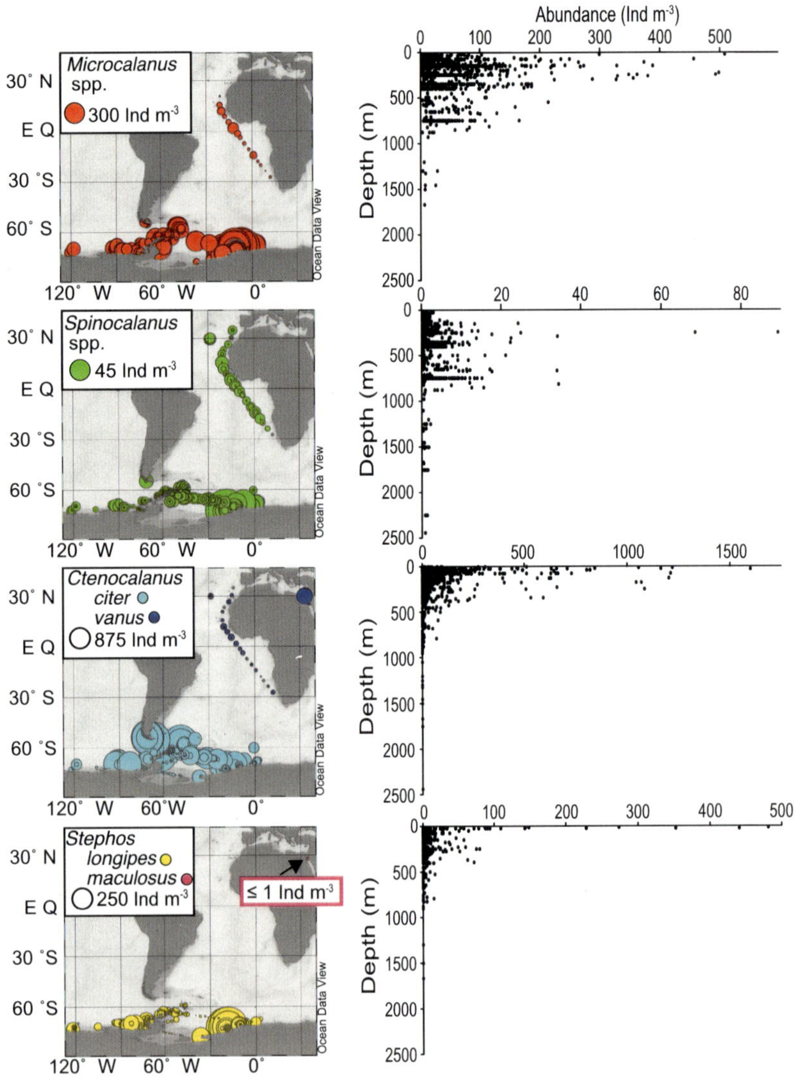

Figure 4. Distribution and abundance as individuals per cubic metre (Ind m^{-3}) of selected genera (*Microcalanus*, *Spinocalanus*, *Ctenocalanus*, *Stephos*). Depth (m) is the mean depth of each sampling depth interval ("Depth water").

and data sets have been published previously in scientific articles. These articles are linked to the respective data sets (under "Related to").

To use the data, they can be downloaded individually as tab-delimited text files or altogether as a .zip file to allow an import to other software, e.g. R (R core team, 2018) or Ocean Data View (Schlitzer, 2015) for further analysis. Due to the consistent taxonomic nomenclature the individual files can be concatenated easily. It should be kept in mind, however, that not all data sets are directly comparable due to differences in net type and mesh sizes (see Sect. 2.2.4). As noted in Sect. 3.2, several species undergo pronounced seasonal and diel vertical migrations. Therefore, nets from surface waters may not always sample the full vertical extent of the populations, particularly of the biomass dominants.

To evaluate the vertical and spatial distribution of marine plankton, hydrographic information such as temperature and salinity profiles is essential. The relevant publications are available at https://doi.org/10.1594/PANGAEA.884619; see "Further details". Recently, a summary of the physical oceanography of R/V *Polarstern* has been published (Driemel et al., 2017) with CTD data archived in PANGAEA as well (Rohardt et al., 2016), except for the cruises PS04 (ANT-II/2), PS14 (ANT-VII/2), PS21 (ANT-X/3), PS63 (ANT-XX/1) and PS65 (ANT-XXII/2) (see Table 1). For these five cruises information on temperature and salinity profiles exist only for PS63 (Schnack-Schiel et al., 2010) and for PS65 the CTD profiles can be downloaded (https://doi.org/10.1594/PANGAEA.742627; Absy et al., 2008). For M11/4 a CTD data set is also available (https://doi.org/10.1594/PANGAEA.742745; Stein, 2010). To connect the CTD data with the corresponding plankton net haul the metadata "Event" and "Date/time" can be used. Furthermore, cruise track and station information are available in the cruise reports as well as on the station tracks for each cruise (https://pangaea.de/expeditions/, last access: 10 February 2018). For the other two R/V *Meteor* cruises hydrographic information is available in scientific articles (M42/3: Beckmann and Mohn, 2002; Mohn and Beckmann, 2002; M44/2: Cornils et al., 2005; Plähn et al., 2002). Metadata information of the cruise JB03 can be downloaded from https://www.bodc.ac.uk/resources/inventories/cruise_inventory/report/5916/ (last access: 10 February 2018). To date, no hydrographic information is publicly available for the cruises DAE79/80 and VH1094.

Additionally, abundances of all other zooplankton organisms in the net samples used for the copepod data sets are available for the four cruises ANT-X/3, ANT-XVIII/5b, M42/3 and M44/2. These can be downloaded at https://doi.org/10.1594/PANGAEA.883833, https://doi.org/10.1594/PANGAEA.884581, https://doi.org/10.1594/PANGAEA.883771 and https://doi.org/10.1594/PANGAEA.883779.

All data presented here are archived in the database PANGAEA. There are, however, other data archiving initiatives that also store data on copepod abundance and distribution such as COPEPOD (https://www.st.nmfs.noaa.gov/copepod/, last access: 10 February 2018), BODC (https://www.bodc.ac.uk, last access: 10 February 2018) or OBIS (http://www.iobis.org, last access: 10 February 2018). The here-presented data, however, have not been published in any other cataloguing initiative before.

4 Data availability

In total 33 data sets with 349 stations were archived in the PANGAEA® (Data Publisher for Earth & Environmental Science, www.pangaea.de) database (Cornils and Schnack-Schiel, 2018). The persistent identifier https://doi.org/10.1594/PANGAEA.884619 links to the splash page of the data compilation. We encourage the users of these data to cite both the DOI of the data collection in PANGAEA as well as the present data publication as a courtesy to Sigrid Schnack-Schiel and the people preparing the data for open access. Metadata include DOIs to cruise reports and related physical oceanography. Data are provided in consistent format as tab-delimited ASCII files and are available through open access under a CC BY license (Creative Commons Attribution 3.0 Unported).

5 Concluding remarks

Pelagic marine ecosystems are threatened by increasing water temperatures due to climate change. These environmental changes are also expected to cause shifts in the community structure of pelagic organisms. Within the pelagic food web copepods have a central role as intermediator between the microbial loop and higher trophic level. Due to their short life cycles and their high diversity copepods offer a unique opportunity to study effects of environmental variables on numerous taxa with different life cycle strategies. It is also known that their species composition and abundance often reflect environmental changes such as temperature, seasonal variability or stratification (Beaugrand et al., 2002). To understand the complexity of ecological niches and ecosystem functioning, but also to investigate the effects of environmental changes, a detailed knowledge of species diversity, distribution and abundance is essential. The present data compilation provides further information on spatial, vertical and temporal distribution of copepod species and may thus be used to obtain a better picture of species biogeographies. Many individual data sets can also be linked to corresponding CTD profiles (Table 1) and may thus be useful for modelling approaches such as species distribution or environmental niche modelling.

Furthermore, for all calanoid copepods females, males and copepodites were enumerated separately, and for selected species a distinction between copepodite stages was made. This detailed resolution of abundance data will also allow future investigations on life cycle strategies as well as show how the different stages interact with the environment (e.g. temperature, currents, depth).

Author contributions. AC and HG were initiators of the data collection and the paper. Data collection and initial analysis was carried out by SiS. Copepod identification (except cyclopoid copepods) and the curation of taxon names was carried out by EM and AC. Data curation was carried out by RS and StS. AC drafted the paper and EM, StS, and HG provided input.

Competing interests. The authors declare that they have no conflict of interest.

Acknowledgements. We would like to thank numerous scientists, technicians and students who helped with the sampling on-board, and the sample processing and analysis in Bremerhaven, in particular Ruth Alheit. We are grateful to the crews of R/Vs *Polarstern*, *Meteor*, *Victor Hensen*, *John Biscoe* and *Polarsirkel*, who helped in many ways during every expedition. We would also like to thank Thomas Brey who strongly supported the effort to archive this data collection. Finally, we thank Angus Atkinson and two anonymous reviewers whose comments improved the paper.

Edited by: Falk Huettmann
Reviewed by: Angus Atkinson and two anonymous referees

References

Absy, J. M., Schröder, M., Muench, R. D., and Hellmer, H. H.: Physical oceanography from 120 CTD stations during POLARSTERN cruise ANT-XXII/2 (ISPOL), PANGAEA, https://doi.org/10.1594/PANGAEA.742627, 2008.

Atkinson, A.: Life cycle strategies of epipelagic copepods in the Southern Ocean, J. Marine Syst., 15, 289–311, https://doi.org/10.1016/S0924-7963(97)00081-X, 1998.

Atkinson, A., Siegel, V., Pakhomov, E., and Rothery, P.: Long-term decline in krill stock and increase in salps within the Southern Ocean, Nature, 432, 100–103, https://doi.org/10.1038/nature02996, 2004.

Atkinson, A., Ward, P., Hunt, B., Pakhomov, E. A., and Hosie, G. W.: An overview of Southern Ocean zooplankton data: Abundance, biomass, feeding and functional relationships, CCAMLR Sci., 19, 171–218, 2013.

Bé, A. W. B., Ewing, M., and Linton, L. W.: A quantitative multiple opening-and-closing plankton sampler for vertical towing, ICES J. Mar. Sci., 25, 36–46, https://doi.org/10.1093/icesjms/25.1.36, 1959.

Beaugrand, G., Reid, P. C., Ibanez, F., and Lindley, J. A.: Reorganization of North Atlantic marine copepod biodiversity and climate, Science, 296, 1692–1694, https://doi.org/10.1126/science.1071329, 2002.

Beckmann, A. and Mohn, C.: The upper ocean circulation at Great Meteor Seamount, Ocean Dynam., 52, 194–204, https://doi.org/10.1007/s10236-002-0018-3, 2002.

Bode, M., Hagen, W., Cornils, A., Kaiser, P., and Auel, H.: Copepod distribution and biodiversity patterns from the surface to the deep sea along a latitudinal transect in the eastern Atlantic Ocean (24N to 21S), Prog. Oceanogr., 161, 66–77, https://doi.org/10.1016/j.pocean.2018.01.010, 2018.

Boysen-Ennen, E. and Piatkowski, U.: Meso-and macrozooplankton communities in the Weddell Sea, Antarctica, Polar Biol., 9, 17–35, https://doi.org/10.1007/BF00441761, 1988.

Boysen-Ennen, E., Hagen, W., Hubold, G., and Piatkowski, U.: Zooplankton biomass in the ice-covered Weddell Sea, Antarctica, Mar. Biol., 111, 227–235, https://doi.org/10.1007/BF01319704, 1991.

Bradford-Grieve, J. M.: Colonization of the pelagic realm by calanoid copepods, Hydrobiologia, 485, 223–244, https://doi.org/10.1023/A:1021373412738, 2002.

Chust, G., Vogt, M., Benedetti, F., Nakov, T., Villéger, S., Aubert, A., Vallina, S. M, Righetti, D., Not, F., Biard, T., Bittner, L., Benoiston, A.-S., Guidi, L., Villarino, E., Gaborit, C., Cornils, A., Buttay, L., Irisson, J.-O., Chiarelo, M., Vallim, A. L., Blanco-Bercial, L., Basconi, L., and Ayata, S.-D.: Mare Incognitum: A Glimpse into Future Plankton Diversity and Ecology Research, Front. Mar. Sci., 4, 68, https://doi.org/10.3389/fmars.2017.00068, 2017.

Conover, R. J.: Comparative life history in the genera *Calanus* and *Neocalanus* in high latitudes of the northern hemisphere, Hydrobiologia, 167, 127–142, https://doi.org/10.1007/BF00026299, 1988.

Cornils, A. and Schnack-Schiel, S. B.: Abundance of planktonic Copepoda (Crustacea) during METEOR cruise M44/2 (Gulf of Aqaba, Red Sea) – additional stations, PANGAEA, https://doi.org/10.1594/PANGAEA.881901, 2017.

Cornils, A. and Schnack-Schiel, S. B.: Abundance and distribution of planktonic Copepoda in the Southern Ocean and other regions from 1980 to 2005, PANGAEA, https://doi.org/10.1594/PANGAEA.884619, 2018.

Cornils, A., Schnack-Schiel, S. B., Hagen, W., Dowidar, M., Stambler, N., Plähn, O., and Richter, C.: Spatial and temporal distribution of mesozooplankton in the Gulf of Aqaba and the northern Red Sea in February/March 1999, J. Plankton Res., 27, 505–518, https://doi.org/10.1093/plankt/fbi023, 2005.

Cornils, A., Metz, C., and Schnack-Schiel, S. B.: Abundance of planktonic Cyclopoida (Copepoda, Crustacea) during POLARSTERN cruise ANT-XI/3 (PS29), PANGAEA, https://doi.org/10.1594/PANGAEA.879718, 2017a.

Cornils, A., Metz, C., and Schnack-Schiel, S. B.: Abundance of selected planktonic Cyclopoida (Copepoda, Crustacea) during POLARSTERN cruise ANT-III/3 (PS06), PANGAEA, https://doi.org/10.1594/PANGAEA.879771, 2017b.

Cornils, A., Metz, C., and Schnack-Schiel, S. B.: Abundance of selected planktonic Cyclopoida (Copepoda, Crustacea) during POLARSTERN cruise ANT-V/3 (PS10), PANGAEA, https://doi.org/10.1594/PANGAEA.879772, 2017c.

Cornils, A., Metz, C., and Schnack-Schiel, S. B.: Abundance of selected planktonic Cyclopoida (Copepoda, Crustacea) during POLARSTERN cruise ANT-XII/4 (PS35), PANGAEA, https://doi.org/10.1594/PANGAEA.879773, 2017d.

Currie, R. and Foxton, P.: The Nansen closing method with vertical plankton nets. J. Mar. Biol. Assoc. UK, 35, 483–492, https://doi.org/10.1017/S002531540001033X, 1956.

Driemel, A., Fahrbach, E., Rohardt, G., Beszczynska-Möller, A., Boetius, A., Budéus, G., Cisewski, B., Engbrodt, R., Gauger, S., Geibert, W., Geprägs, P., Gerdes, D., Gersonde, R., Gordon, A. L., Grobe, H., Hellmer, H. H., Isla, E., Jacobs, S. S., Janout, M., Jokat, W., Klages, M., Kuhn, G., Meincke, J., Ober, S., Østerhus, S., Peterson, R. G., Rabe, B., Rudels, B., Schauer, U., Schröder, M., Schumacher, S., Sieger, R., Sildam, J., Soltwedel, T., Stangeew, E., Stein, M., Strass, V. H., Thiede, J., Tippenhauer, S., Veth, C., von Appen, W.-J., Weirig, M.-F., Wisotzki, A., Wolf-Gladrow, D. A., and Kanzow, T.: From pole to pole: 33 years of physical oceanography onboard R/V *Polarstern*, Earth Syst. Sci. Data, 9, 211–220, https://doi.org/10.5194/essd-9-211-2017, 2017.

Dubischar, C. D., Lopes, R. M., and Bathmann, U. V.: High summer abundances of small pelagic copepods at the Antarctic Polar Front – implications for ecosystem dynamics, Deep-Sea Res. Pt. II, 49, 3871–3887, https://doi.org/10.1016/S0967-0645(02)00115-7, 2002.

Edwards, M. and Richardson, A. J.: The impact of climate change on the phenology of the plankton community and trophic mismatch, Nature, 430, 881–884, https://doi.org/10.1038/nature02808, 2004.

Hernroth, L.: Sampling and filtration efficiency of two commonly used plankton nets. A comparative study of the Nansen net and the Unesco WP 2 net, J. Plankton Res., 9, 719–728, https://doi.org/10.1093/plankt/9.4.719, 1987.

Hopcroft, R. R., Roff, J. C., and Chavez, F. P.: Size paradigms in copepod communities: a re-examination, Hydrobiologia, 453, 133–141, https://doi.org/10.1023/A:101316791, 2001.

Hopkins, T. L.: The zooplankton community of Croker passage, Antarctic Peninsula, Polar Biol., 4, 161–170, https://doi.org/10.1007/BF00263879, 1985.

Hopkins, T. L. and Torres, J. J.: The zooplankton community in the vicinity of the ice edge, western Weddell Sea, March 1986, Polar Biol., 9, 79–87, https://doi.org/10.1007/BF00442033, 1988.

Hosie, G. W., Fukuchi, M., and Kawaguchi, S.: Development of the Southern Ocean continuous plankton recorder survey, Prog. Oceanogr., 58, 263–284, https://doi.org/10.1016/j.pocean.2003.08.007, 2003.

Humes, A. G.: How many copepods?, Hydrobiologia, 292, 1–7, https://doi.org/10.1007/BF00229916, 1994.

Huys, R. and Boxshall, G. A.: Copepod evolution, The Ray Society, London, England, 468 pp., 1991.

Kouwenberg, J. H. M., Razouls, C., and Desreumaux, N.: 6.6. Southern Ocean Pelagic Copepods, in: The Biogeographic Atlas of the Southern Ocean, edited by: De Broyer, C., Koubbi, P., Griffith H. J., Raymond, B., d'Udekem d'Acoz, C., Van de Putte, A. D., Danis, B., David, B., Grant, S., Gutt, J., Held, C., Hosie, G., Huettmann, F., and Post, A., The Scientific Committee on Antarctic Research, Cambridge, 209–296, 2014.

Longhurst, A. R.: Relationships between diversity and the vertical structure of the upper ocean, Deep-Sea Res., 32, 1535–1570, https://doi.org/10.1016/0079-6611(85)90036-9, 1985.

Longhurst, A. R. and Harrison, W. G.: The biological pump: profiles of plankton production and consumption in the upper ocean, Prog. Oceanogr., 22, 47–123, https://doi.org/10.1016/0079-6611(89)90010-4, 1989.

McGowan, J. A. and Brown, D. M.: A new opening-closing paired zooplankton net, Scripps Inst. Ocean., 66–23, 54 pp., 1966.

McLeod, D. J., Hosie, G. W., Kitchener, J. A., Takahashi, K. T., and Hunt, B. P. V.: Zooplankton Atlas of the Southern Ocean: The SCAR SO-CPR Survey (1991–2008), Polar Sci., 4, 353–385, https://doi.org/10.1016/j.polar.2010.03.004, 2010.

Michels, J., Schnack-Schiel, S. B., Pasternak, A. F., Mizdalski, E., Isla, E., and Gerdes, D.: Abundance of copepods during POLARSTERN cruise ANT-XXI/2 (BENDIX), PANGAEA, https://doi.org/10.1594/PANGAEA.754015, 2012.

Mohn, C. and Beckmann, A.: The upper ocean circulation at Great Meteor Seamount, Ocean Dynam., 52, 179–193, https://doi.org/10.1007/s10236-002-0017-4, 2002.

Motoda, S.: Devices of simple plankton apparatus, Mem. Fac. Fish., Hokkaido Univ., 7, 73–94, http://hdl.handle.net/2115/21829 (last access: 10 February 2018), 1959.

Nansen, F.: Closing-nets for vertical hauls and for horizontal towing, ICES J. Mar. Sci., s1, 3–8, https://doi.org/10.1093/icesjms/s1.67.3, 1915.

Paffenhöfer, G. A. and Mazzocchi, M. G.: Vertical distribution of subtropical epiplanktonic copepods, J. Plankton Res., 25, 1139–1156, https://doi.org/10.1093/plankt/25.9.1139, 2003.

Pakhomov, E. A., Perissinotto, R., and McQuaid, C. D.: Zooplankton structure and grazing in the Atlantic sector of the Southern Ocean in late austral summer 1993: Part 1. Ecological zonation, Deep-Sea Res. Pt. I, 47, 1663–1686, https://doi.org/10.1016/S0967-0637(99)00122-3, 2000.

Plähn, O., Baschek, B., Badewien, T. H., Walter, M., and Rhein, M.: Importance of the Gulf of Aqaba for the formation of bottom water in the Red Sea, J. Geophys. Res., 107, 3108, https://doi.org/10.1029/2000JC000342, 2002.

Postel, L., Fock, H., and Hagen, W.: 4 – Biomass and Abundance, in: ICES Zooplankton Methodology Manual, edited by: Harris, R., Wiebe, P., Lenz, J., Skjoldal, H. R., and Huntley, M, Academic Press, London, 83–192, https://doi.org/10.1016/B978-012327645-2/50005-0, 2000.

Razouls, C., de Bovee, F., Kouwenberg, J., and Desreumaux, N.: Diversity and geographic distribution of marine planktonic copepods, Sorbonne Universite, CNRS, available at: http://copepodes.obs-banyuls.fr/ (last access: 10 February 2018), 2005–2018.

R Core Team: R: A Language and Environment for Statistical Computing, Foundation for Statistical Computing, Vienna, Austria, available at: https://www.R-project.org, last access: 10 February 2018.

Reid, P. C. and Edwards, M.: Long-term changes in the pelagos, benthos and fisheries of the North Sea, Senck. Marit., 31, 107–115, https://doi.org/10.1007/BF03043021, 2001.

Rivero-Calle, S., Gnanadesikan, A., Del Castillo, C. E., Balch, W. M., and Guikema, S. D.: Multidecadal increase in North Atlantic coccolithophores and the potential role of rising CO_2, Science, 350, 1533–1537, https://doi.org/10.1126/science.aaa9942, 2015.

Rohardt, G., Fahrbach, E., Beszczynska-Möller, A., Boetius, A., Brunßen, J., Budéus, G., Cisewski, B., Engbrodt R., Gauger, S., Geibert, W., Geprägs, P., Gerdes, D., Gersonde, R., Gordon, A. L., Hellmer, H. H., Isla, E., Jacobs, S. S., Janout, M., Jokat, W., Klages, M., Kuhn, G., Meincke, J., Ober, S., Østerhus, S., Peterson, R. G., Rabe, B., Rudels, B., Schauer, U., Schröder, M., Sildam, J., Soltwedel, T., Stangeew, E., Stein, M., Strass, V. H., Thiede, J., Tippenhauer, S., Veth, C., von Appen, W.-J., Weirig, M.-F., Wisotzki, A., Wolf-Gladrow, D. A., and Kanzow, T.: Physical oceanography on board of POLARSTERN (1983-11-22 to 2016-02-14), PANGAEA, https://doi.org/10.1594/PANGAEA.860066, 2016.

Rutherford, S., D'Hondt, S., and Prell, W.: Environmental controls on the geographic distribution of zooplankton diversity, Nature, 400, 749–753, https://doi.org/10.1038/23449, 1999.

Schlitzer, R.: Ocean Data View, available at: http://odv.awi.de (last access: 2 January 2018), 2015.

Schminke, H. K.: Entomology for the copepodologist, J. Plankton Res., 29, i149–i162, https://doi.org/10.1093/plankt/fbl073, 2007.

Schnack-Schiel, S. B.: Aspects of the study of the life cycles of Antarctic copepods, Hydrobiologia, 453/454, 9–24, https://doi.org/10.1023/A:1013195329066, 2001.

Schnack-Schiel, S. B.: Abundance of copepods during POLARSTERN cruise ANT-VII/2 (EPOS I), PANGAEA, https://doi.org/10.1594/PANGAEA.754736, 2010.

Schnack-Schiel, S. B. and Mizdalski, E.: Seasonal variations in distribution and population structure of *Microcalanus pygmaeus* and *Ctenocalanus citer* (Copepoda: Calanoida) in the eastern Weddell Sea, Antarctica, Mar. Biol., 119, 357–366, https://doi.org/10.1007/BF00347532, 1994.

Schnack-Schiel, S. B., Thomas, D., Dieckmann, G. S., Eicken, H., Gradinger, R., Spindler, M., Weissenberger, J., Mizdalski, E., and Beyer, K.: Life cycle strategy of the Antarctic calanoid copepod *Stephos longipes*, Prog. Oceanogr., 36, 45–75, https://doi.org/10.1016/0079-6611(95)00014-3, 1995.

Schnack-Schiel, S. B., Hagen, W., and Mizdalski, E.: Seasonal carbon distribution of copepods in the eastern

Weddell Sea, Antarctica, J. Marine Syst., 17, 305–311, https://doi.org/10.1016/S0924-7963(98)00045-1, 1998.

Schnack-Schiel, S. B., Michels, J., Mizdalski, E., Schodlok, M. P., and Schröder, M.: Abundance of copepods from multinet samples during POLARSTERN cruise ANT-XXII/2 (ISPOL), PANGAEA, https://doi.org/10.1594/PANGAEA.646297, 2007.

Schnack-Schiel, S. B., Mizdalski, E. and Cornils, A.: Abundance of copepods from multinet samples during POLARSTERN cruise ANT-XX/1, version 1, PANGAEA, https://doi.org/10.1594/PANGAEA.753644, 2010a.

Schnack-Schiel, S. B., Mizdalski, E., and Cornils, A.: Copepod abundance and species composition in the Eastern subtropical/tropical Atlantic, Deep-Sea Res. Pt. II, 57, 2064–2075, https://doi.org/10.1016/j.dsr2.2010.09.010, 2010b.

Shreeve, R. S., Tarling, G. A., Atkinson, A., Ward, P., Goss, C., and Watkins, J.: Relative production of *Calanoides acutus* (Copepoda: Calanoida) and *Euphausia superba* (Antarctic krill) at South Georgia, and its implications at wider scales, Mar. Ecol. Prog. Ser., 298, 229–239, https://doi.org/10.3354/meps298229, 2005.

Skjoldal, H. R., Wiebe, P. H., Postel, L., Knutsen, T., Kaartvedt, S., and Sameoto, D. D.: Intercomparison of zooplankton (net sampling systems: Results from the ICES/GLOBEC sea-going workshop, Prog. Oceanogr., 108, 1–42, https://doi.org/10.1016/j.pocean.2012.10.006, 2013.

Smetacek, V. and Nicol, S.: Polar ocean ecosystems in a changing world, Nature, 437, 362–368, https://doi.org/10.1038/nature04161, 2005.

Smith, J. N., De'ath, G., Richter, C., Cornils, A., Hall-Spencer, J. M., and Fabricius, K. E.: Ocean acidification reduces demersal zooplankton that reside in tropical coral reefs, Nat. Clim. Change, 6, 1124–1129, https://doi.org/10.1038/nclimate3122, 2016.

Stein, M.: Physical oceanography during METEOR cruise M11/4, Bundesforschungsanstalt für Fischerei, Hamburg, PANGAEA, https://doi.org/10.1594/PANGAEA.742745, 2010.

Tarling, G. A., Ward, P., and Thorpe, S. E.: Spatial distributions of Southern Ocean mesozooplankton communities have been resilient to long-term surface warming, Glob. Change Biol., 24, 132–142, https://doi.org/10.1111/gcb.13834, 2017.

Tittensor, D. P., Mora, C., Jetz, W., Lotze, H. K., Ricard, D., Berghe, E. V., and Worm, B.: Global patterns and predictors of marine biodiversity across taxa, Nature, 466, 1098–1101, https://doi.org/10.1038/nature09329, 2010.

Van Guelpen, L., Markle, D. F., and Duggan, D. J.: An evaluation of accuracy, precision, and speed of several zooplankton subsampling techniques, J. Cons. Int. Explor. Mer., 40, 226–236, https://doi.org/10.1093/icesjms/40.3.226, 1982.

Ward, P., Atkinson, A., Schnack-Schiel, S. B., and Murray, A. W. A.: Regional variation in the life cycle of *Rhincalanus gigas* (Copepoda: Calanoida) in the Atlantic sector of the Southern Ocean – re-examination of existing data (1928 to 1993), Mar. Ecol. Prog. Ser., 157, 261–275, https://doi.org/10.3354/meps157261, 1997.

Ward, P., Tarling, G. A., and Thorpe, S. E.: Mesozooplankton in the Southern Ocean: Spatial and temporal patterns from Discovery Investigations, Prog. Oceanogr., 120, 305–319, https://doi.org/10.1016/j.pocean.2013.10.011, 2014.

Weikert, H. and John, H.-C.: Experiences with a modified Bé multiple opening-closing plankton net, J. Plankton Res., 3, 167–176, https://doi.org/10.1093/plankt/3.2.167, 1981.

Wiebe, P. H.: A computer model study of zooplankton patchiness and ist effects on sampling error, Limnol. Oceanogr., 16, 29–38, https://doi.org/10.4319/lo.1971.16.1.0029, 1971.

Wiebe, P. H. and Benfield, M. C.: From the Hensen net toward four-dimensional biological oceanography, Prog. Oceanogr., 56, 7–136, https://doi.org/10.1016/S0079-6611(02)00140-4, 2003.

WoRMS Editorial Board: World Register of Marine Species, available at: http://www.marinespecies.org at VLIZ (last accessed 25 January 2018), https://doi.org/10.14284/170, 2018.

Zwally, H. J.: Detection of Change in Antarctica, in: Antarctic Science, edited by: Hempel G., Springer, Berlin, Heidelberg, 126–143, https://doi.org/10.1007/978-3-642-78711-9_10, 1994.

6

Anthropogenic land use estimates for the Holocene –HYDE 3.2

Kees Klein Goldewijk[1,3], Arthur Beusen[2,3], Jonathan Doelman[3], and Elke Stehfest[3]

[1]Copernicus Institute of Sustainable Development, Utrecht University, Utrecht, P.O. Box 80115, the Netherlands

[2]Department of Earth Sciences, Utrecht University, Utrecht, P.O. Box 80115, the Netherlands

[3]PBL Netherlands Environmental Assessment Agency, The Hague, the Netherlands

Correspondence to: Kees Klein Goldewijk (c.g.m.kleingoldewijk@uu.nl)

Abstract. This paper presents an update and extension of HYDE, the History Database of the Global Environment (HYDE version 3.2). HYDE is an internally consistent combination of historical population estimates and allocation algorithms with time-dependent weighting maps for land use. Categories include cropland, with new distinctions for irrigated and rain-fed crops (other than rice) and irrigated and rain-fed rice. Grazing lands are also provided, divided into more intensively used pasture and less intensively used rangeland, and further specified with respect to conversion of natural vegetation to facilitate global change modellers. Population is represented by maps of total, urban, rural population, population density and built-up area. The period covered is 10 000 before Common Era (BCE) to 2015 Common Era (CE). All data can be downloaded from https://doi.org/10.17026/dans-25g-gez3.

We estimate that global population increased from 4.4 million people (we also estimate a lower range < 0.01 and an upper range of 8.9 million) in 10 000 BCE to 7.257 billion in 2015 CE, resulting in a global population density increase from 0.03 persons (or capita, in short cap) km^{-2} (range 0–0.07) to almost 56 cap km^{-2} respectively. The urban built-up area evolved from almost zero to roughly 58 Mha in 2015 CE, still only less than 0.5 % of the total land surface of the globe.

Cropland occupied approximately less than 1 % of the global land area (13 037 Mha, excluding Antarctica) for a long time period until 1 CE, quite similar to the grazing land area. In the following centuries the share of global cropland slowly grew to 2.2 % in 1700 CE (ca. 293 Mha, uncertainty range 220–367 Mha), 4.4 % in 1850 CE (578 Mha, range 522–637 Mha) and 12.2 % in 2015 CE (ca. 1591 Mha, range 1572–1604 Mha). Cropland can be further divided into rain-fed and irrigated land, and these categories can be further separated into rice and non-rice. Rain-fed croplands were much more common, with 2.2 % in 1700 CE (289 Mha, range 217–361 Mha), 4.2 % (549 Mha, range 496–606 Mha) in 1850 CE and 10.1 % (1316 Mha, range 1298–1325 Mha) in 2015 CE, while irrigated croplands used less than 0.05 % (4.3 Mha, range 3.1–5.5 Mha), 0.2 % (28 Mha, range 25–31 Mha) and 2.1 % (277 Mha, range 273–278 Mha) in 1700, 1850 and 2015 CE, respectively. We estimate the irrigated rice area (paddy) to be 0.1 % (13 Mha, range 9–16 Mha) in 1700 CE, 0.2 % (28 Mha, range 26–31 Mha) in 1850 CE and 0.9 % (118 Mha, range 117–120 Mha) in 2015 CE.

The estimates for land used for grazing are much more uncertain. We estimate that the share of grazing land grew from 5.1 % in 1700 CE (667 Mha, range 507–820 Mha) to 9.6 % in 1850 CE (1192 Mha, range 1068–1304 Mha) and 24.9 % in 2015 CE (3241 Mha, range 3211–3270 Mha). To aid the modelling community we have divided land used for grazing into more intensively used pasture, less intensively used converted rangeland and less or unmanaged natural unconverted rangeland. Pasture occupied 1.1 % in 1700 CE (145 Mha, range 79–175 Mha), 1.9 % in 1850 CE (253 Mha, range 218–287 Mha) and 6.0 % (787 Mha, range 779–795 Mha) in 2015 CE, while rangelands usually occupied more space due to their occurrence in more arid regions and thus lower yields to sustain livestock. We estimate converted rangeland at 0.6 % in 1700 CE (82 Mha range 66–93 Mha), 1 % in 1850 CE (129 Mha range 118–136 Mha) and 2.4 % in 2015 CE (310 Mha range 306–312 Mha), while the unconverted natural rangelands occupied approximately 3.4 % in 1700 CE (437 Mha, range 334–533 Mha), 6.2 % in 1850 CE (810 Mha, range 733–881 Mha) and 16.5 % in 2015 CE (2145 Mha, range 2126–2164 Mha).

1 Introduction

Humans have emerged as the most important driving force of landscape transformation (Butzer, 1964; Ellis, 2015; Ellis et al., 2013; Kaplan et al., 2011). Long before the start of the Holocene, humans used fire as a tool to open up the landscape for grazing of wildlife to facilitate easier hunting. Then, roughly 12 000 years ago, the domestication of plants and animals slowly started and sedentary agriculture became more and more widespread. This led to conversion of natural ecosystems to other types of land use such as cropland, grazing land for livestock and built-up areas, a process that has accelerated greatly during the past few hundred years. This has led to the current situation in which more than 37 % of the ice-free land in the world is used for agriculture or settlements and another 30 % is more or less under the influence of humans, causing many natural resources to be heavily used or even near depletion (Ellis et al., 2013; Foley et al., 2005).

Land use plays an important role in the climate system (Feddema et al., 2005; Le Quéré et al., 2015). Many ecosystem processes are directly or indirectly climate driven, and together with anthropogenic land cover changes (ALCCs), they determine how the land surface will evolve over time (Betts, 2006; Brovkin et al., 2006; Claussen et al., 2001). The climate system is influenced in a biophysical manner by affecting radiative forcing through a changing albedo or heat fluxes (Matthews et al., 2003; Myhre and Myhre, 2003; Davin et al., 2007) but also through biogeochemical processes by exchanging greenhouse gases (GHGs) and other gases with the atmosphere (Betts, 2006; Brovkin et al., 2006; Betts et al., 2007). All these feedbacks on land use can be either positive or negative (Claussen et al., 2001).

ALCCs, mainly through conversion of undisturbed ecosystems to other land uses (e.g. deforestation for agriculture or grazing, forestry, or infrastructure), result in GHG emissions and have contributed considerably in the past to the cumulative carbon dioxide (CO_2) increase in the atmosphere. This part of the global carbon cycle is an important factor in the climate system and has been the focus of many studies (Feddema et al., 2005; Friedlingstein et al., 2006; Findell et al., 2007; Plattner et al., 2008; Strassmann et al., 2008). However, these carbon fluxes are still not well quantified in the global carbon budget (Le Quéré et al., 2015; Pacala et al., 2001). Uncertainties in quantifying the impact of land use changes on the global carbon cycle lead to uncertainties in projections of atmospheric CO_2 and climate and consequently affect policy makers in establishing reasonable emission mitigation strategies.

The existing global estimates of historical land use are rare and rather uncertain (Klein Goldewijk and Verburg, 2013). This can partly be explained by the sheer lack of reliable data, and hence by the different approaches used to reconstruct the trends, which vary from simple bookkeeping methods, to "hindcast" modelling techniques, to simulations of anthropogenic deforestation based on population densities (Houghton et al., 1983; Kaplan et al., 2011; Klein Goldewijk et al., 2011; Pongratz et al., 2008). The first analyses of carbon emissions from land use change were based on an approach that has come to be known as the bookkeeping approach (Houghton and Hackler, 2002; Houghton et al., 1983). This approach refers to the model used to calculate the changes in carbon following a change in land use. The model uses defined response curves (for four carbon pools: living biomass, woody debris, harvested wood products and soils) to determine the annual sources and sinks of carbon from each hectare undergoing a change related to management. The various uses of land in the first studies included croplands, pastures and the harvest of wood. Ramankutty and Foley (1999) present a study on global croplands only for the 1700–1992 CE period. They created a global representation of permanent croplands in 1992 CE, at 5 min spatial resolution, by calibrating a remotely sensed land cover classification data set against cropland inventory data. Next, they hindcasted the 1992 CE data with a simple land use change model back to 1700 CE. This was updated and expanded with pasture estimates for the 1700–2007 CE period (Ramankutty, 2012). Klein Goldewijk (2001) presented a reconstruction of croplands and pasture for the 1700–2000 CE period at a 30 arcmin resolution, which was updated to a 5 arcmin resolution and expanded to the 10 000 BCE to 2010 CE period (Klein Goldewijk et al., 2010, 2011). It primarily assumed a nearly constant per capita land use hindcasting approach, with changing allocation algorithms over time. Pongratz et al. (2008) presented a reconstruction of global agricultural areas for the 800–1992 CE period, the Millennium Land Cover Reconstruction (ML08). They aimed to provide spatially explicit maps at 0.5° resolution that reach further back in history than the data sets available at that time. The authors relied on existing land cover data where possible and used a wealth of literature on agricultural practices and agrotechnological innovations for all regions in the world; with this, per capita land use estimates could be accounted for as evolving over time and at the same time changing differently in the different regions of the globe. They also used seven different population databases, and the upper and lower ends of uncertainties of per capita land use based on literature were applied. This led to two different trajectories of land cover change that reflect the upper and lower bounds within the uncertainty range of historical land cover changes, around a third, medium estimate that used a specific population data set and assumed constant per capita land use changes over time, which was a key shortcoming. The Kaplan and Krumhardt (KK10) scenario was developed to provide a basis for calculating the effects of anthropogenic land cover change on the global carbon cycle over the Holocene (Kaplan et al., 2011). The scenario is based on a non-linear empirical relationship between forest cover on land suitable for crop or pasture and country-level population density observed in Europe over the past 2 millennia on the basis of demographer estimates of population density and reconstructions of forest cover (Kaplan et al., 2011, 2009). The population density–forest cover relationship is approximated as a logistic function, with the most rapid rates of deforestation at intermediate aggregate population densities of ca. 50–100 persons per km^{-2}. This function results in an implied per capita land use that is a function of population density, which is further modulated by the potential suitability of the landscape for rain-fed agriculture, and pasture ALCC is still not successfully implemented in many models and studies of global change. As a result, climate modelling in paleo-mode or projection mode that tries to take ALCC into account is seriously hindered (Gaillard et al., 2010). The few scenarios of past ALCC that exist, e.g. ML08 (Pongratz et al., 2009), HYDE (Klein Goldewijk et al., 2011) and KK10 (Kaplan et al., 2011, 2009), show very large differences indeed (Gaillard et al., 2010). Therefore, improved descriptions of past anthropogenic land cover change on the global spatial scale are urgently needed.

This study presents a key update and extension of the former HYDE 3.1 historical land use database (Klein Goldewijk et al., 2010, 2011). This recent version, HYDE 3.2, is an improved and internally consistent combination of new historical population data and land use allocation algorithms, which vary over time. Categories include cropland, with a new distinction between irrigated and rain-fed crops (other than rice) and irrigated and rain-fed rice. In addition, grazing lands are provided and categorized into more intensively used pasture and less intensively used rangeland. Population is also represented by maps of total, urban, rural population and population density as well as built-up area. The period covered now is 10 000 BCE to 2015 CE.

An overview of the new features of HYDE 3.2 include

- improved allocation algorithms (a smoother transition from current "satellite-based" allocation towards "rule-based" allocation in the past);
- updated historical estimates of many land use categories (e.g. improved census data of 55 countries);
- a distinction between rain-fed and irrigated cropland;
- an introduction of the land use category rice (rain-fed and irrigated);
- the use of additional data (cropland layer map of Waldner et al., 2016; irrigation data of Siebert, 2008, and Siebert et al., 2015, MIRCA data set) and rice data from You et al., 2014);
- the use of the most recent and higher-resolution satellite information on land cover from the ESA consortium for 2010 reference maps;
- a new distinction between pasture and rangeland according to the intensity of use, and a further classification of rangeland into rangeland under natural vegetation and rangeland involving conversion of natural vegetation;
- an extension of the database to the year 2015.

2 Methodology and data

2.1 Input data for population

The basis for our population data is the United Nations World Populations Prospects (2008 Revision) for the 1950–2015 period. The pre-1950 historical estimates were largely taken from McEvedy and Jones (1978), Livi-Bacci (2007), and Maddison (2001). We supplemented them with the sub-national population numbers of Populstat (Lahmeyer, personal communication, 2004) and many other country-specific sources (see the Supplement A). Time series were constructed for each province or state of every country of the world. We used the current administrative units and kept them constant over time for simplicity reasons. Other historical sources were adjusted to match the current boundaries of HYDE 3.2 if needed by taking fractions of those former, often larger, administrative units (e.g. Roman Empire). The methodology is similar to that described in detail in Klein Goldewijk et al. (2010).

2.2 Spatial coverage for population

For the present, we used the spatial patterns from LandScan (2014) for the year 2012 in our weighting maps to allocate our total population for each administrative unit. LandScan, developed at the Department of Energy's Oak Ridge National Laboratory, USA, is a global population database showing population totals at 1 km resolution. The modelling process uses subnational-level census counts for each country and primary geospatial input or ancillary data sets, including land cover, roads, slope, urban areas, village locations and high-resolution imagery analysis.

2.3 Input data for land use

In this section, we describe the national and subnational data for all land use categories, from 10 000 BCE until present day, and the construction of land use maps for the present day. The allocation method to derive land use maps for the historic period back to 10 000 BCE is described in Sect. 3.2.

2.3.1 Cropland and grazing land statistics

We started with country totals for the FAO categories of "arable land and permanent crops" and "permanent meadows and pastures", further referred to here as "cropland" and "grazing land" respectively (FAO, 2015). The FAO data reach back to 1960, and thus for the period 1960–2015 CE, the FAO data were followed exactly, though complemented in some countries by subnational statistics. For the pre-1960 CE period another method had to be used, and we combined population estimates with per capita land use estimates. For 1960, per capita cropland and grazing land was derived from FAO data and then modelled back in time, assuming that they were not constant but followed a curved trajectory. The curve can differ between countries and often resembles either a concave-shaped, Bell-shaped or convex-shaped curve, depending on the (limited) historical sources found. The 1960 CE per capita land use value is generally lower than for the present day since population numbers have exploded since World War II. Land use was, however, limited by the lack of technology and this restricted the maximum amount of land that subsistent farmers could handle. We first estimated land use per capita for each country, and by multiplying this with the total population, we computed the total areas for cropland and grazing.

Further specific input statistics on a subnational level were taken for some of the larger countries in the world, such as the USA (USDA, 2006), Canada (Urquhart and Buckley, 1965), Mexico (World Atlas of Agriculture, 1969), Argentina (Vazquez-Presedo, 1988), Brazil (N. Ramankutty, personal communication, 2001), India (Flint and Richards, 1991; Indiastat, 2009), China (China National Bureau of Statistics, 2008; Ge et al., 2008; He et al., 2012, 2013) and Australia (Australian Bureau of Statistics, 2001). Furthermore, other country-specific sources were used when available for specific information per country. The methodology has also been described in detail in Klein Goldewijk et al. (2011); see also Supplement A for country-specific land use data sets.

For a spatially explicit depiction of present land cover we used the ESA Land Cover consortium maps (ESA, 2016; Hollman et al., 2013). The consortium produced a time series of three global 300 m spatial resolution land cover data sets predominantly based on MERIS satellite data, representative for the 1998–2002 CE, 2003–2007 CE and 2008–2012 CE periods. To produce the satellite reference map of HYDE 3.2, the most recent epoch has been used, representing 2010 CE. The ESA classes are aggregated to 5 arcmin grid cell fractions consistent with the HYDE land mask. Cropland and grazing land are allocated to the grid using the various ESA land cover classes, which are used in two ways: classes are grouped according to their probability to contain cropland and pasture, and then allocation takes place accordingly. First, the procedure tries to allocate as much cropland (or pasture) to the land cover classes in the probability group 1 (allocation step 1). Then in the second and third rounds as much of the remaining land is allocated to probability groups 2 and 3 (allocation steps 2 and 3), and finally, in a fourth round, priority group 4 is used to allocate the remaining land. In most cases, FAO areas are already allocated in steps 1 or 2. Several classes are defined to never contain cropland (or pasture). Additionally, we define the fraction of an ESA land cover class that is actually covered by cropland and pasture (numbers in Table 1): on cropland and grassland land cover types this is assumed to be 90 % to account for small areas of infrastructure, wetlands, unsuitable terrain, steep slopes or small patches of vegetation that are not explicitly identified in the original land cover product (Verburg et al., 2009). The mosaic cropland land cover types are defined to have 60 and 40 % of cropland or pasture following the managed grass definition of Poulter et al. (2015). First, cropland is allocated, then grazing land.

The allocation procedure first allocates FAO cropland areas to land cover types dominated by cropland (see probability group 1 in Table 1). Within the area that can be allocated a preference is given to cropland locations from the unified cropland layer map (Waldner et al., 2016). This is done in order to make use of the high-quality empirical data that are available and to correct for the characteristic that the ESA cropland land cover types are most likely a mix of cropland and pasture lands in many regions. If the land cover types dominated by cropland (probability group 1) do not cover sufficient area to allocate all FAO cropland area, land cover areas within the second category (shrubland cover, open forest or grassland) are allocated. If these areas are also insufficient, cropland is allocated to sparse vegetation, finally followed by bare lands. Priority is always given to cropland locations from the unified cropland layer map.

A similar approach is used to allocate grazing land in each country, following an almost identical order of allocation (see Table 1). In addition to a preferred allocation to the land cover types dominated by cropland, which were also used to allocate cropland; grazing land is also preferably allocated to the grassland land cover type. Cropland is given priority over grazing land and allocated before grassland, and thus the cropland that was allocated in the previous step reduces the area available for pasture. In addition, various region-specific rules are implemented. First, in Canada, Russia and the USA, various subnational regions known to have very little to no pasture area are excluded from allocation, e.g. inaccessible tundra areas. Second, in countries with substantial grazing land that can be allocated in large homogeneous areas (e.g. the Sahara) a preference is given to locations in close proxim-

Table 1. Percentages of cropland and grazing land allocated in four steps from the original ESA CCI land cover classes (between 0 and 100 %).

LCCS Class	UNLCCS land cover class description	Cropland A	Cropland B	Pasture C	Pasture B
10	Cropland, rain-fed	90	1	90	1
11	Herbaceous cover	90	1	90	1
12	Tree or shrub cover	90	1	90	1
20	Cropland, irrigated or post-flooding	90	1	90	1
30	Mosaic cropland (> 50 %)/natural vegetation (tree, shrub, herbaceous)	60	1	60	1
40	Mosaic natural vegetation (tree, shrub, herbaceous cover) (> 50 %)	40	1	40	1
50	Tree cover, broadleaved, evergreen, closed to open (> 15 %)	–	–	–	–
60	Tree cover, broadleaved, deciduous, closed to open (> 15 %)	–	–	–	–
61	Tree cover, broadleaved, deciduous, closed (> 40 %)	–	–	–	–
62	Tree cover, broadleaved, deciduous, open (15–40 %)	30	2	30	2
70	Tree cover, needleleaf, evergreen, closed to open (> 15 %)	–	–	–	–
71	Tree cover, needleleaf, evergreen, closed (> 40 %)	–	–	–	–
72	Tree cover, needleleaf, evergreen, open (15–40 %)	30	2	30	2
80	Tree cover, needleleaf, deciduous, closed to open (> 15 %)	–	–	–	–
81	Tree cover, needleleaf, deciduous, closed (> 40 %)	–	–	–	–
82	Tree cover, needleleaf, deciduous, open (15–40 %)	30	2	30	2
90	Tree cover, mixed leaf type (broadleaved and needleleaf)	–	–	–	–
100	Mosaic tree and shrub (> 50 %)/herbaceous cover (> 50 %)	40	2	40	2
110	Mosaic herbaceous cover (> 50 %)/tree and shrub (> 50 %)	60	2	60	2
120	Shrubland	90	2	90	2
121	Evergreen shrubland	90	2	90	2
122	Deciduous shrubland	90	2	90	2
130	Grassland	90	2	90	1
140	Lichens and mosses	–	–	–	–
150	Sparse vegetation (tree, shrub, herbaceous cover) (< 15 %)	90	3	–	3
152	Sparse shrub (< 15 %)	90	3	–	3
153	Sparse herbaceous cover (< 15 %)	90	3	–	3
160	Tree cover, flooded, fresh or brackish water	–	–	–	–
170	Tree cover, flooded, saline water	–	–	–	–
180	Shrub or herbaceous cover, flooded, fresh/saline/brackish water	–	–	90	3
190	Urban areas	–	–	–	–
200	Bare areas	90	4	90	4
201	Consolidated bare areas	90	4	90	4
202	Unconsolidated bare areas	90	4	90	4
210	Water bodies	–	–	–	–
220	Permanent snow and ice	–	–	–	–

A: fraction available for cropland. B: allocation step. C: fraction available for grazing land.

ity to previously allocated grazing area. The resulting maps for current cropland and grazing land are used as reference maps Wcrop_satellite and Wgrazing_satellite in the HYDE allocation procedure (see Sect. 3.2)

2.3.2 Rice

Total rice area statistics per country were taken from FAO for the post-1960 CE period. For the pre-1960 CE period, data for some rice-growing countries were taken from Mitchell (2007a, b, c) for the 1890–1950 CE period. The rest were hindcasted on a per capita basis for countries for which no historical data could be found. Global totals were compared with literature where possible to tune the per capita numbers. A correction was made for harvested area versus physical area. In China, for example this ratio of physical area to harvested area is around 0.6 due to multiple and triple cropping (Frolking et al., 2002), in Bangladesh 0.67 in 1980 CE and 0.55 in 2010 CE (Ara et al., 2016), in Vietnam 0.63 (range 1.25–0.48) (Wiens, 1998), and India 0.71–0.73 (GYGA, 2017; Kalaiselvi and Sundar, 2011). The 0.6 ratio was also applied to other Asian rice-producing countries; the rest of the world is assumed to have a ratio of 1.0. Ricepedia (2015) presents estimates for the fraction of rice that is irrigated per country for the present, and an assump-

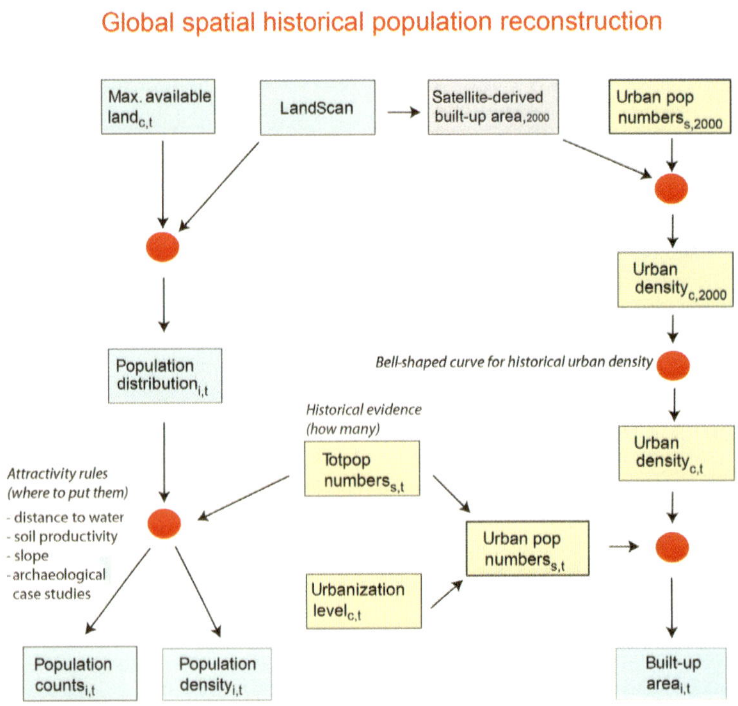

Figure 1. Population allocation scheme.

tion was made for that fraction at the start of agriculture. For the years in between, data were linearly interpolated towards the 2010 CE value (see Table S1 in the Supplement, "faostat_rice" sheet, columns CN : FM). In general, most arid regions were assumed to be irrigated, others rain-fed. Thus, for each time step the total amount of rice could be computed. Rain-fed rice was allocated first (most cost effective), irrigated rice second. The reference maps for rice were derived from MAP SPAM (You et al., 2014) and are used for the allocation for the present. They gradually change into combined weighting maps of the HYDE rules; see Sect. 3.

2.3.3 Irrigation

The input data for the category "areas equipped for irrigation" for the post-1960 CE period were taken from FAO (2015) and for the 1900–1960 CE period from Siebert et al. (2015). For the 1700–1900 CE period estimates were taken from Siebert (2008). For the pre-1700 CE period a per capita estimate was used to match global estimates of irrigated area found in literature. The statistics for the actual irrigated area for the 1960–2010 CE period were derived by multiplying FAO's category "area equipped for irrigation" with the fraction of "equipped/actual irrigated area" from the GMIA_v5 database of Siebert et al. (2015). For many countries this fraction is 1.0, but not for all countries. Therefore, we used the last known ratio and applied this to the complete series of "areas equipped for irrigation" to compute the historical time series of "actual irrigated area" (see Table S2). For the spatial representation of current global areas equipped for irrigation and actual irrigated areas, we used the Siebert et al. (2015) global data set of monthly irrigated and rain-fed crop areas around the year 2000 (MIRCA2000).

3 The HYDE allocation of population and land use

3.1 Allocation of population and built-up area

For the present, the spatial patterns from LandScan (2014) are used in our weighting maps to allocate our total population for each administrative unit. These patterns are gradually replaced with combined weighting maps based on various proxies such as soil suitability, slope, distance to water and case studies from several disciplines (e.g. history, archaeology) when going back in time (Fig. 1). Built-up areas (Uarea) are computed by dividing total urban population in a country at time t by average urban densities in a country at time t. These curves are computed per country and are assumed to follow a Bell-shaped curve trajectory, a specific pattern for increase and decrease in urban density in cities and towns over time but on different scales in various regions. See Fig. 1 and for a full description of the methodology Klein Goldewijk et al. (2010) and for the built-up area curve Supplement B.

3.2 Allocation of land use

3.2.1 Non-usable areas

We excluded specific areas from allocation. Protected areas as derived from the database of protected areas from the United Nations Environment World Conservation Monitoring Centre (UNEP/WCMC, 2013) were excluded from agricultural use. However, as protected areas were only established relatively recently, they are only excluded after 1900 CE until present. Before 1900 CE, they are potentially available for agriculture, but due to the tendency of the allocation model to not shift, keeping the location of agricultural areas when moving back in time, and due to their often low accessibility, these areas are not likely to be used before 1900 CE. Additionally, so-called "non-used areas" in Australia were excluded from the allocation of agriculture, based on a map from the Australian National Land and Water Resources Audit (NLWRA, 2001).

3.2.2 Allocation of land use

The order of allocation is as follows: first, we allocate cropland, then rice, then irrigation and finally grazing land.

Cropland

The method to allocate historical cropland is followed for each grid cell of 5′ by 5′ (ca. 85 km^2 around the Equator). The maximum area of a grid cell that is available for allocation of cropland ($Garea_{max}$) is calculated as follows:

$$Garea_{max} = [Garea_{cell} - Uarea_t - Parea_t - NLarea_t], \quad (1)$$

where $Garea_{cell}$ is the land area of a 5′ grid cell (spherical Earth) except water bodies, snow and ice; $Uarea_t$ is the urban built-up area for year t; $Parea_t$ is protected area, but only valid for the post-1900 period (they did not exist before); and NLarea is the no-land-use area in central Australia (see Table S3 for the original input data and Fig. 2 for the cropland allocation scheme).

The land use statistics are allocated to grid cells according to a mix of two weighting maps: a reference map of 2010 CE for cropland ($Wcrop\_satellite_{2010}$), derived from satellite imagery of ESA (2016) for the present (Eq. 2), and a historical map ($Wcrop\_historic_t$) that is constructed on the basis of the six rules as described in the next section (Eq. 3).

$$Wcrop\_reference_t = Wcrop\_satellite_{2010} \quad (2)$$

For allocating historical cropland, six major assumptions were made. (i) Urban built-up areas (Uarea) were excluded for allocation (see Eq. 1). (ii) A population density (Wpopd) less than 0.1 cap km^{-2} does not allow permanent agriculture. (iii) Areas with better soil suitability according to the Global Agro-Ecological Zones map (GAEZ, 2000) FAO IIASA are used first (Wsuit). (iv) Easily accessible areas such as coastlines and river plains derived from Natural Earth (2015) are more promising for early settlement (Wcoast and Wriver). (v) Inaccessible terrains with steep slopes derived from NOAA-NGDC 5-Minute Gridded Global Relief Data (ETOPO5, 2005) are less promising for settlement (Wslope). (vi) Below an annual mean temperature of 0 °C no agriculture is possible (Wtemp_crop). The temperature map is derived from the Climate Research Unit (CRU) database of the University of East Anglia, UK, as an average for the 1960–1990 period (New et al., 1997). We normalized all weighting maps of these assumptions between 0 and 1 and multiplied them into a final, unique weighting map for each time step. This methodology is similar to the one already described in Klein Goldewijk et al. (2011).

$$Wcrop\_historic_t = Wpop_t \cdot Wsuit \cdot Wriver \cdot Wcoast \cdot Wslope \cdot Wtemp\_crop \quad (3)$$

The final allocation is a combination of $Wcrop\_reference_t$ and $Wcrop\_historic_t$. We assume that the influence of $Wcrop\_reference_t$ is zero in the year 1500 CE and 100 % in 2010 CE. Alternatively, the influence of $Wcrop\_historic_t$ is zero in 2010 CE and 100 % in 1500 CE (and the pre-1500 CE period as well); see Supplement D for the technical description.

Rice

All rice-producing area is assumed to fall within the area defined earlier as cropland. For the spatial representation of global rice areas we used the International Food Policy Research Institute (IFPRI) Spatial Production Allocation Model (SPAM) maps from You et al. (2014). They present maps of the irrigated and rain-fed harvested area of rice.

Current reference maps for irrigated and rain-fed rice allocation:

$$Wir\_rice_{2010} = Wir\_rice\_satellite_{2010} \quad (4)$$

$$Wrf\_rice_{2010} = Wrf\_rice\_satellite_{2010}. \quad (5)$$

Extra weighting for the past is also determined by a historical geography map for rice-growing areas for the year 1000 CE (M. Widgren, personal communication, 2015). Also, the current reference maps for irrigated and rain-fed rice remain important for the allocation since the assumption is made that most rice-growing areas are very old, implying that current patterns are, to a great extent, representative for ancient patterns as well (see Table S4 for original input data and Fig. 2 for the cropland allocation scheme).

Historical weighting maps for rice allocation:

$$Wir\_rice_t = Wrice_{1000} \cdot Wir\_rice_{2010} \quad (6)$$

$$Wrf\_rice_t = Wrice_{1000} \cdot Wrf\_rice_{2010}. \quad (7)$$

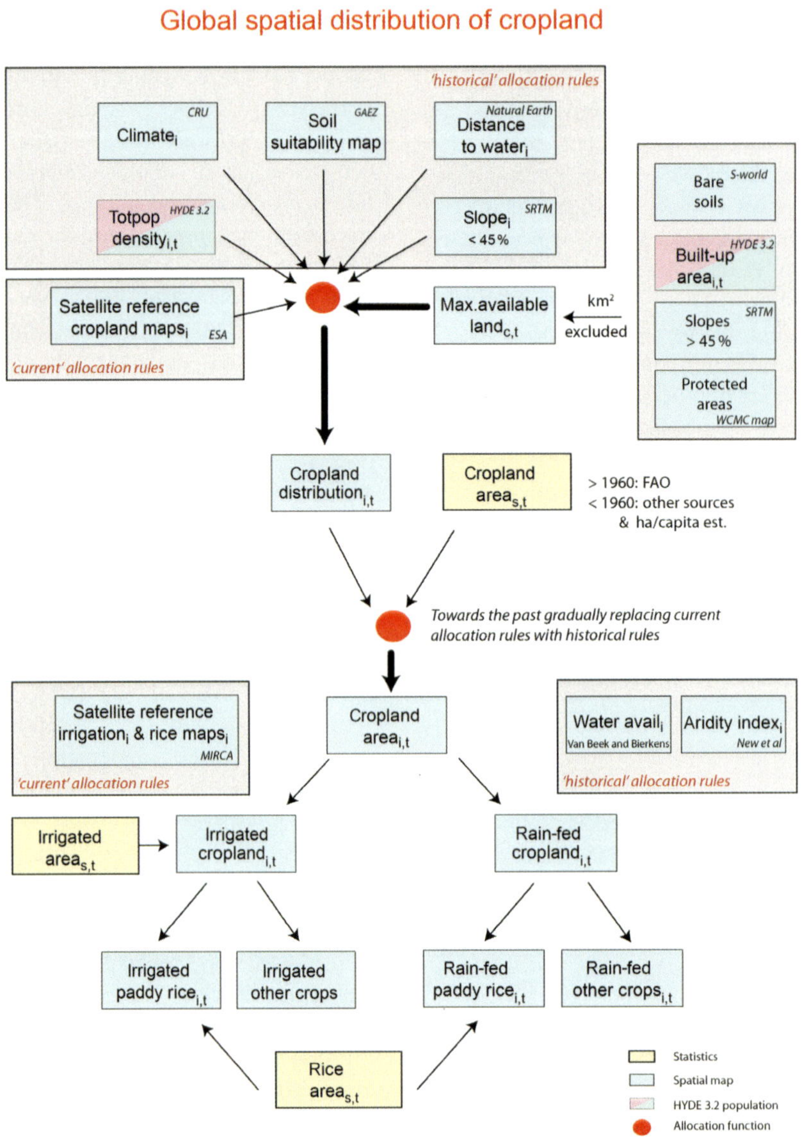

Figure 2. Cropland allocation scheme.

Irrigated area

The allocation of irrigated areas is comparable with that of cropland. First we use the reference map for 2000 CE for irrigated areas derived from the MIRCA (monthly irrigated and rainfed crop areas around the year 2000) database (Siebert et al., 2015). When hindcasting, we apply the HYDE rules for allocation based on the following assumptions. All irrigated area must fall within the computed cropland area. Next, we check whether there is enough water available to irrigate (Wwav), for which we use, as a proxy, a discharge map derived from Van Beek and Bierkens (2008). Furthermore, we assume that when the aridity index (Waridity) is low, the need for irrigation is higher. The aridity index is computed as annual precipitation divided by annual evapotranspiration, both derived from New et al. (1997). See Table S5 for original input data and Fig. 2 for the allocation scheme.

Current reference map for irrigated land allocation:

$$\mathrm{Wirri}_{2010} = \mathrm{Wirri\_satellite}_{2010}. \quad (8)$$

Historical weighting maps for irrigated land allocation:

$$\mathrm{Wirri}_t = \mathrm{Waridity} \cdot \mathrm{Wwav} \cdot \mathrm{Wirri}_{2010}. \quad (9)$$

Grazing land

As described above (Sect. 2.3), we define grazing land as land used for mowing or grazing livestock, based on the FAO category "permanent meadows and pastures". Grazing land can be a variety of ecosystems, ranging from managed irrigated grasslands to unmanaged open savannah woodlands to semi-shrub/scrub, almost desert, lands. The method to allocate grazing land (to $Garea_{max}$ minus the already allocated built-up and cropland area) is comparable to the procedure for cropland. Natural grassland ecosystems as computed by the BIOME model were supplementary to the weighting maps. Biomes are a combination of plant functional types, defined by climatic variables and soil properties; see Prentice et al. (1992). Natural grasslands are assumed to be more attractive for use of grazing then other types (Wbiome); areas with a higher net primary production are also more favourable for grazing (Wnpp), and grazing is assumed to be impossible at annual average temperatures below $-10\,^{\circ}$C (Wtemp_grazing).

Current reference map for grazing land allocation:

$$\text{Wgrazing}_{2010} = \text{Wgrazing\_satellite}_{2010}. \quad (10)$$

We multiply the historical weighting maps for grazing land allocation:

$$\text{Wgrazing}_t = \text{Wpopd}_t \cdot \text{Wbiome} \cdot \text{Wnpp} \cdot \text{Wtemp\_grazing}. \quad (11)$$

See Table S6 for original input data and Fig. 3 for the grazing land allocation scheme.

Disaggregating grazing land into pasture and rangeland

For grazing lands, a distinction is made between more intensively managed pasture and extensively managed rangelands. The main difference between these two types of grassland is that rangelands comprise natural grasslands, shrublands, woodlands, wetlands, and deserts and grow primarily native vegetation, rather than plants established by humans, and typically have low livestock densities. The distinction, however, is often less straightforward, and transitions between the two forms exist in both time and space. To distinguish between these types of grasslands in a simple and transparent way, also historically, we apply – based on expert judgement – a population density and an aridity index, as low animal densities can be related to low population density, or to low productivity of the natural vegetation, which is approximated via the aridity index. When the aridity index (defined as annual precipitation divided by annual evapotranspiration) of a grid cell defined as grazing land is less than 0.5, or when the aridity index is higher than 0.5 but the population density is less than 5 inhabitants km^{-2}, then it is defined as rangeland.

Whether or not natural vegetation has been converted to establish grazing land is a very relevant question for studying the impacts of land use change. While most rangelands occur on land with mostly natural vegetation, low-intensity livestock grazing is also located in former forest or woodland areas, e.g. in Brazil. Therefore, after consultation with the Land Use Model Intercomparison Project (LUMIP), we recommend that our maps of pasture and rangeland be used for modelling purposes in the following way: for pastures, all natural vegetation is cleared and replaced by grass species. For rangeland, the natural vegetation remains intact if it is non-forest, but is cleared if it is forest. HYDE includes this distinction by providing two types of rangeland: (1) rangeland-natural is located in non-forest biomes (terrestrial ecoregions; Olson et al., 2001) and is therefore assumed not to have undergone conversion of natural vegetation. (2) Rangeland-converted is located in forest biomes (terrestrial ecoregions; Olson et al., 2001) and is assumed to have undergone conversion of natural vegetation. We also test this simple method against a more advanced, satellite-based approach to distinguish pasture and rangeland for the present day and find mostly a good agreement, except for some areas in Africa and central Brazil (see Supplement C).

4 Results

4.1 Population

We have constructed historical population maps for a 12 000-year period at a $5'$ by $5'$ grid resolution. Population numbers were very low at the start of the Neolithic era. We estimate 4.4 million in 10 000 BCE, which is in line with ranges of 1–20 million found in literature, with most estimates below 6 million; see also Klein Goldewijk et al. (2010) and Fig. 4. Our world population estimate for 5000 BCE is 19 million, in range with values between 5 and 20 million found in literature (Klein Goldewijk et al., 2010). Furthermore, we also estimate a lower estimate of 2 million and an upper estimate of 36 million in 5000 BCE (see Sect. 5). These lower and upper population estimates are eventually multiplied with the lower and upper per capita land use estimates to illustrate the bandwidth of uncertainty around our best estimate. It shows that especially for the pre-1700 CE period we have been very cautious in determining the lower and upper estimates, often outside the literature estimates, in order to capture any possible (extreme) population scenario in the past. We further estimate a global population in 1 CE of 232 million people (literature range 170–330 million, our uncertainty range 58–406 million).

Demographic and socioeconomic changes such as the rise and fall of the Egyptian, Greek, Roman, Indian, Chinese and American empires led to a global population estimate of 253 million in 500 CE (literature estimate 190 million, our range 101–406) and 323 million in 1000 CE (literature range 253–345, our range 176–568). The Columbian Exchange was the onset of true globalization. Shortly after Columbus landed in the Americas, the Industrial Revolution fuelled by colonization by Europeans of the Americas, Australia and later Africa, accompanied by a huge agricultural expansion, in the temperate regions and a bit later in the tropics as well. After 1700 CE, population growth accelerated, with an estimate of 592 million people in 1700 CE (literature range 410–680, our range 444–740), 943 million in 1800 CE (literature range 890–1000, our range 802–1086), 1643 million in 1900 CE (literature range 1571–1710, our range 1561–1725), 2531

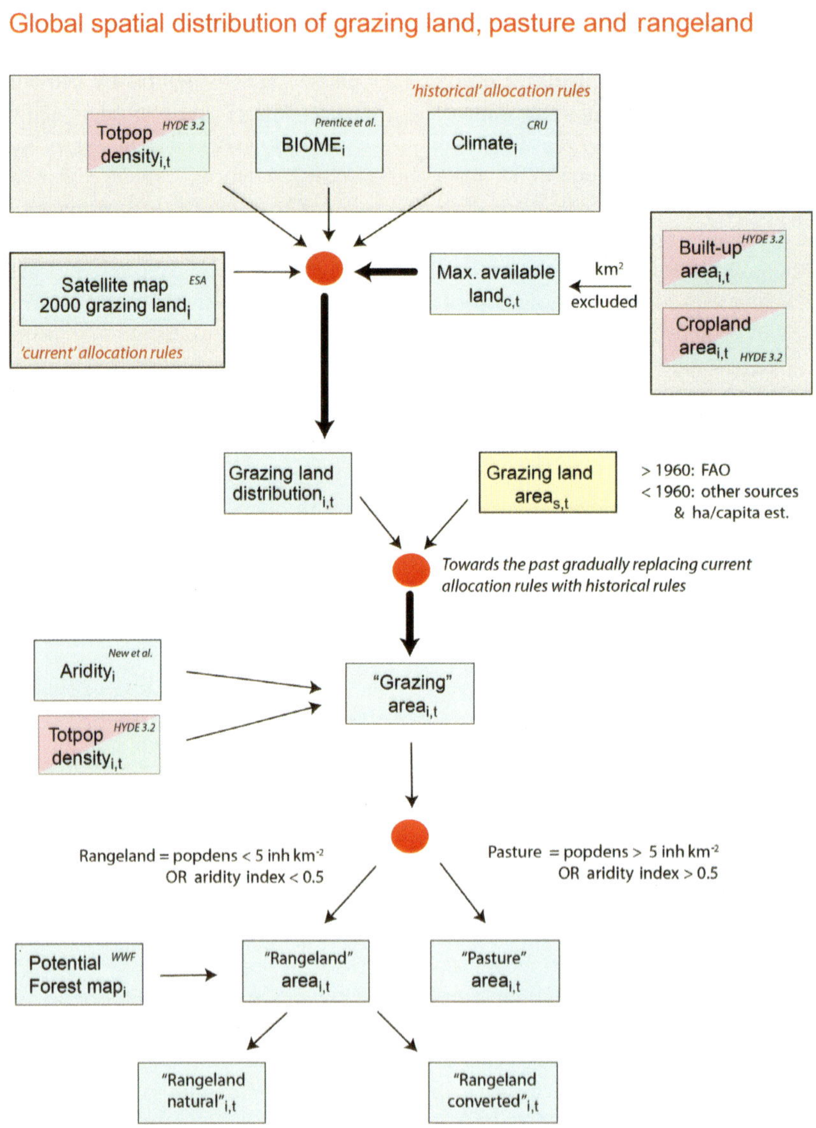

Figure 3. Grazing land allocation scheme.

million in 1950 CE, 4489 million in 1980 CE, 6113 million in 2000 CE and 7260 million in 2015 CE (see Table S7). The area occupied by built-up area (housing, building, etc.) is still very modest compared to the total land surface available: less than 0.10 % in 1900 CE and still less than 0.5 % in 2015 CE.

4.2 Croplands

Agriculture developed very slowly at the start of the Holocene after the domestication of plants and animals, in place as well as over time. We estimate the corresponding global cropland area extent in 5000 BCE at a very modest 6 million ha (range 1–18), which corresponds to 0.31 ha cropland cap^{-1} (range 0.05–0.57). Technology was limited and agriculture was dependent on local climate (environment) conditions. Agriculture was more prevalent throughout the Mediterranean, northern India and in eastern China during Greek and Roman times, most likely due to the existence of highly developed irrigation schemes. We estimate the global cropland area to reach 146 million ha at 1 CE (range 50–344), corresponding to 0.63 ha cropland cap^{-1} (range 0.16–1.10); see also Fig. 4.

The cropland area in 500 CE is estimated to be 133 Mha (67–269), 140 Mha (82–251) in 800 CE, 209 Mha (124–309) in 1100 CE and 219 Mha (148–299) in 1400 CE, which translates globally to around the same number of 0.50 (0.20–0.84) ha cropland cap^{-1} over that period because of population fluctuations. Our cropland estimate of 140 Mha for 800 CE is somewhat higher than the 136 Mha (range 80–220) of Pongratz et al. (2008) and close to their 197 Mha estimate

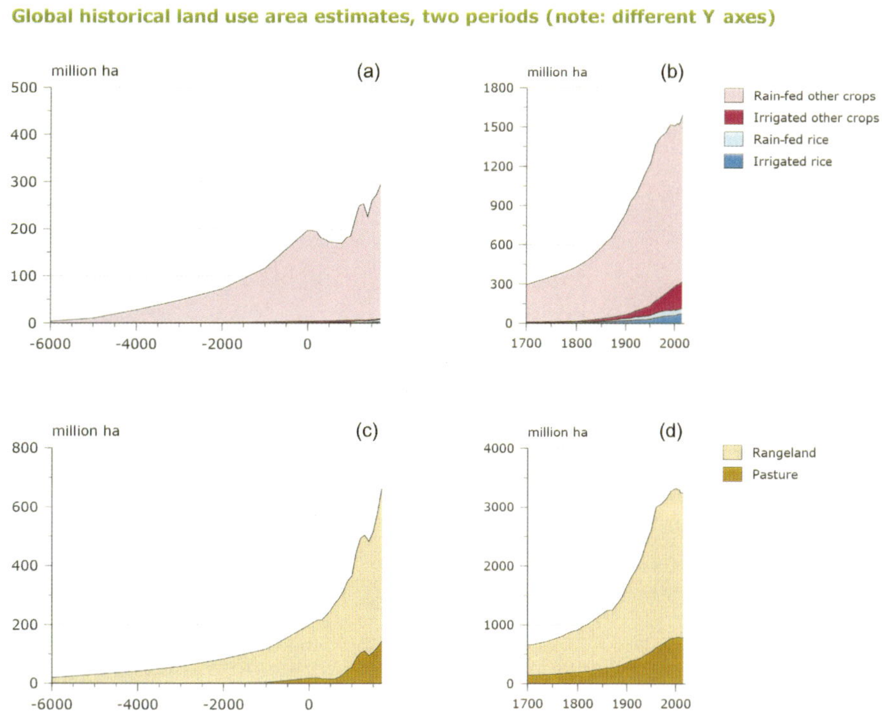

Figure 4. Summary of historical land use area estimates for different time periods. Cropland area for 6000 BCE–1700 CE **(a)** and 1700–2010 CE **(b)**. Grazing land area for 6000 BCE–1700 CE **(c)** and 1700–2010 CE **(d)**.

for 1100 CE but slightly lower than the 233 Mha for 1400 CE (see Tables 2 and S7).

As result of large population growth and technological developments, the total global area of cropland has doubled at a rapid pace since the 16th century from 293 Mha in 1700 CE to 578 Mha in 1850 CE, 1223 Mha in 1950 CE, 1532 Mha in 2000 CE and 1593 Mha in 2015 CE. The increase in global cropland area appears to have levelled off during the early 2000s but recently we have see an increase again. Ramankutty and Foley (1999) and Pongratz et al. (2008) estimated a value of around 400 Mha for 1700 CE. The latter can probably be explained by the fact that their hindcasting starting point in 1990 CE was already higher than that of the FAO because they also used non-FAO data (Pongratz et al., 2008); see Table 2.

Globally, the area of cropland per person increased slowly to a maximum of 0.8 ha cap^{-1} until 3000 BCE. Then it slowly decreased again to less than 0.5 ha cap^{-1} at the end of the 18th century, with a small temporary increase to 0.53 ha cap^{-1} in the early 20th century. However, after 1950 CE it decreased again to 0.22 ha cap^{-1} in 2015 CE due to the strong population growth. Apparently, technology alone could not compensate entirely for the explosive population growth after World War II, and since the best soils are already occupied, this trend will continue.

4.3 Irrigated land

Irrigation played a vital role in the existence and spread of agriculture. In general, irrigation can be defined as applying additional water (in addition to natural rainfall) to the soil in order to enhance crop yield. First, surface water was diverted from lakes, streams and rivers to other places in the landscape. Later, various types of pumps were used, driven by either livestock (oxen, mules, horses), manpower or, eventually, machines. The earliest archaeological evidence for irrigation can be dated to around 6000 BCE in Jordan and Egypt (Sojka et al., 2002). In the millennia that followed, irrigation was diffused across the Levant region and the Mediterranean. At the same time, irrigation emerged independently in India, Pakistan and China. In addition, in the Americas the Inca, Maya and Aztec already had irrigation schemes in the first millennium. A little later in time, the Hohokam practiced irrigation in the dry southwestern USA only to disappear mysteriously in the 14th century (Sojka et al., 2002). We estimate the global irrigated area to be less than 0.2 Mha in 5000 BCE, 2.6 million ha in 1 CE, very slowly increasing to 4.1 Mha in 1000 CE and levelling off at 4.2 Mha in 1500 CE; see Table 3.

The 19th century marked a significant change in technology in many areas of science. Ancient irrigation technology worked properly in already favourable environmental conditions (rainfall, terrain, slope), but when population grew and agriculture expanded into areas with less favourable condi-

Table 2. Regional cropland area estimates (million ha).

Regional cropland area estimates (million ha)																	
	4000 BCE	3000 BCE	2000 BCE	1000 BCE	1 CE	800 CE	1100 CE	1400 CE	1700 CE	1800 CE	1850 CE	1900 CE	1920 CE	1950 CE	1960 CE	2000 CE	2015 CE
North America	0.0	0.0	0.0	0.2	0.4	0.6	0.8	1.1	0.8	8.3	44	153	199	232	237	230	208
Latin America	1.1	3.8	5.3	6.1	6.9	11.1	15.2	19.5	9.0	16.0	20	32	51	84	101	161	198
Europe	1.0	3.3	7.2	14.4	33.6	26.9	43.1	50.2	78.0	100.7	123	148	151	155	157	136	127
USSR	0.5	1.4	1.6	2.8	3.1	5.3	8.3	12.0	21.6	46.9	70	130	154	196	235	212	204
N.Africa_M.East	13.4	16.4	16.9	20.8	25.4	23.6	21.3	18.0	21.3	24.0	32	41	46	63	78	87	86
Tropical Africa	0.3	1.2	1.9	2.8	4.4	13.8	19.4	27.1	39.1	42.2	48	61	70	112	144	202	249
China	2.7	6.0	12.2	24.1	47.8	24.4	61.5	46.9	62.3	80.7	89	94	103	110	110	133	128
South Asia	1.5	3.6	6.4	11.3	21.6	30.4	33.9	36.7	49.6	94.8	132	143	149	185	210	220	222
SE Asia	0.1	0.5	0.7	1.1	1.7	3.7	4.9	7.1	11.6	15.3	18	34	47	62	73	100	120
Pacific dev.	0.0	0.0	0.0	0.0	0.0	0.0	0.0	0.0	0.0	0.1	0.3	4.3	10	24	34	50	50
World	21	36	52	84	145	140	209	219	293	429	20	840	51	1223	1377	1531	1591
– lower estimate	5	9	14	26	48	80	124	148	220	366	522	800	940	1186	1342	1512	1572
– upper estimate	49	84	123	197	333	251	309	299	367	494	637	884	1022	1259	1413	1542	1604
Other global estimates																	
Houghton et al. (1983)									264								
Esser (1991)												1390	1570	1910			
Richards (1990)									265		537		913	1170			
Ramankutty and Foley (1999)									405	678	821	1144	1301	1528	1688		
Klein Goldewijk (2001), HYDE 2.0									266	402	537	813	944	1230			
Pongratz et al. (2008)						136	197	233	401								
Klein Goldewijk (2001), HYDE 3.1					130				300	418	562	849	995	1214		1531	
FAO (2015)															1372	1519	1591

Table 3. Regional irrigated area estimates (million ha).

Regional actual irrigated area estimates (million ha)													
	4000 BCE	3000 BCE	2000 BCE	1000 BCE	1 CE	1000 CE	1500 CE	1700 CE	1800 CE	1900 CE	1960 CE	2000 CE	2015 CE
North America	0.00	0.00	0.00	0.00	0.00	0.0	0.0	0.0	0.0	3.2	15.7	22.6	22.1
Europe	0.00	0.00	0.00	0.03	0.11	0.2	0.4	0.1	0.3	2.2	5.8	14.9	21.4
USSR	0.00	0.00	0.00	0.00	0.03	0.1	0.1	0.2	0.3	2.9	7.0	12.0	11.4
Pacific dev.	0.00	0.01	0.01	0.01	0.05	0.1	0.2	0.3	0.7	3.2	6.4	13.2	13.2
China	0.86	0.99	1.05	1.14	1.28	1.9	1.0	0.9	0.9	2.8	10.1	20.7	23.8
Latin America	0.00	0.00	0.00	0.00	0.01	0.0	0.1	0.1	0.2	0.8	2.4	5.4	6.8
N.Africa_M.East	0.07	0.15	0.30	0.60	0.91	1.2	1.7	1.8	4.1	11.3	39.7	48.8	64.4
Tropical Africa	0.02	0.04	0.06	0.11	0.20	0.4	0.5	0.7	2.0	16.4	35.0	73.4	86.4
South Asia	0.00	0.00	0.00	0.00	0.00	0.1	0.2	0.4	0.8	4.6	10.4	20.2	23.8
SE Asia	0.00	0.00	0.00	0.00	0.00	0.0	0.0	0.0	0.0	0.1	0.7	1.8	2.4
World	0.95	1.19	1.42	1.90	2.61	4.1	4.2	4.5	9.3	47.5	133.2	233.0	275.6
Other global estimates													
Siebert (2008)								4.5	10.5	53.2		278.9	
Siebert et al. (2014)												255.2	
Framji et al. (1982)										40.0			
Michael (2008)										40.0			
Li et al. (2009)										50.0			
Siebert et al. (2015)										63.0	144.5	262.9	
Freydank and Siebert (2008)								5.1	10.6	53.2			
FAO (2015)											163.5	288.9	

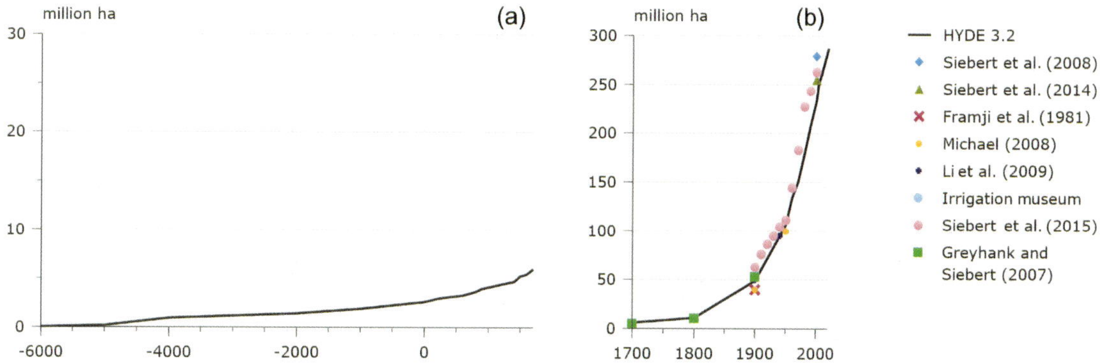

Figure 5. Historical (actual) irrigated area estimates for two periods: 6000 BCE–1700 CE **(a)** and 1700–2010 CE **(b)**.

tions, things had to change. Electrical, steam and internal combustion engines became available for pumping water – even in remote areas from large groundwater reservoirs many metres deep. This enabled farmers to grow crops where agriculture had previously not been possible; hence, the area in which irrigation was used increased on a large scale.

We estimate the global irrigated area in 1700 CE to be 4.5 Mha, which is similar to the estimate of Siebert (2008) and close to the 5.1 Mha of Freydank and Siebert (2008); see Fig. 5. Our estimate for 1800 CE of 9.3 Mha is close to the 10.5 Mha of Siebert (2008) and the 10.6 Mha of Freydank and Siebert (2008). In addition, our estimate of 48 Mha for 1900 CE is well within the literature range of 40–63 Mha (Framji et al., 1982; Freydank and Siebert, 2008; Li et al., 2009; Michael, 2008; Siebert, 2008; Siebert et al., 2015). Finally, we estimate the global irrigated area in 2015 CE at 276 Mha, with a large share in China (63 Mha) South Asia (88 Mha) and South East Asia (24 Mha).

4.4 Rice

Many studies agree that the origins of rice agriculture began in the lower Yangtze River valley in eastern China (Barker, 2006; Bellwood, 2001; Fuller et al., 2007; Zhang and Wang, 1998). These findings form the basis for the hypothesis that rice cultivation that led to domestication began in 6000 BCE. During the period from 6000 BCE until around 4000 BCE, systematic cultivation of rice species had become well established (Opferkuch, 2016).

We cautiously estimate rice area in China to be less than 0.0004 Mha in 8000 BCE, 0.03 Mha in 6000 BCE, 0.02 Mha in 4000 BCE, 0.11 Mha in 2000 BCE, 1 Mha in 1 CE and 1.2 Mha in 1000 CE (see Table 4). However, these numbers are highly uncertain and must be treated with care. Numbers for South East and South Asia show a similar pattern but that might be incorrect for the distant past (before 1 CE). Fuller and Qin (2011) suggest that rice arrived later in South East and South Asia and was fully domesticated around 4000 BCE. On a global scale, we estimate a rice area of 0.004 Mha in 8000 BCE, 0.02 Mha in 6000 BCE, 0.11 Mha in 4000 BCE, 0.49 Mha in 2000 BCE, 2.6 Mha in 1 CE and 4.8 Mha in 1000 CE. This is in good agreement with the study of Fuller et al. (2011), who estimated a total rice area of 0.10 Mha in 4000 BCE, 0.20 Mha in 3000 BCE, 0.50 Mha in 2000 BCE, 0.80 Mha in 1000 BCE and 2.50 Mha in 1 CE (see Table 4 and Fig. 6). We did not find any global rice area estimates for the 1000 CE–1900 CE period in literature.

4.5 Grazing land

We estimate grazing land in 5000 BCE to be around 30 million ha (range 5–58), corresponding to 1.58 ha cap^{-1} of grazing land (range 0.23–2.87). Grazing land is estimated to be 199 million ha (range 50–353) in 1 CE and 1.05 ha grazing land cap^{-1} (range 0.22–1.54). The area for grazing land was higher than for cropland, namely 314 Mha (range 155–526) in 800 CE, 444 Mha (range 260–645) in 1100 CE and 483 Mha (range 326–660) in 1400 CE, which corresponds globally to 1.10 ha of grazing land cap^{-1} and 1.12 and 1.09 ha cap^{-1}. Our grazing land estimates are, in general, higher than the Pongratz et al. (2008) estimates; they present 140 Mha for 800 CE (range 80–210), 198 Mha for 1100 CE and 227 Mha for 1400 CE (see Tables 5 and S7).

We compute a global area used for grazing of 664 Mha in 1700 CE, which is higher than the 526 Mha of Houghton et al. (1983). The largest differences appear to be in tropical Africa, but the largest differences could arise from the fact that we distinguish between pasture and rangeland (145 Mha of pasture and 519 Mha of rangeland in 1700 CE). The latter is assumed to only involve a conversion from natural ecosystem to grazing land if it is located on forest land, whereas Houghton by design wanted to exclude grasslands, which did not involve conversion of natural shrubs. In 1800 CE the area occupied for grazing had grown to 915 Mha (189 Mha pasture and 726 Mha rangeland), and 1657 Mha in 1900 CE (351 Mha pasture and 1306 Mha rangeland). The area of grazing land reached a peak around 2000 CE with 3322 Mha (784 Mha pasture and 2538 Mha rangeland) and has been slowly decreasing towards the present day. The largest increase in pastures appears to have occurred in the second

Table 4. Regional rice area estimates (million ha).

Regional rice area estimates (million ha)	4000 BCE	3000 BCE	2000 BCE	1000 BCE	1 CE	1000 CE	1500 CE	1850 CE	1900 CE	1930 CE	1960 CE	2000 CE	2015 CE
North America	0.00	0.00	0.00	0.00	0.00	0.00	0.00	0.00	0.15	0.39	0.86	1.23	0.88
Latin America	0.00	0.00	0.00	0.00	0.00	0.00	0.00	0.10	0.40	1.14	6.19	6.40	5.36
Europe	0.00	0.00	0.00	0.00	0.00	0.03	0.06	0.17	0.28	0.30	0.39	0.43	0.47
USSR	0.00	0.00	0.00	0.00	0.00	0.02	0.04	0.16	0.22	0.27	0.33	0.51	0.43
N.Africa_M.East	0.00	0.00	0.01	0.02	0.07	0.13	0.14	0.30	0.34	0.49	1.04	1.36	1.33
Tropical Africa	0.00	0.00	0.00	0.00	0.00	0.09	0.42	1.14	1.33	1.71	3.20	7.05	11.46
China	0.02	0.05	0.11	0.16	1.00	1.18	3.19	16.22	15.39	19.33	21.08	20.51	20.78
South Asia	0.08	0.16	0.31	0.60	1.18	1.95	2.35	4.65	10.26	13.30	36.79	42.19	42.84
SE Asia	0.01	0.02	0.06	0.15	0.36	1.40	2.49	5.64	12.13	18.99	23.91	30.06	34.57
Pacific dev.	0.00	0.00	0.00	0.00	0.00	0.00	0.00	0.00	0.01	0.01	0.04	0.15	0.17
World	0.11	0.22	0.49	0.93	2.62	4.80	8.70	28.38	40.50	55.94	93.83	109.89	118.29
Other global estimates													
Fuller et al. (2011), wet rice	0.05	0.20	0.40	0.70	2.30								
Fuller et al. (2011), total rice	0.10	0.20	0.50	0.80	2.50								
Barker et al. (1985), harvested are										81.3			
FAO (2015), harvested area											115.4	132.9	

FAO: area harvested. HYDE: standing crop area.

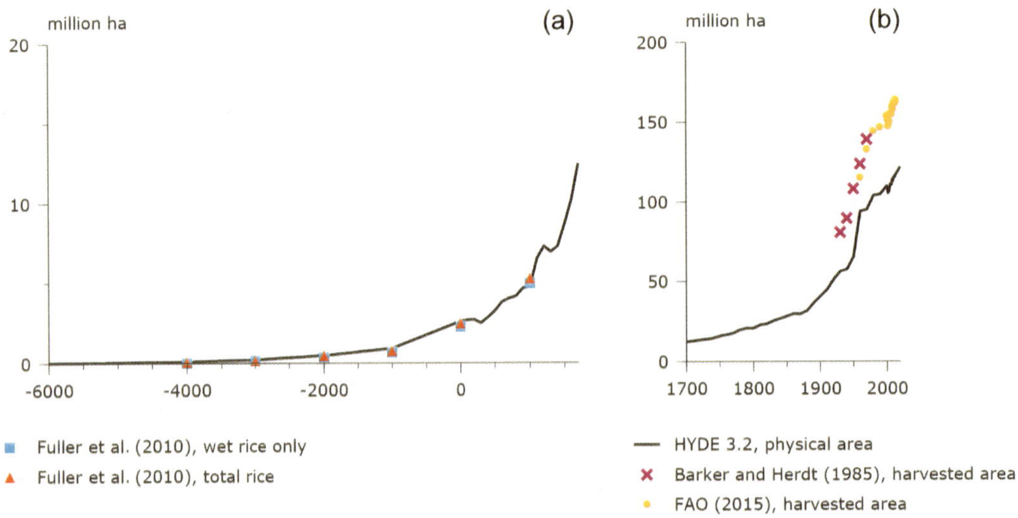

Figure 6. Historical (physical) rice area estimates for two periods: 6000 BCE–1700 CE **(a)** and 1700–2010 CE **(b)**.

half of the 20th century in tropical Africa, followed by Latin America, and to a lesser extent in China. In North America and China, it seems to have stabilized in the last 10 to 20 years, while in Africa it is still increasing. Europe already had its peak around 1900 CE and is still declining slowly. Rangeland areas also increased in tropical Africa and Pacific developed countries (Japan, Australia, and New Zealand) up to World War II, but then declined slowly or stabilized at current levels.

Grazing land had a global low per capita average number of around 1.0 ha cap^{-1} until 1 CE, then a moderate increase to 1.13 ha cap^{-1} towards 1000 CE and then a gradually decrease again to 0.45 ha cap^{-1} in 2015 CE (see Table S7). However, regionally there are large differences. Sparsely populated countries such as Australia, Botswana, Mauritania and Namibia have more than 40 ha of grazing land per capita available today and Mongolia and Western Sahara even have around 150 ha cap^{-1}.

Table 5. Regional grazing land area estimates (million ha).

Regional grazing and area estimates (million ha)	4000 BCE	3000 BCE	2000 BCE	1000 BCE	1 CE	800 CE	1100 CE	1400 CE	1700 CE	1800 CE	1850 CE	1900 CE	1920 CE	1950 CE	1960 CE	2000 CE	2015 CE
North America	0.0	0.0	0.0	0.0	0.0	0.0	0.0	0.0	4.2	18.2	36.1	158.0	181.3	278.1	280.8	251.8	266.7
Latin America	0.0	0.0	0.0	0.0	0.0	0.0	0.0	0.0	29.5	45.4	73.1	152.7	216.7	370.7	457.2	553.5	559.2
Europe	0.9	1.7	3.9	7.6	19.0	31.4	75.3	74.2	87.5	78.5	88.6	97.2	99.0	98.8	90.7	78.2	70.2
USSR	13.0	15.1	17.7	20.6	24.0	38.5	47.2	58.6	79.8	118.9	148.5	225.7	263.3	274.0	301.0	361.7	369.5
N.Africa_M.East	4.5	6.9	10.7	15.8	26.1	38.1	36.8	33.3	41.0	62.2	107.9	140.9	156.2	196.0	250.6	307.8	292.8
Tropical Africa	15.4	22.7	34.0	51.6	79.0	147.2	180.1	222.4	301.8	352.6	400.9	458.5	500.1	697.0	799.6	804.9	766.3
China	7.4	10.9	16.1	19.7	48.1	45.2	90.1	80.0	98.5	214.5	298.0	288.1	317.0	311.4	367.3	520.1	506.1
South Asia	0.0	0.1	0.3	0.6	1.4	10.4	11.0	9.8	13.2	16.1	18.8	24.8	28.1	38.6	51.7	49.1	48.5
SE Asia	0.2	0.2	0.5	0.9	1.2	2.9	3.3	4.4	7.1	7.1	8.0	12.5	15.5	18.3	17.3	17.2	17.5
Pacific dev.	0.0	0.0	0.0	0.0	0.0	0.0	0.0	0.2	1.2	1.4	11.6	98.8	175.5	328.2	393.5	378.3	344.6
World	41.4	57.7	83.1	116.6	198.8	313.8	443.8	483.0	663.5	914.9	1191.6	1657	1953	2611	3010	3323	3241
Other global estimates																	
Houghton et al. (1983)									526								
Pongratz et al. (2008)						144	198	227	370								
Klein Goldewijk (2001), HYDE 2.0									524	942	1310	1955	2282	2930			
Klein Goldewijk et al. (2011), HYDE 3.1					106				324	513	721	1294	1769	2464		3429	
FAO (2015)															3095	3424	3359

4.6 Regional trends

There are large differences in the land use history of the world's regions. Figure 7 depicts the land use development for selected time periods for the different world regions. Europe, the Middle East–North Africa–Turkey (MENAT), China and South Asia started agriculture much earlier than most other regions. North and Latin America and Oceania witnessed a relative late expansion of agriculture, but – once started – the expansion of cropland has been massive. In most regions, now and historically, cropland has been largely rain-fed, while South Asia, South East Asia and China have also had a significant share of irrigated crop area. The share of pasture in the total grazing land area is relatively high in Europe and South East Asia, while the share of rangeland is more dominant in North America, Latin America, the Middle East, the former USSR, tropical Africa and Pacific developed countries.

The share of irrigated land was very modest compared to the total amount of agricultural land for a long time. Only during the 20th century did this share increase, especially in South Asia and South East Asia.

Spatial patterns of agricultural land use are depicted in Fig. 8. The maps represent land use for the year 2010 for the different HYDE categories. Irrigated croplands are mainly concentrated in India, China, central Asia and North America. The main rain-fed cropland areas are more widespread all over the world in and concentrate in central and eastern North America, southeast Latin America, Europe, central Asia, sub-Saharan Africa, India, East and South East Asia, and the southwest and southeast parts of Australia. Irrigated rice is concentrated in eastern China, northern and eastern India, South East Asia, Japan, the Philippines, and Indonesia, while rain-fed rice can mostly be found in India, South East Asia, Madagascar, western Africa and Brazil.

The pasture areas are mostly located outside the arid regions such as the eastern USA, central America, eastern Brazil and Argentina, Europe, and the northern ranges of central Asia, sub-Saharan Africa, Madagascar, and eastern China. The rangelands can be found in the more arid and remote regions such as the western part of the USA and Mexico, southwest Brazil and the southern part of Argentina, the Sahel region of Africa, and the dry regions of southern Africa, North Africa, and the Arabian peninsula, the dryer parts of central Asia, Mongolia, and China, and the remote parts of Australia, except for the deep interior.

5 Uncertainties

There are large and many uncertainties that come with the hindcast methods applied in this study. We start with good and reliable data from the United Nations World Population Prospects for the post-1950 period, but before 1950 there is a strong dependency on a few historical population data sources such as McEvedy and Jones (1978), Madisson (2001) and Livi-Bacci (2007), which are for some regions and time periods not undisputed and probably on the low side of estimates (Klein Goldewijk and Verburg, 2013). Therefore, especially deeper into the past (pre-1500 CE), the numbers must be treated with caution. However, as already stated in Klein Goldewijk and Verburg (2013), the resulting demographic growth rates seem plausible, and thus the data are acceptable for the purpose of this study.

A similar statement can be made for the different land use estimates. The FAO data for post-1960 are quite reliable, al-

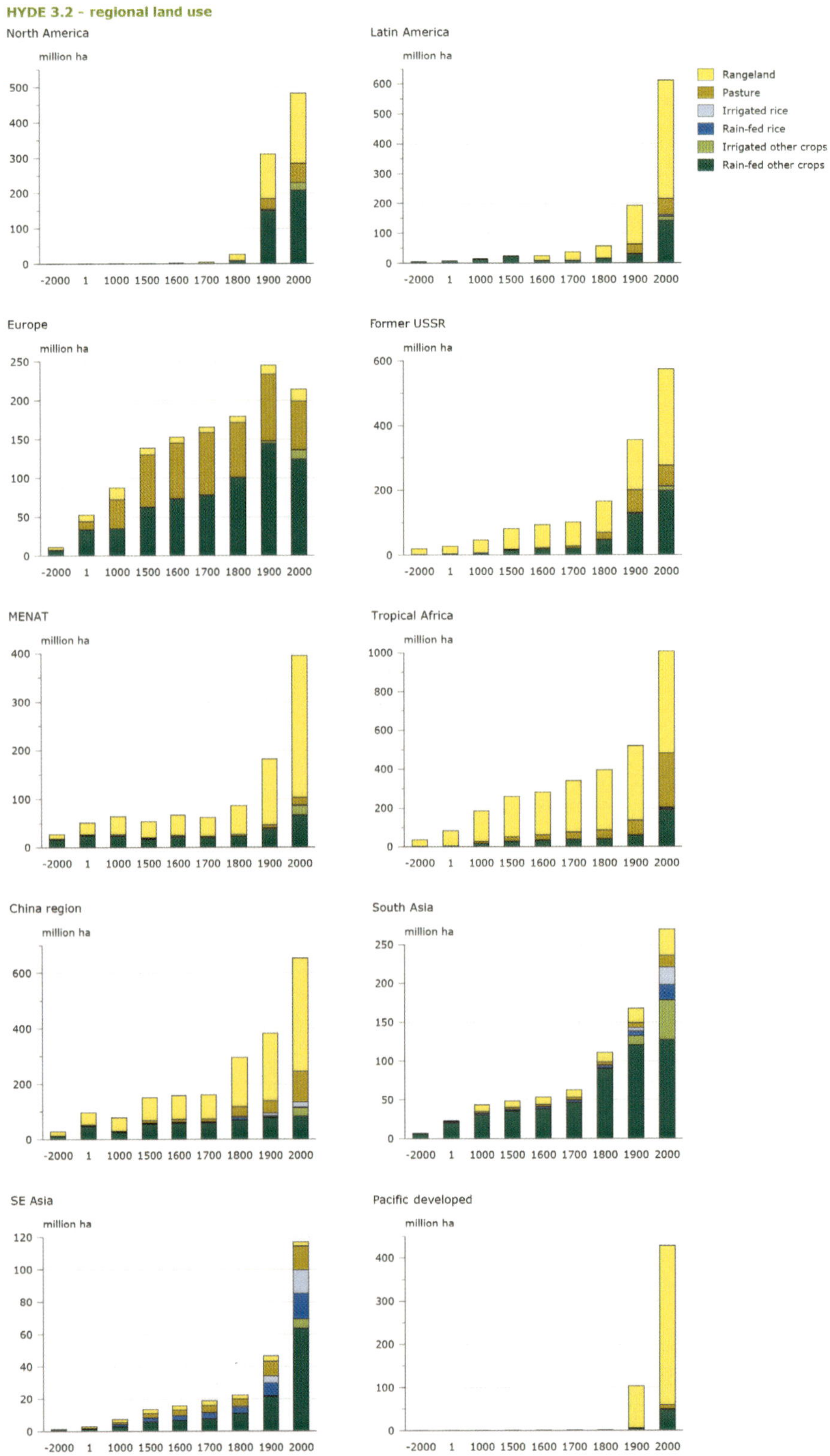

Figure 7. Historical land use estimates by region, for irrigated and rain-fed rice and other crops, rangeland and pasture. Please note the rangeland and pasture distinction is purely an indication of the intensity of use, not of the land use conversions involved (see Sect. 3.2.2 "Grazing land").

though for some countries they can be questioned, even for the present. The hindcasting approach of using land use per capita is a pragmatic approach, but it is very sensitive to the shape of the curve into the past. The minimum and maximum values per curve are very uncertain and vary substantially by country and in time. The study of Klein Goldewijk and Verburg (2013) clearly showed that the shape of the per capita curve profoundly determines the total agricultural area, especially in the pre-industrial era. Basic assumptions are that it is not zero, but there is also an absolute limit to the amount of work a person can do per day or year (Williams, 2000).

Making the distinction between pasture and rangeland is difficult and uncertain, also in the recent past, as this distinction is not covered by FAO statistics, though a distinction between permanent meadows and pastures into "cultivated" and "natural growing" is already available for some countries (FAOSTAT, 2017). We therefore applied a combined population density and aridity index to reflect livestock densities and an additional rule that pastures involve conversion of natural vegetation, while rangeland only involves conversion of natural vegetation if it was forest or woodland. The method has been compared to a more advanced approach and shows good agreement with satellite data (see Supplement C).

On top of the default estimates of population, cropland and grazing land, we also estimated lower and upper uncertainty ranges. These uncertainty ranges were partly based the ranges we could find in literature and partly on our own expert judgement and should be treated with care. The uncertainty range A is cautiously estimated at 1 % in 2000 CE, 5 % in 1900 CE, 10 % in 1800 CE, 25 % in 1700 CE, 50 % in 850 CE, 75 % in 1 CE and 95 % in 10 000 BCE. The uncertainty range B is twice the uncertainty range of A and should be considered as such since in our opinion it is highly unlikely that areas of cropland or grazing land have been outside this range in the past. The years in between were linearly interpolated (the method is similar to that described in Klein Goldewijk et al., 2010), and the resulting ranges are depicted in Fig. 8. A regional summary of cropland, pasture, population, per capita cropland area and per capita pasture area is presented in Table S7.

Figure 9a and b depict the range in total population estimates that we could find in literature and the HYDE baseline estimate (solid black line) for the period 10 000 BCE to 1700 CE (plate A) and 1700 CE to 2000 CE (plate B). The range in literature estimates for the last 300 years are not so large, meaning that apparently the uncertainty is consequently also rather small. This is also reflected in the range we applied to our own estimates, on top of the baseline, varying from 1 % in 2000 CE to 25 % in 1700 CE. Before 1700 CE the literature estimates show much more variation. For example, the lower literature estimate found for 1 CE is 27 % lower than our baseline estimate, while the upper estimate found in literature is 42 % higher than ours. Our range of plus and minus 75 % for 1 CE is therefore on the cautious side but expresses our idea that values outside our uncertainty bands are very unlikely to have occurred in the past. Figure 9c and d are the resulting total population estimates of HYDE, with their uncertainty bands A and B. Please note the different y axes. Figure 9e and f depict the HYDE baseline estimate of cropland with their uncertainties. Figure 9g and h show the HYDE estimates for grazing land.

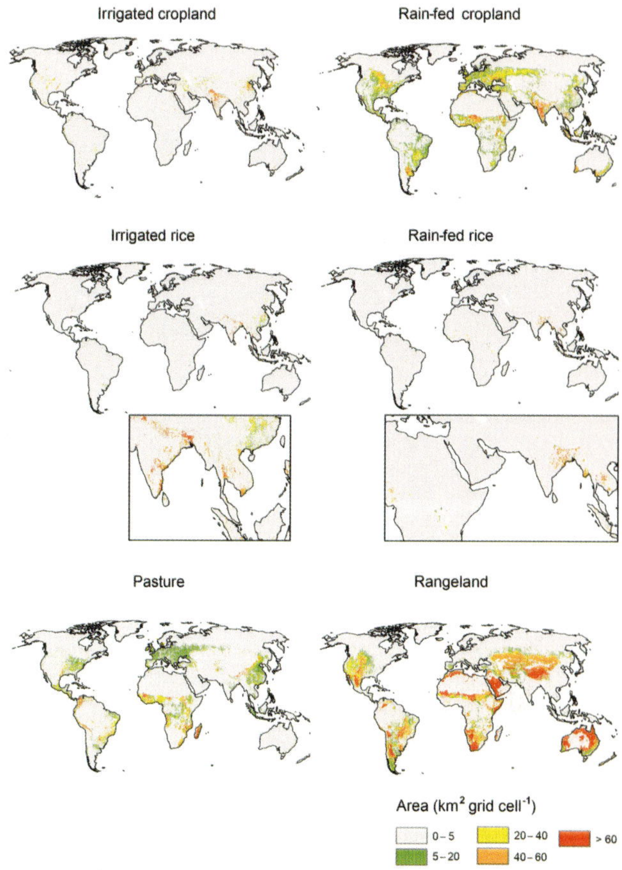

Figure 8. Spatial maps of cropland (irrigated and rain-fed), rice (irrigated and rain-fed), and pasture and rangeland for 2010 CE (six panels).

The final estimate of cropland and grazing land in HYDE is thus computed by multiplying population numbers and the per capita land use estimate for each scenario; a baseline, a first lower and upper estimate scenario (range A) and the second lower and upper estimate scenario (range B); see Fig. 9. The estimates for HYDE 3.2 are in general higher than the older estimates of HYDE 3.1. This is partly because of the use of new (census) data sources from the Clio Infra project (https://www.clio-infra.eu/), and partly because of new insights into long-term per capita land use (see also Klein Goldewijk et al., 2017). Especially grazing land has now been estimated to be much higher than before, reflecting that livestock grazing has been a long-term and widespread activity.

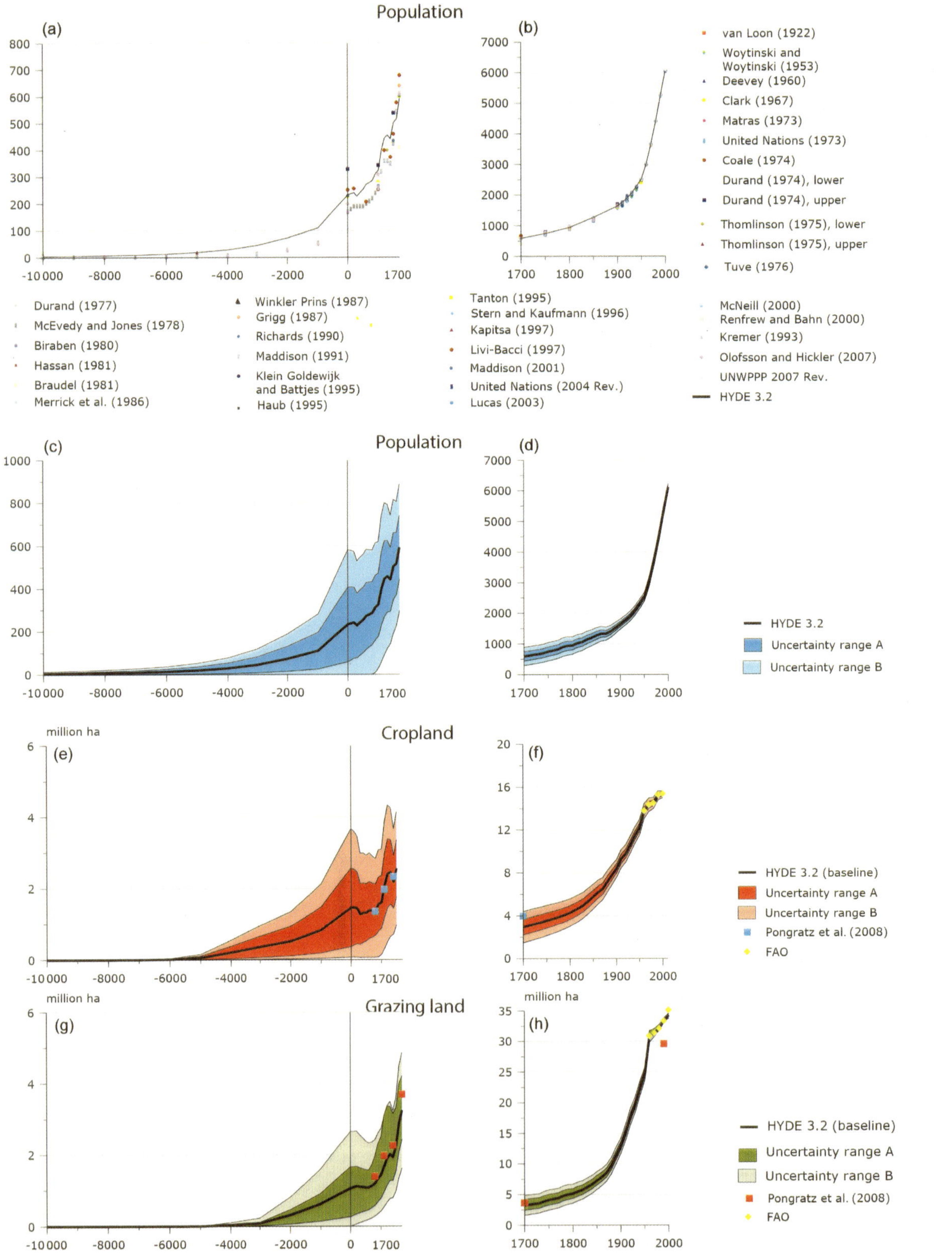

Figure 9. Available sources for population estimates **(a, b)**, and uncertainty ranges for global population **(c, d)**, cropland **(e, f)** and grazing land **(g, h)**, split into recent past (1700 CE onwards) and before 1700 CE to allow for different scales.

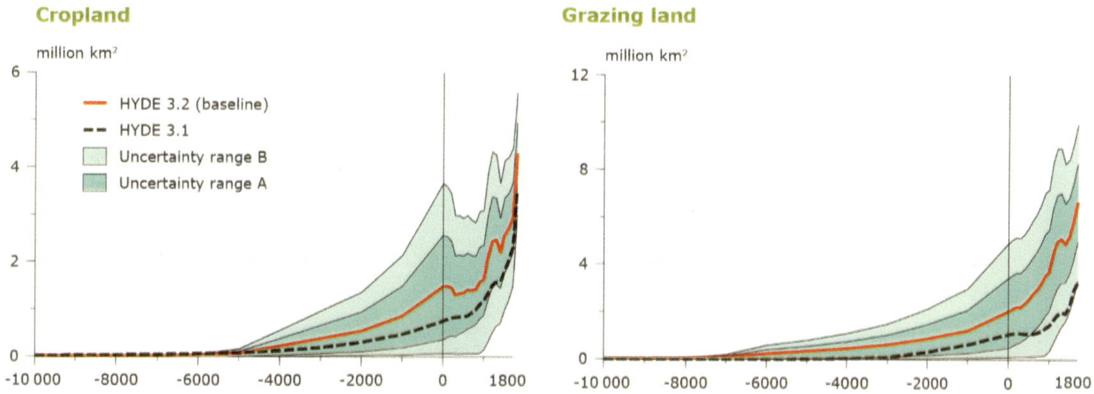

Figure 10. Comparison of HYDE 3.1 and HYDE 3.2, with uncertainty ranges.

We also quantified the amount of variation or dispersion in the input parameters in the per capita land use categories, used for computation of the total areas. Figure 10 depicts the results. The standard deviation (function STDEV.P in Excel) for cropland per capita remains around plus and minus 0.5 ha cap^{-1} for a long time. Only in the distant past, before 1 CE, does it increase to 1 ha cap^{-1}. The variation in the per capita grazing land values is much larger, up to 25 ha cap^{-1}, especially in the 18th and 19th centuries during the period of colonization. Before 1600 CE it is around 5 ha cap^{-1} (see Fig. 11).

The variation in irrigated land per capita is relatively high in recent years due to several irrigation projects worldwide. The peak in 4000 BCE is due to the high value for Egypt and some Middle Eastern countries. The per capita rice variation has been plus and minus 0.03 ha cap^{-1} (roughly around $\pm 50\,\%$) during the last decades, but then increases towards more than 100 % before 1960 CE (see Fig. 11).

Another issue is that due to the absence of transient Holocene climate and vegetation maps, we simplified things by using weighting maps for current climate and biome over the whole time period. Although the climate in 10 000 BCE differs from the present day (Armesto et al., 2009; Bertrand and Van Ypersele, 2002; Kropelin et al., 2008; Tett et al., 2005; Verschuren et al., 2002), we believe that the lower temperature thresholds we used are still valid, partially since they are just one of the factors involved. The same applies to the biome map. The Sahara desert was more like a grassland/savanna type during the pre-5000 BCE period (Verschuren et al., 2000) and changed into the current desert state later on, but it has been hardly populated (ample agriculture); thus, allocation procedure remained unchanged during that era.

Analysis by Prestele et al. (2016) show that it is quite crucial for future outcomes to determine at what spatial starting point global change models and integrated assessment models start their simulations. Therefore, we have also experimented with a different choice for the cropland and grazing land reference maps to see whether it would have an influence, and if so, to what extent the spatial patterns of land use would change. In other words, how important is it to use the most recent satellite information for input land use data series for integrated global modelling exercises. We took the old HYDE 3.1 reference maps of DISCover/GLC 2000 (Loveland and Belward, 1997; Loveland et al., 2000) valid for the reference year 2000, the IIASA/IFPRI cropland map (Fritz et al., 2015) valid for the reference year 2005 and the 2010 reference map (ESA, 2016) valid for the year 2010. Next, we ran a fuzzy numerical kappa analysis with the software from the Map Comparison Kit (Visser and de Nijs, 2006). The fuzzy numerical kappa analysis between the 2000 reference map (DISCover/GLC2000) and the 2010 reference map (ESA) yielded a substantial degree of agreement of 0.744 (1.00 = very high degree of agreement, 0.00 = very low). The comparison of the 2000 reference map (DISCover/GLC2000) with the 2005 reference map (IIASA/IFPRI) gave a fuzzy numerical kappa value of 0.727. Furthermore, the comparison between the 2005 reference map (IIASA/IFPRI) and the 2010 reference map (ESA) yielded a fuzzy numerical kappa value of 0.810. This means that there is overall a substantial agreement (or in other words relatively little difference) between the three reference maps for cropland. The fuzzy numerical kappa analysis between the 2000 reference map (DISCover/GLC2000) and the 2010 reference map (ESA) was 0.635. The comparison between the 2005 reference map (IIASA/IFPRI) and the 2010 reference map (ESA) yielded a kappa fuzzy numerical of 0.643. There is no grazing land map of 2005 from IIASA/IFPRI.

All this is in agreement with a study of 43 simulations from 11 global-scale land use–land change models of Prestele et al. (2016), showing that cropland is more consistent among the different reference starting conditions than grazing land. However, differences do exist, usually on the local scale. First, the original satellite input is different between the three data sets, with different sensors. Second, different algorithms were used to classify the land cover

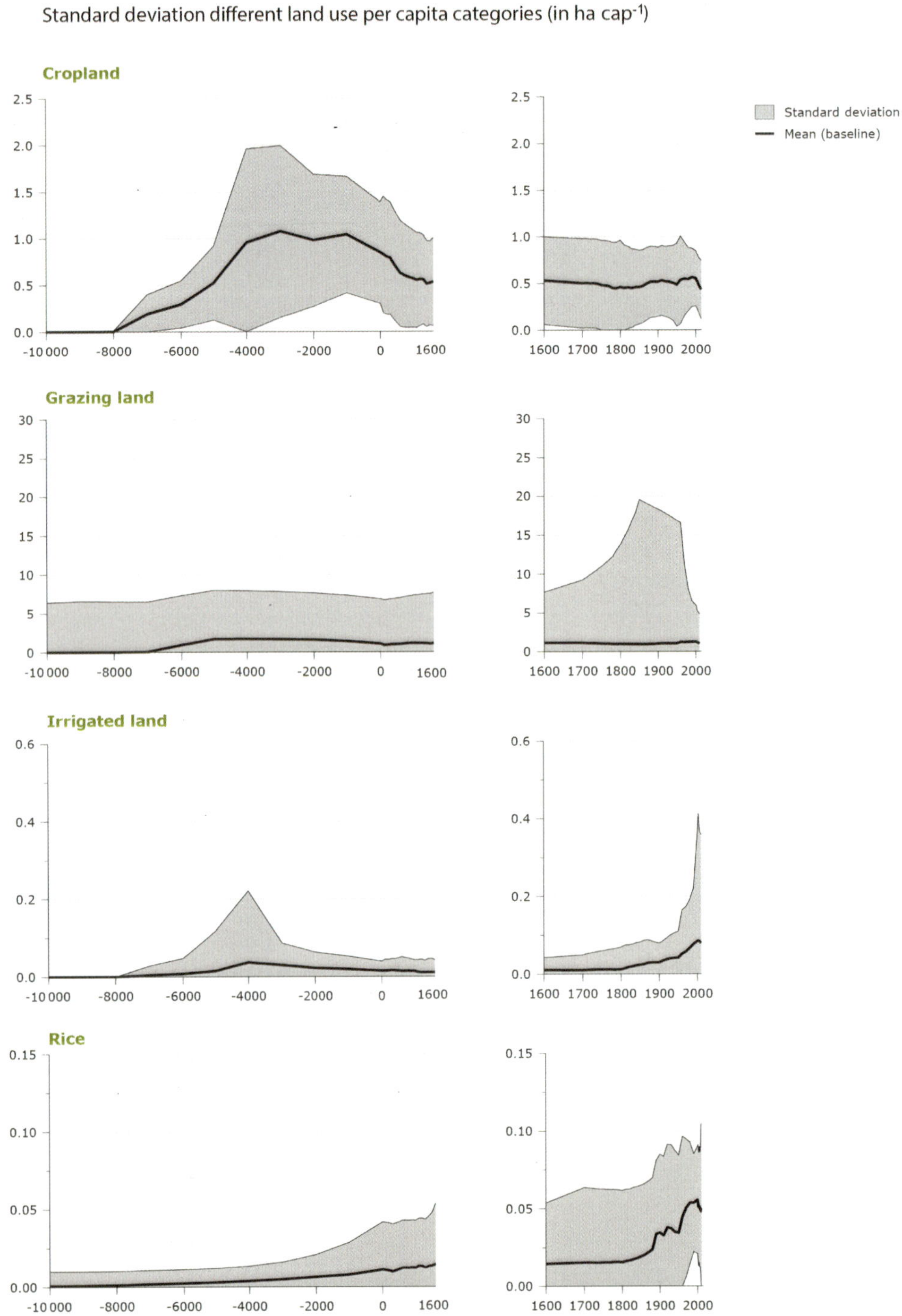

Figure 11. Standard deviation of input data per capita.

classes, which also led to different results. Third, when time progresses, the absolute areas of cropland have sometimes changed over time, either by expansion or abandonment, leading to different maps. We believe that for this hindcasting study, it does not play an important role which reference map we used, but it can be important for future modelling studies.

Finally, an important point to be made is that in this HYDE 3.2 version no empirical data are systematically used (yet) to improve the historical land use reconstructions. All allocation in the distant past is performed following general, globally applicable HYDE allocation rules. As Morrison (2015) rightly pointed out, these rules are often made with a Eurocentric point of view. We fully acknowledge this

Table 6. Summary of global population and land use estimates.

Historical global population (million) land use estimates (million ha).								
	5000 BCE	1 CE	500 CE	1000 CE	1500 CE	1700 CE	1850 CE	2015 CE
Population	19.0	232	253	323	503	592	1271	7301
Cropland	5.8	145	133	162	256	293	578	1591
Rain-fed area	5.6	142	129	157	252	289	549	1316
Net irrigated area	0.2	3	3	4	4	4	28	276
Net rice area	0.1	3	3	5	9	12	28	118
Paddy rice	0.0	1	1	2	2	3	12	75
Rain-fed rice	0.1	2	2	3	6	10	16	43
Grazing	30.1	199	249	366	515	664	1192	3241
Pasture	0.0	17	15	55	105	145	253	787
Rangeland	30.0	181	234	310	410	519	939	2454
% agric/total land area	0.3 %	2.6 %	2.9 %	4.0 %	5.9 %	7.3 %	13.6 %	37.1 %

since these rules were used as a first simple attempt to allocate land use in the distant past. We are aware that there is a need for a much more regionalized approach. Very promising work is underway in the Past Global Change (PAGES, http://www.pages.org/) LandCover6K working group initiative, for which archaeologists, historians, geographers, paleoecologists and land use reconstructors join efforts for the first time to collect and provide rich data from all different disciplines in order to improve the ALCC time series. In return, this is beneficial for several model intercomparison projects such as the Coupled Model Intercomparisons Project (CMIP) and the Land Use Model Intercomparison Project (LUMIP), which aim to further advance understanding of the impacts of land use and land cover change on climate (Lawrence et al., 2016).

6 Data availability

This HYDE version replaces earlier beta releases as provided at DANS DOI https://doi.org/10.17026/dans-25g-gez3. All data can be downloaded from ftp://ftp.pbl.nl/hyde/hyde3.2/ and supplementary data can be downloaded from ftp://ftp.pbl.nl/hyde/supplementary/Klein_Goldewijk_et_al_2017_HYDE32_paper/.

7 Discussion

We have estimated population and specific land use categories for the Holocene. Population has grown exponentially; however, while growth rates were initially very low, only during the last century did they dramatically increase to more than 1 %. Grazing land and cropland also increased but at a slower rate. The majority of cropland is rain-fed, but irrigated croplands and rice areas – despite their relatively small share – have been a very important factor in the increase in yield and production. We have estimated that total agricultural land occupied less than 0.5 % of the total land area in 5000 BCE, 3.4 % in 1 CE, 13.6 % in 1850 CE and 37.4 % in 2015 CE. Table 6 summarizes our estimates for population and land use for the Holocene.

This and the few other existing global estimates of historical land use are rather uncertain (Klein Goldewijk and Verburg, 2013). This can partly be explained by the sheer lack of reliable data, and hence by the large difference in methods and assumptions used for reconstruction. Most estimates of historic land use cover the period from today back to 1700; only two go further back in time (Pongratz et al., 2008, Kaplan et al., 2010). In general, the core of the methodology is regional and national or subnational estimates of historic cropland and pasture areas, which are allocated on a grid in a hindcasting mode starting from today's land cover. Similar to the HYDE methodology, the estimates of cropland and pasture areas are a mix of original data on agricultural area and calculations based on population data and assumed cropland per capita factors in time.

For example, Ramankutty and Foley (1999) present a simple approach to derive geographically explicit changes in global croplands from 1700 to 1992 CE. They created a global representation of permanent croplands in 1992 CE at 5′ spatial resolution by calibrating a remotely sensed land cover classification data set against cropland inventory data. Then, from a variety of sources, they compiled an extensive database of historical cropland inventory data, at the national and subnational level. Next, they hindcasted the 1992 CE data with a simple land use change model back to 1700 CE. Recently, the data set was updated to 2007 and revised. The range in cropland estimates (see Table 2) mostly originates from different definitions of cropland area. Pongratz et al. (2008), who covered the time period 800–1992 CE, already show higher cropland areas in the 20th century, and this is maintained for the entire historic period.

Table 7. Comparison total agricultural land estimates: HYDE 3.2 and Kaplan et al. (2011).

Total agricultural land (Mha)									
	4000 BCE	1000 BCE	1 CE	100 CE	500 CE	1000 CE	1500 CE	1600 CE	1850 CE
HYDE 3.2	62	200	344	355	382	527	771	850	1769
Kaplan et al. (2010)	186	871	1360	1440	1530	1800	2300	2160	2940

On a grid cell level, further difference between historic land use products can arise. Most methods use simple land use change models with specific, partly socio-economic suitability factors. Additionally, various assumptions are made as to which natural vegetation type is preferentially converted to agriculture (see also Prestele et al., 2016), i.e. whether agricultural land is taken preferentially from the forested, grass-covered or shrub-covered parts of the grid cell; this decision can cause large differences in the reconstructed extent of natural vegetation types (Reick et al., 2013). For the Pongratz et al. (2008) reconstruction, allocation rules for cropland were derived from the existing high-resolution (5 arcmin) maps of cropland and potential vegetation, while pasture is first allocated to grass as far as possible, then to the area of the woody vegetation types. This procedure reflects the human behaviour of minimizing effort: clearing of forest is generally not performed if sufficient natural grassland is available for grazing (Houghton, 1999).

The hindcasting prior to 1700 CE is even more uncertain than the period thereafter, as data become increasingly scarce. Therefore, the only practical approach is to use population data and assumptions about per capita needs for agricultural land. A key factor in explaining differences between these estimates is temporal and spatial changes in per capita land requirements: earlier versions of the HYDE database kept per capita land use constant over time or, in the uncertainty estimates, homogeneously varied it with time across the globe; the KK10 (Kaplan et al., 2010) data set assumed that land use intensification depends on the development stage of a society, which was represented by population density (this relationship was derived from data for five European countries and applied to the globe). By contrast, Pongratz et al. (2008) used a wealth of literature on agricultural practices and agro-technological innovations for all regions in the world (note that such literature exists more readily for the last millennium as compared to the longer time periods HYDE and KK10 covered); with this, per capita land use estimates could be accounted for as evolving over time and at the same time changing differently in the different regions of the globe. This approach accounts for, for example, the spread of wet rice farming in China that drastically lowered per capita land needs over time, or the importance of hunting, gathering, and fishing in some of the traditional societies in the Americas.

In addition to the per capita need for agricultural land, population data are the second determining factor in our analysis. To account for their uncertainties, different population databases were used and the upper and lower ends of an uncertainty range were assessed. Together with the uncertainty range in per capita cropland, this led to two different trajectories of land cover change that reflect the upper and lower bounds within the uncertainty range of historical land cover changes, around the default estimate that used a specific population data set and assumed constant per capita land use changes over time.

A key shortcoming for application of the reconstruction on a fine spatial scale is the implicit assumption of constant land use patterns over time. While relative changes between countries are captured by the regionally specific population and per capita land use data, within a given country the pattern of agricultural extent from present day is kept fixed and simply scaled with national totals. Some regional adjustments have been made where this assumption has been known to not apply, e.g. the differential agricultural expansion of different parts of Russia or the arrival of Europeans to only specific parts of Australasian and North and South American countries.

The total amount of agricultural area computed by our study is significantly smaller than the estimate of Kaplan et al. (2011). This can largely be explained by differences in methodology; Kaplan et al. (2011) used a correlation between deforestation and population density, varying over time with changes in technology and development. It was calibrated for Europe and then extrapolated to the rest of the world, which very likely caused an overestimation of land use change estimates. Since HYDE 3.2 does not estimate anthropogenic activities such as wood harvesting or shifting cultivation, our resulting estimates of deforestation are also necessarily lower; see Table 7 for an overview of the total agricultural areas from both studies.

Despite all these uncertainties, there is a great need for historic land use reconstruction, especially in the climate change modelling community. Earlier versions of the HYDE database have been used in the Land-Use Harmonization (LUH) data sets (Hurtt et al., 2011), and this most recent HYDE 3.2 version is again applied in the follow-up product LUH2, complemented with future scenarios of land use change. These LUH data sets are specifically developed to meet the needs of integrated assessment models (IAMs) and Earth system models (ESMs) developing future scenarios and simulations as part of international assessment (IPCC's 5th Assessment Report and soon CMIP6 and the 6th Assessment Report). IAMs produce estimates of future regional or gridded land use as part of their future scenario projections, while ESMs require spatio-temporal information on land use activities from at least 800 CE onwards, including estimates of secondary (recovering) land age, area and biomass density to quantify the impacts of human land use on the Earth system. The LUH data sets make sure that IAM future projections do smoothly transition from the end of historical reconstructions. In addition to the LUH data sets that are provided to the scientific community, in which model inputs and decisions are systematically varied and the resulting changes in model outputs are examined, large-scale sensitivity analyses are also performed. The analysis shows that our model is most sensitive to the historical start date, the choice of primary vs. secondary priority for land use conversions, and the inclusion of wood harvesting and shifting cultivation.

The LUH data, of which the HYDE database is an essential part, have now been used by almost all ESMs producing simulations for the CMIP5 experiments and the IPCC AR5 process (Brovkin et al., 2013; Jones et al., 2011; Shevliakova et al., 2009). The LUH data have also been used and highlighted in several other studies, e.g. Thomson et al. (2010), Pereira et al. (2010) and Jones et al. (2013). These studies have made use of the detailed land use transition information in the LUH products to quantify land conversion events and track the resulting demographic effects of land disturbance and recovery, ultimately helping to close regional and global carbon budgets as well as impacts on future biodiversity.

Competing interests. The authors declare that they have no conflict of interest.

Acknowledgements. We thank the reviewers for their constructive comments. This research was performed with the support of NWO grant no. 016.158.021. This study was also undertaken as part of the Past Global Changes (PAGES) LandCover6K working group project, which in turn received support from the US National Science Foundation and the Swiss Academy of Sciences.

Edited by: David Carlson
Reviewed by: two anonymous referees

References

Ara, I., Lewis, M., and Ostendorf, B.: Spatio-temporal analysis of the impact of climate, cropping intensity and means of irrigation: an assessment on rice yield determinants in Bangladesh, Agriculture & Food Security, 5, 12, https://doi.org/10.1186/s40066-016-0061-9, 2016.

Armesto, J. J., Manuschevich, D., Mora, A., Smith-Ramirez, C., Rozzi, R., Abarzua, A. M., and Marquet, P. A.: From the Holocene to the Anthropocene: A historical framework for land cover change in southwestern South America in the past 15,000 years, Land Use Policy, 27, 148–160, 2009.

Australian Bureau of Statistics: 7124.0 – Historical Selected Agriculture Commodities, by State (1861 to Present), Australian Goverment, Canberra office, Canberra, Australia, 2001.

Barker, G.: The archaeology of foraging and farming at Niah Cave, Sarawak, Asian Perspect., 44, 90–106, 2006.

Barker, R., Herdt, R. W., and Rose, B.: The Rice enconomy of Asia, Resources for the Future, John Hopkins Press, Washington D.C., 1985.

Bellwood, P.: Early Agriculturalist Population Diasporas? Farming, Languages, and Genes, Annu. Rev. Anthropol., 30, 181–207, 2001.

Bertrand, C. and Van Ypersele, J. P.: Transient climate simulation forced by natural and anthropogenic climate forcings, Int. J. Climatol., 22, 623–648, 2002.

Betts, R. A.: Forcings and feedbacks by land ecosystem changes on climate change, J. Phys. IV, 139, 119–142, https://doi.org/10.1051/jp4:2006139009, 2006.

Betts, R. A., Falloon, P. D., Klein Goldewijk, K., and Ramankutty, N.: Biogeophysical effects of land use on climate: Model simulations of radiative forcing and large-scale temperature change, Agr. Forest Meteorol., 142, 216–233, 2007.

Biraben: Essai sur l'évolution du nombre des hommes, Population, French Edn., 34, 13–25, 1979.

Biraben, J.-N.: An Essay Concerning Mankind's Evolution, Population, Springer USA, 1980.

Bogue, D. J.: The Population of the United States, Historical Trends and Future Projections, The Free Press, New York, 1985.

Braudel, F.: The structure of everyday life: The limits of the possible, University of California Press, Berkely CA, 1981.

Brovkin, V., Claussen, M., Driesschaert, E., Fichefet, T., Kicklighter, D., Loutre, M. F., Matthews, H. D., Ramankutty, N., Schaeffer, M., and Sokolov, A.: Biogeophysical effects of historical land cover changes simulated by six Earth system models of intermediate complexity, Clim. Dynam., 26, 587–600, 2006.

Brovkin, V. L., Boysen, L., Arora, V. K., Boisier, J. P., Cadule, P., Chini, L., Claussen, M., Friedlingstein, P., Gayler, V., van den Hurk, B., Hurtt, G., Jones, C., Kato, E., de Noblet-Ducoudré, N., Pacifico, F., Pongratz, J., and Weiss, M.: Effect of anthropogenic land-use and land cover changes on climate and land carbon storage in CMIP5 projections for the 21st century, J. Climate, 26, 6859–6881, 2013.

Butzer, K. W.: Environment and archeology, an introduction to Pleistocene geography, Aldine Publishing Co., Chicago, 1964.

Caldwell, J. C. and Schindlmayr, T.: Historical population estimates: unraveling the Consensus, Popul. Dev. Rev., 28, 183–204, 2002.

China National Bureau of Statistics: http://www.stats.gov.cn/english, last access: 31 May 2008.

Clark, W. C.: Population geography, Prog. Hum. Geog., 1, 136–141, 1977.

Claussen, M., Brovkin, V., and Ganopolski, A.: Biophysical versus biogeochemical feedbacks of large-scale land cover change, Geophys. Res. Lett., 28, 1011–1014, 2001.

Coale, A. J.: Growth and Structure of Human Populations: A Mathematical Investigation, Princeton University Press, Princeton Legacy Library, New Jersey, USA, 1974.

Deevey, E. S.: The human population, Scientific American, CCIII, 195–204, 1960.

Denevan, W. M.: The native population of the Americas in 1492, University of Winsconsin Press, Madison, USA, 1992.

Durand, J. D.: Historical Estimates of World Population: An Evaluation, Popul. Dev. Rev., 3, 253–296, 1977.

Durand, J. D.: Historical Estimates of World Population: An Evaluation, PSC Analytical and Technical Reports Number 10, Table 2, University of Pennsylvania, Population Studies Center, Philadelphia, USA, 1974.

Ellis, E. C.: Ecology in an anthropogenic biosphere, Ecol. Monogr., 85, 287–331, 2015.

Ellis, E. C., Kaplan, J. O., Fuller, D. Q., Vavrus, S., Klein Goldewijk, K., and Verburg, P. H.: Used Planet: A Global History, P. Natl. Acad. Sci. USA, 110, 7978–7985, 2013.

Esser, G.: Osnabrück Biosphere Model: Structure, Construction, Results, in: Modern Ecology, Basic and Applied Aspects, edited by: Esser, G. and Overdieck, D., Elsevier Sci Publ., New York, 1991.

ESA: Three global LC maps for the 2000, 2005 and 2010 epochs, European Space Agency, available at: http://maps.elie.ucl.ac.be/CCI/viewer/index.php (last access: May 2017), 2016.

ETOPO5: 5-Minute Gridded Global Relief Data Collection (ETOPO5), NOAA-NGDC (Ed.), Boulder, Colorado, USA, 2005.

FAO: FAOSTAT, Food and Agriculture Organization of the United Nations, Rome, Italy, available at: http://www.fao.org, last access: June 2015.

Feddema, J. J., Oleson, K. W., Bonan, G. B., Mearns, L. O., Buja, L. E., Meehl, G. A., and Washington, W. M.: Atmospheric science: The importance of land-cover change in simulating future climates, Science, 310, 1674–1678, 2005.

Flint, E. P. and Richards, J. F.: Historical analysis of changes in land use and carbon stock of vegetation in South and Southeast Asia, Can. J. Forest Res., 21, 91–110, 1991.

Foley, J. A., DeFries, R., Asner, G. P., Barford, C., Bonan, G., Carpenter, S. R., Chapin, F. S., Coe, M. T., Daily, G. C., Gibbs, H. K., Helkowski, J. H., Holloway, T., Howard, E. A., Kucharik, C. J., Monfreda, C., Patz, J. A., Prentice, I. C., Ramankutty, N., and Snyder, P. K.: Global consequences of land use, Science, 309, 570–574, 2005.

Framji, K., Garg, B., and Lutrha, S.: Irrigation and drainage in the world – a global review, International Commission on Irrigation and Drainage, New Dehli, India, 1982.

Freydank, K. and Siebert, S.: Towards mapping the extent of irrigation in the last century: a time series of irrigated area per country, Universität Frankfurt am Main, Frankfurt, Germany, 2008.

Fritz, S., See, L., McCallum, I., You, L., Bun, A., Moltchanova, E., Duerauer, M., Albrecht, F., Schill, C., Perger, C., Havlik, P., Mosnier, A., Thornton, P., Wood-Sichra, U., Herrero, M., Becker-Reshef, I., Justice, C., Hansen, M., Gong, P., Abdel Aziz, S., Cipriani, A., Cumani, R., Cecchi, G., Conchedda, G., Ferreira, S., Gomez, A., Haffani, M., Kayitakire, F., Malanding, J., Mueller, R., Newby, T., Nonguierma, A., Olusegun, A., Ortner, S., Rajak, D. R., Rocha, J., Schepaschenko, D., Schepaschenko, M., Terekhov, A., Tiangwa, A., Vancutsem, C., Vintrou, E., Wenbin, W., van der Velde, M., Dunwoody, A., Kraxner, F., and Obersteiner, M.: Mapping global cropland and field size, Glob. Change Biol., 21, 1980–1992, 2015.

Frolking, S., Qiu, J., Boles, S., Xiao, X., Liu, J., Zhuang, Y., Li, C., and Qin, X.: Combining remote sensing and ground census data to develop new maps of the distribution of rice agriculture in China, Global Biogeochem. Cy., 16, 38–31, 2002.

Fuller, D. and Qin, L.: Pathways to Asian civilizations: Tracing the origins and spread of rice and rice cultures, Rice, 4, 78–92, 2011.

Fuller, D., Qin, Q., and Harvey, E.: A Critical Assessment of Early Agriculture in East Asia, with emphasis on Lower Yangtze Rice Domestication, Pradghara (Journal of the Uttar Pradesh State Archaeology Department), 18, 18–20, 2007.

Fuller, D., van Etten, J., Manning, K., Castillo, C., Kingwell-Banham, E., Weisskopf, A., Qin, L., Sato, Y.-I., and Hijmans, R. J.: The contribution of rice agriculture and livestock pastoralism to prehistoric methane levels: An archaeological assessment, The Holocene, 21, 743–759, 2011.

GAEZ: Joined Global Agro-Ecological Zones Project (Global-AEZ), FAO-IIASA, Laxenburg, Austria, available at: http://www.fao.org/ag/agl/agll/gaez/index.htm (last access: May 2015), 2000.

Gaillard, M.-J., Sugita, S., Mazier, F., Trondman, A.-K., Broström, A., Hickler, T., Kaplan, J. O., Kjellström, E., Kokfelt, U., Kuneš, P., Lemmen, C., Miller, P., Olofsson, J., Poska, A., Rundgren, M., Smith, B., Strandberg, G., Fyfe, R., Nielsen, A. B., Alenius, T., Balakauskas, L., Barnekow, L., Birks, H. J. B., Bjune, A., Björkman, L., Giesecke, T., Hjelle, K., Kalnina, L., Kangur, M., van der Knaap, W. O., Koff, T., Lagerås, P., Latalowa, M., Leydet, M., Lechterbeck, J., Lindbladh, M., Odgaard, B., Peglar, S., Segerström, U., von Stedingk, H., and Seppä, H.: Holocene land-cover reconstructions for studies on land cover-climate feedbacks, Clim. Past, 6, 483–499, https://doi.org/10.5194/cp-6-483-2010, 2010.

Ge, Q. S., Dai, J. H., He, F. N., Pan, Y., and Wang, M. M.: Land use changes and their relations with carbon cycles over the past 300 a in China, Sci. China Ser. D, 51, 871–884, 2008.

Grigg, D. B. (Ed.): The industrial revolution and land transformation, John Wiley, New York, 1987.

GYGA: Global Yield Gap and Water Productivity Atlas, Wageningen University, University of Nebraska, Water for Food, Daugherty Global Institute, Wageningen, 2017.

Hassan, F. A.: Demographic Archaeology: Studies in Archaeology, Academic Press, London, 1981.

Haub, C.: How many people have ever lived on Earth?, Population Today, 1995. 1995.

He, F., Li, S., and Zhang, X.: Reconstruction of cropland area and spatial distribution in the mid-Northern Song Dynasty (AD1004–1085), J. Geogr. Sci., 22, 359–370, 2012.

He, F., Li, S., Zhang, X., Ge, Q., and Dai, J.: Comparisons of cropland area from multiple datasets over the past 300 years in the traditional cultivated region of China, J. Geogr. Sci., 23, 978–990, 2013.

Hollman, R., Merchant, C. J., Saunders, R., Downy, C., Buchwitz, M., Cazenave, A., Chuvieco, E., Defourny, P., de Leeuw, G., Forsberg, R., Holzer-Popp, T., Paul, F., Sandven, S., Sathyendranath, S., van Roozendael, M., and Wagner, W.: The ESA climate change initiative, B. Am. Meteorol. Soc., 94, 1541–1552, https://doi.org/10.1175/BAMS-D-11-00254.1, 2013.

Houghton, R. A.: The annual net flux of carbon to the atmosphere from changes in land use 1850–1990, Tellus B, 51, 298–313, 1999.

Houghton, R. A. and Hackler, J. L.: Carbon Flux to the Atmosphere from Land-Use Changes: 1850 to 1990, Carbon Dioxide Information Analysis Center, U.S. Department of Energy, Oak Ridge National Laboratory, Oak Ridge, Tennessee, USA, 86 pp., 2002.

Houghton, R. A., Hobbie, J. E., Melillo, J. M., Moore, B., Peterson, B. J., Shaver, G. R., and Woodwell, G. M.: Changes in the Carbon Content of Terrestrial Biota and Soils between 1860 and 1980: A Net Release of CO_2 to the Atmosphere, Ecol. Monogr., 53, 236–262, 1983.

Hurtt, G. C., Chini, L. P., Frolking, S., Betts, R., Feddema, J. J., Fischer, G., Hibbard, K. A., Janetos, A. C., Jones, C., Klein Goldewijk, K., Kindermann, G., Kinoshita, T., Riahi, K., Shevliakova, E., Smith, S., Stehfest, E., Thomson, A., Thornton, P., van Vuuren, D., and Wang, Y. P.: Harmonization of land-use scenarios for the period 1500–2100: 600 years of global gridded annual land-use transitions, wood harvest, and resulting secondary lands, Climatic Change, 109, 117–161, 2011.

Indiastat: Statistics India, Ministry of Agriculture, Government of India, Datanet India Pvt. Ltd. D-100, First Floor, Okhla Industrial Area, Phase-I, New Delhi, 2009.

Jones, C. D., Hughes, J. K., Bellouin, N., Hardiman, S. C., Jones, G. S., Knight, J., Liddicoat, S., O'Connor, F. M., Andres, R. J., Bell, C., Boo, K.-O., Bozzo, A., Butchart, N., Cadule, P., Corbin, K. D., Doutriaux-Boucher, M., Friedlingstein, P., Gornall, J., Gray, L., Halloran, P. R., Hurtt, G., Ingram, W. J., Lamarque, J.-F., Law, R. M., Meinshausen, M., Osprey, S., Palin, E. J., Parsons Chini, L., Raddatz, T., Sanderson, M. G., Sellar, A. A., Schurer, A., Valdes, P., Wood, N., Woodward, S., Yoshioka, M., and Zerroukat, M.: The HadGEM2-ES implementation of CMIP5 centennial simulations, Geosci. Model Dev., 4, 543–570, https://doi.org/10.5194/gmd-4-543-2011, 2011.

Jones, A. D., W. Collins, W., Edmonds, J., Torn, M., Janetos, A., Calvin, K., Thomson, A., Chini, L., Mao, J., Shi, X., Thornton, P., Hurtt, G., and Wise, M.: Greenhouse gas policy influences climate via direct effects of land-use change, J. Climate, 26, 3657–3670, https://doi.org/10.1175/JCLI-D-12-00377.1, 2013.

Kalaiselvi, S. and Sundar, I.: Interstate Disparity in Cropping Intensity in India, International Journal of Business Management, Economics and Information Technology, 3, 269–273, 2011.

Kapitsa, S.: A model of world population growth as an experiment in systematic research (Model' rosta naseleniya zemli kak opyt sistemnogo issledovaniya), Voprosy Statistiki, 8, 46–57, 1997.

Kaplan, J. O., Krumhardt, K. M., Ellis, E. C., Ruddiman, W. F., Lemmen, C., and Klein Goldewijk, K.: Holocene carbon emissions as a result of anthropogenic land cover change, The Holocene, 21, 775–791, 2011.

Kaplan, J. O., Krunhardt, K. M., and Zimmermann, N.: The prehistoric and preindustrial deforestation of Europe, Quaternary Sci. Rev., 28, 3016–3034, 2009.

Klein Goldewijk, K.: Estimating global land use change over the past 300 years: The HYDE database, Global Biogeochem. Cy., 15, 417–433, 2001.

Klein Goldewijk, K. and Battjes, J. J.: The IMAGE Hundred Year (1890–1990) Database for Integrated Environmental Assessments (HYDE Version 1.0), National Institute of Public Health and the Environment (RIVM), Bilthoven, Report 482523001, 170 pp., 1995.

Klein Goldewijk, K. and Verburg, P. H.: Uncertainties in global-scale reconstructions of historical land use: an illustration using the HYDE data set, Landscape Ecol., 28, 861–877, 2013.

Klein Goldewijk, K., Beusen, A., and Janssen, P.: Long term dynamic modeling of global population and built-up area in a spatially explicit way: HYDE 3.1, The Holocene, 20, 565–573, 2010.

Klein Goldewijk, K., Beusen, A., van Drecht, G., and de Vos, M.: The HYDE 3.1 spatially explicit database of human induced land use change over the past 12,000 years, Global Ecol. Biogeogr., 20, 73—86, 2011.

Klein Goldewijk, K., Dekker, S. C., and van Zanden, J. L.: Per-capita estimations of long-term historical land use and the consequences for global change research, Journal of Land Use Science, 12, 313–337, https://doi.org/10.1080/1747423X.2017.1354938, 2017.

Kropelin, S., Verschuren, D., Lezine, A. M., Eggermont, H., Cocquyt, C., Francus, P., Cazet, J. P., Fagot, M., Rumes, B., Russell, J. M., Darius, F., Conley, D. J., Schuster, M., Von Suchodoletz, H., and Engstrom, D. R.: Climate-driven ecosystem succession in the Sahara: The past 6000 years, Science, 320, 765–768, 2008.

LandScan: Landscan Global Population Database, The 2012 Revision, Oak Ridge National Laboratory, Oak Ridge, Tennessee, available at: http://www.ornl.gov/landscan (last access: January 2015), 2014.

Lawrence, D. M., Hurtt, G. C., Arneth, A., Brovkin, V., Calvin, K. V., Jones, A. D., Jones, C. D., Lawrence, P. J., de Noblet-Ducoudré, N., Pongratz, J., Seneviratne, S. I., and Shevliakova, E.: The Land Use Model Intercomparison Project (LUMIP) contribution to CMIP6: rationale and experimental design, Geosci. Model Dev., 9, 2973–2998, https://doi.org/10.5194/gmd-9-2973-2016, 2016.

Le Quéré, C., Moriarty, R., Andrew, R. M., Peters, G. P., Ciais, P., Friedlingstein, P., Jones, S. D., Sitch, S., Tans, P., Arneth, A., Boden, T. A., Bopp, L., Bozec, Y., Canadell, J. G., Chini, L. P., Chevallier, F., Cosca, C. E., Harris, I., Hoppema, M., Houghton, R. A., House, J. I., Jain, A. K., Johannessen, T., Kato, E., Keeling, R. F., Kitidis, V., Klein Goldewijk, K., Koven, C., Landa, C. S., Landschützer, P., Lenton, A., Lima, I. D., Marland, G., Mathis, J. T., Metzl, N., Nojiri, Y., Olsen, A., Ono, T., Peng, S., Peters, W., Pfeil, B., Poulter, B., Raupach, M. R., Regnier, P., Rödenbeck, C., Saito, S., Salisbury, J. E., Schuster, U., Schwinger, J., Séférian, R., Segschneider, J., Steinhoff, T., Stocker, B. D., Sutton, A. J., Takahashi, T., Tilbrook, B., van der Werf, G. R., Viovy, N., Wang, Y.-P., Wanninkhof, R., Wiltshire, A., and Zeng, N.: Global carbon budget 2014, Earth Syst. Sci. Data, 7, 47–85, https://doi.org/10.5194/essd-7-47-2015, 2015.

Li, Y., Thenkabail, P. S., Biradar, C. M., Noojipady, P., Dheeravath, V., Velpuri, M., Gangalakunta, O., and Cai, X. L.: History of irrigated areas in the world, in: Global croplands for food security, edited by: Thenkabail, P. S., Lyon, J. G., Turral, H., and Biradar, C. M., CRC Press, Taylor and Francis Group, Boca Raton, FL, USA, 2009.

Livi-Bacci, M.: A concise history of world population, 4th Edn., Blackwell Publishing, Oxford, UK, 2007.

Loveland, T. R. and Belward, A. S.: IGBP-DIS global 1 km land cover data set, DISCover: First results, Int. J. Remote Sens., 18, 3289–3295, 1997.

Loveland, T. R., Reed, B. C., Brown, J. F., Ohlen, D. O., Zhu, Z., Yang, L., and Merchant, J. W.: Development of a global land cover characteristics database and IGBP DISCover from 1 km AVHRR data, Int. J. Remote Sens., 21, 1303–1330, 2000.

Lucas, D.: Chapter three, World Population Growth, The Australian National University,Canberra, New South Wales, Australia, 2003.

Maddison, A.: The world economy: a millennial perspective, OECD, Paris, France, 2001.

McEvedy, C. and Jones, R.: World Atlas of population history, Penguin Books Ltd., Hammondsworth, UK, 1978.

Merrick, T. W.: World population in transition, Population Bulletin, 41, 53 pp., 1986.

Michael, A. M.: Irrigation: Theory and Practice, Vikas Publishing House PVT ltd, New Delhi, 2008.

Mitchell, B. R.: International Historical Statistics, Africa, Asia & Oceania: 1750–2005, 5th Edn., Palcrave MacMillan, New York, USA, 2007a.

Mitchell, B. R.: International Historical Statistics, Europe: 1750–2005, 6th Edn., Palcrave MacMillan, New York, USA, 2007b.

Mitchell, B. R.: International Historical Statistics, The Americas: 1750–2005, 6th Edn., Palcrave MacMillan, New York, USA, 2007c.

Morrison, K.: Provincializing the Anthropocene, Seminar, 673, 75–80, 2015.

Natural Earth: ne_10m_rivers_lake_centerlines, edited by: Natural Earth, http://www.naturalearthdata.com, last access: January 2015.

New, M., Hulme, M., and Jones, P.: A 1961–1990 mean monthly climatology of global land areas. Climate Research Unit, edited by: University of East Anglia, Norwich, 1997.

NLWRA: National Land and Water Resources Audit 2001, Australian Natural Resources Atlas, V 2.0., Land, Land Use Australia, 2001.

Olofsson, J. and Hickler, T.: Effects of human land-use on the global carbon cycle during the last 6,000 years, Veg. Hist. Archaeobot., 17, 605–615, 2008.

Olson, D. M., Dinerstein, E., Wikramanayake, E. D., Burgess, N. D., Powell, G. V. N., Underwood, E. C., D'Amico, J. A., Itoua, I., Strand, H. E., Morrison, J. C., Loucks, C. J., Allnutt, T. F., Ricketts, T. H., Kura, Y., Lamoreux, J. F., Wettengel, W. W., Hedao, P., and Kassem, K. R.: Terrestrial Ecoregions of the World: A New Map of Life on Earth, BioScience, 51, 933–938, 2001.

Opferkuch, K.: Quantifying Historical Rice Patterns. An evaluation and development of the current HYDE total rice agricultural area maps for the Southeast Asian region, Utrecht University, Utrecht, 2016.

Pacala, S. W., Hurtt, G. C., Baker, D., Peylin, P., Houghton, R. A., Birdsey, R. A., Heath, L., Sundquist, E. T., Stallard, R. F., Ciais, P., Moorcroft, P., Caspersen, J. P., Shevliakova, E., Moore, B., Kohlmaier, G., Holland, E., Gloor, H., Harmon, M. E., Fan, S. M., Sarmiento, J. L., Goodale, C. L., Schimel, D., and Field, C. B.: Consistent land- and atmosphere-based U.S. carbon sink estimates, Science, 292, 2316–2320, 2001.

Pereira, H. M., Leadley, P. W., Proença, V., Alkemade, R., Scharlemann, J. P. W., Fernandez-Manjarrés, J. F., Araújo, M. B., Balvanera, P., Biggs, R., Cheung, W. W. L., Chini, L., Cooper, H. D., Gilman, E. L., Guénette, S., Hurtt, G. C., Huntington, H. P., Mace, G. M., Oberdorff, T., Revenga, C., Rodrigues, P., Scholes, R. J., Sumaila, U. R., and Walpole, M.: Scenarios for Global Biodiversity in the 21st Century, Science, 2010. 1496–1501, 2010.

Pongratz, J., Reick, C., Raddatz, T., and Claussen, M.: A reconstruction of global agricultural areas and land cover for the last millennium, Global Biogeochem. Cy., 22, GB3018, https://doi.org/10.1029/2007GB003153, 2008.

Pongratz, J., Raddatz, T., Reick, C. H., Esch, M., and Claussen, M.: Radiative forcing from anthropogenic land cover change since A.D. 800, Geophys. Res. Lett., 36, L02709, https://doi.org/10.1029/2008GL036394, 2009.

Poulter, B., MacBean, N., Hartley, A., Khlystova, I., Arino, O., Betts, R., Bontemps, S., Boettcher, M., Brockmann, C., Defourny, P., Hagemann, S., Herold, M., Kirches, G., Lamarche, C., Lederer, D., Ottlé, C., Peters, M., and Peylin, P.: Plant functional type classification for earth system models: results from the European Space Agency's Land Cover Climate Change Initiative, Geosci. Model Dev., 8, 2315–2328, https://doi.org/10.5194/gmd-8-2315-2015, 2015.

Prentice, I. C., Cramer, W., Harrison, S. P., Leemans, R., Monserud, R. A., and Solomon, A. M.: A global biome model based on plant physiology and dominance, soil properties and climate, J. Biogeogr., 19, 117–134, 1992.

Prestele, R., Alexander, P., Rounsevell, M. D. A., Arneth, A., Calvin, K., Doelman, J., Eitelberg, D. A., Engström, K., Fujimori, S., Hasegawa, T., Havlik, P., Humpenöder, F., Jain, A. K., Krisztin, T., Kyle, P., Meiyappan, P., Popp, A., Sands, R. D., Schaldach, R., Schüngel, J., Stehfest, E., Tabeau, A., Van Meijl, H., Van Vliet, J., and Verburg, P. H.: Hotspots of uncertainty in land-use and land-cover change projections: a global-scale model comparison, Glob. Change Biol., 22, 3967–3983, https://doi.org/10.1111/gcb.13337, 2016.

Ramankutty, N.: Global Cropland and Pasture Data from 1700–2007, McGill University (Ed.), Montreal, Canada, 2012.

Ramankutty, N. and Foley, J. A.: Estimating historical changes in global land cover: Croplands from 1700 to 1992, Global Biogeochem. Cy., 13, 997–1027, 1999.

Reick, C. H., Raddatz, T., Brovkin, V., and Gayler, V.: Representation of natural and anthropogenic land cover change in MPI-ESM, Journal of Advances in Modelling Earth Systems, 5, 459–482, 2013.

Renfrew, C. and Bahn, P.: Archaeology: Theories, Methods & Practices, Thames & Hudson Ltd., 2008.

Richards, J. F.: Land transformation, Chapter 10, in: The Earth as transformed by human action, Turner, B. L., Clark, W. C., Kates, R. W., Richards, J. F., Mathews, J. T., Meyer, W. B., Cambridge University Press, New York, 1990.

Ricepedia: http://ricepedia.org/rice-around-the-world, last access: February 2015.

Siebert, S.: Technical Report documentation irrigation 17900–1900, University of Bonn, Bonn, 2008.

Siebert, S., Kummu, M., Porkka, M., Döll, P., Ramankutty, N., and Scanlon, B. R.: A global data set of the extent of irrigated land from 1900 to 2005, Hydrol. Earth Syst. Sci., 19, 1521–1545, https://doi.org/10.5194/hess-19-1521-2015, 2015.

Stern, D. I. and Kaufmann, R. K.: Estimates of global anthropogenic sulphate emissions 1860–1993, Center for Energy and Environmental Studies, Boston, Mass, 1996.

Sojka, R. E., Bjorneberg, D. L., and Entry, J. A.: Irrigation: an historical perspective, in: Encyclopedia of Soil Science, Marcel Dekker Inc., 2002. Tanton, J. H.: End of the migration epoch, The Social Contract, IV and V, 1995.

Tett, S., Betts, R., Crowley, T. J., Jones, A., Gregory, J., Oestroem, E., Roberts, D. L., and Woodage, M. J.: Simulating the recent holocene, American Meteorological Society, San Diego, CA, 2021–2029, 2005.

Thomlinson, R.: Demographic problems, controversy over population control, 2nd Edn., Dickenson Publishing Company, CA, 1975.

Tuve, G. L.: Energy, environment, population, and food, Wiley-Interscience, New York, 1976.

UN: Long-range world population projections: Two centuries of population growth, 1950–2150, United Nations Population Division, New York, 1992.

UN: World Population Prospects, The 2003 Revision, United Nations Population Division, New York, 2004.

UN: World Population Prospects, The 2007 Revision, United Nations Population Division, New York, 2008.

UNEP/WCMC: World Database on Protected Areas (WDPA). Protected Areas Programme, Cambridge, UK, 2013.

Urquhart, M. and Buckley, K.: Historical statistics of Canada, University Press, Toronto, Canada, 1965.

USDA: National Agricultural Statistics Service (NASS), http://www.nass.usda.gov (last access: June 2009), 2006.

Van Beek, L. H. P. and Bierkens, M. F. P.: The global hydrological model PCR-GLOBWB: conceptualization, parameterization and verification, Department of Physical Geography, Faculty of Earth Sciences, Utrecht University, Utrecht, 2008.

van Loon, H.: The story of mankind, New York, Liveright, 1922.

Vazquez-Presedo, V.: Estadisticas Historicas Argentinas (Comparadas), Buenos Aires Acad. Nacional de Ciencias Econoìmicas, Buenos Aires, Argentina, 1988.

Verburg, P. H., van de Steeg, J., Veldkamp, A., and Willemen, L.: From land cover change to land function dynamics: A major challenge to improve land characterization, J. Environ. Manage., 90, 1327–1335, 2009.

Verschuren, D., Lalrd, K. R., and Cumming, B. F.: Rainfall and drought in equatorial east Africa during the past 1,100 years, Nature, 403, 410–414, 2000.

Verschuren, D., Johnson, T. C., Kling, H. J., Edgington, D. N., Leavitt, P. R., Brown, E. T., Talbot, M. R., and Hecky, R. E.: History and timing of human impact on Lake Victoria, East Africa, P. Roy. Soc. B-Biol. Sci., 269, 289–294, 2002.

Visser, H. and de Nijs, T.: The Map Comparison Kit, Environ. Modell. Softw., 21, 346–358, 2006.

Waldner, F., Fritz, S., Di Gregorio, A., Plotnikov, D., Bartalev, S., Kussul, N., Gong, P., Thenkabil, P., Hazeu, G., Klein , I., Loew, F., Miettinen, J., Dadhwal, V. K., Lamarche, C., Bontemps, S., and Defourny, P.: A Unified Cropland Layer at 250 m for Global Agriculture Monitoring, Data, 1, 3, https://doi.org/10.3390/data1010003, 2016.

Wiens, T.: Agriculture and Rural Poverty in Vietnam, in: Household Wellfare and Vietnam's Transition, edited by: Dollar, D., Glewwe, P., and Litvack, J. I., World Bank, Washington, USA, 1998.

Williams, M.: Dark ages and dark areas: Global deforestation in the deep past, J. Hist. Geogr., 26, 28–46, 2000.

World Atlas of Agriculture: World Atlas of Agriculture, Istituto Geografico De Agostini S.p.A., Novara, Italy, 1969.

Woytinsky, W. S. and Woytinsky, E. S.: World Population and Production, The Lord Baltimore Press, New York, 1953.

You, L., Wood-Sichra, U., Fritz, S., Guo, Z., See, L., and Koo, J.: Spatial Production Allocation Model (SPAM) 2005 v2.0, available at: http://mapspam.info (last access: May 2015), 2014.

Zhang, J. and Wang, X.: Notes on the recent discovery of ancient cultivated rice at Jiahu, Henan Province: a new theory concerning the origin of Oryza japonica in China, Antiquity, 72, 897–901, 1998.

7

Freshwater fish fauna of rivers of the southern Western Ghats

Anbu Aravazhi Arunkumar[1] **and Arunachalam Manimekalan**[2]

[1]Department of Biotechnology, Karpagam Academy of Higher Education, Coimbatore 641021, Tamil Nadu, India

[2]Biodiversity and Molecular Lab, Department of Environmental Sciences, Bharathiar University, Coimbatore 641046, Tamil Nadu, India

Correspondence: Anbu Aravazhi Arunkumar (anbu.arunkumar@gmail.com)

Abstract. The Western Ghats of India is a UNESCO World Heritage Site and is one of the eight "hotspots" of biological diversity in the world. It is also referred to as the "Great Escarpment of India". This paper provides information on the diversity of freshwater fish fauna of six river systems of the southern Western Ghats. The study area has been identified geographically using a GPS, and the respective topographic map has been digitized using ArcGIS software. The fish fauna were collected from various streams and rivers using cast nets, dip nets, gill nets and drag nets. Among the 31 georeferenced sites sampled from the rivers of the southern Western Ghats, a total of 64 species, belonging to 6 orders, 14 families and 31 genera, were recorded. Among them, the order Cypriniformes was dominant, with 3 families, 18 genera and 49 species (76.6 %). Principal component analysis and cluster analysis were performed to express the contribution of the variables and their influence on the species diversity. Interestingly, of the 31 sites, Thunakadavu stream, Gulithuraipatti, Athirapally, Naduthotam, Nadathittu, Mullaithodu, Thonanthikla, Noolpuzha and Sinnaru exhibited high variation in species diversity. Nearly 15 species were found to be threatened in the Western Ghats. *Garra periyarensis* and *Cirrhinus cirrhosus* are known to be vulnerable and *Hemibagrus punctatus* is critically endangered because of various anthropogenic activities. The study clearly indicates that certain timely measures have to be taken immediately to protect the fish fauna in the southern Western Ghats (https://doi.org/10.1594/PANGAEA.882214).

1 Introduction

The Western Ghats of India is a UNESCO World Heritage Site and is one of the eight "hotspots" of biological diversity in the world. It is also known as the "Great Escarpment of India" (Myers et al., 2000). The range of the Western Ghats runs from north to south along the western edge of the Deccan Plateau, and separates the plateau from a narrow coastal plain, called Konkan, along the Arabian Sea (Fig. 1, Bhavani river system). Of a total of 39 World Heritage Sites in India, which include national parks, wildlife sanctuaries and reserve forests, 20 are in Kerala, 10 in Karnataka, 5 in Tamil Nadu and 4 in Maharashtra; these have to be protected and conserved. The Western Ghats of India has rich freshwater fish fauna with a high level of endemism (Dahanukar et al., 2004). At present, a major part of the faunal diversity of the Western Ghats is threatened by human activities and invasive species (Dahanukar et al., 2004). Hence, knowledge of the diversity and distribution of the fish fauna is essential for designing and implementing conservation strategies. However, data on the fish fauna of the Western Ghats have limitations as most of the rivers have not been surveyed extensively with georeferenced data, and checklists for individual rivers are not available.

The history of Indian freshwater fish fauna goes back to Hamilton (1822), who studied fish fauna found in the river Ganges and its tributaries. Documentation and listing of fish fauna from different parts of India was mainly carried out by Jerdon (1848). A comprehensive and authori-

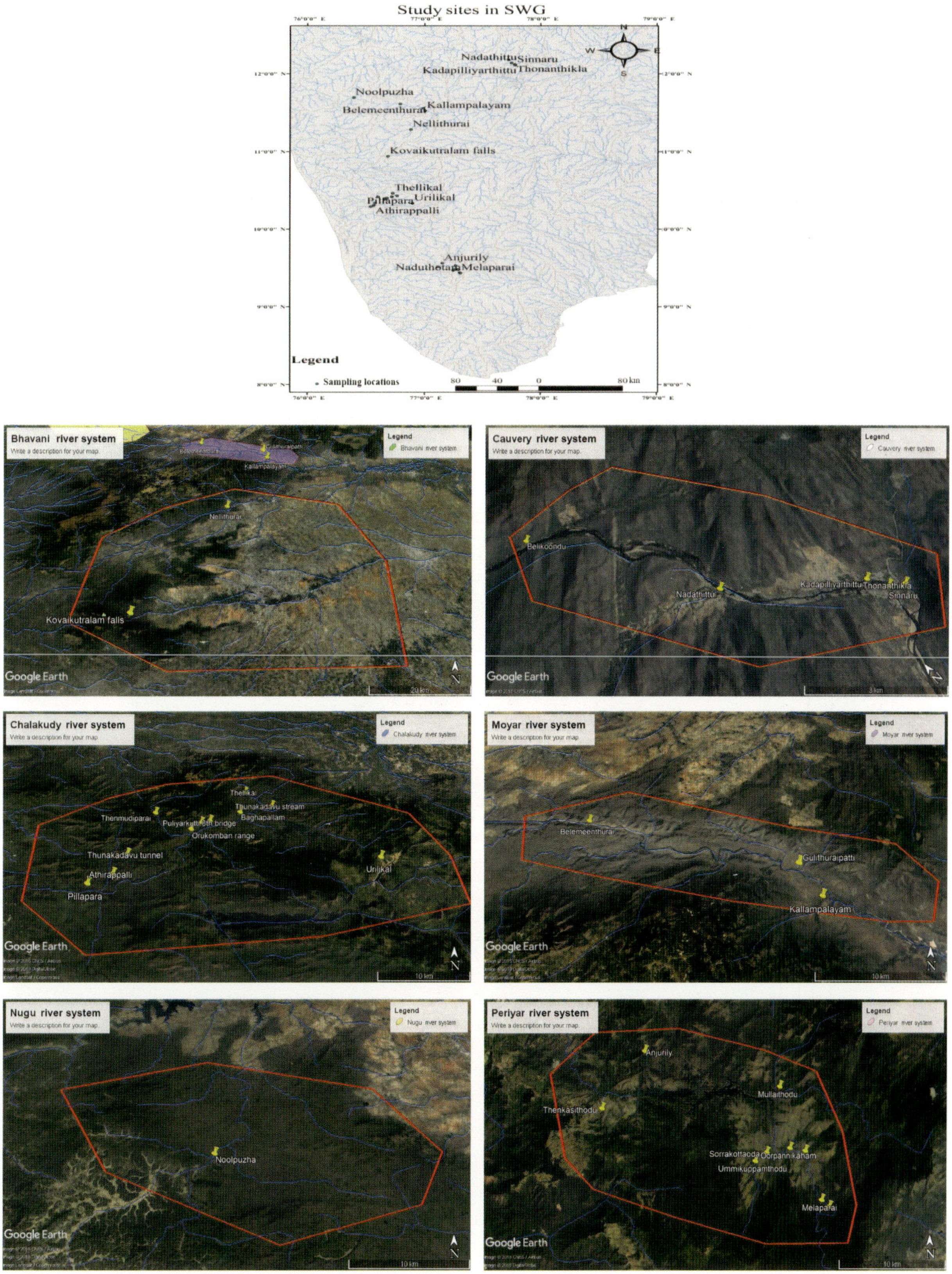

Figure 1. Location of collection sites in six river systems.

tative account on the freshwater fish fauna has been provided by Day (1865, 1871, 1878). Further investigations on the freshwater fish fauna of India, especially in the Western Ghats, were initiated by Hora (1921, 1937, 1941, 1949) and Hora and Misra (1938), and Hora also enunciated the Satpura Hypothesis. Silas (1951) listed 25 fish species from the Anamalai Hills and 10 species from the Nelliampathi Hills. His study extended the distribution of several species earlier known only from the central division of the Western Ghats to the southern division beyond the Palghat Gap. These led to new descriptions, initiating elaborate discussions on the endemism and other zoogeographical relevance of the area, and several new taxa were added from Kerala during this period. Studies on the endemic fish fauna from various streams and rivers in the Western Ghats mountain range have been compiled. Fish diversity in selected streams in northern Karnataka (Arunachalam et al., 1997) and the central Western Ghats (Arunachalam et al., 2000) has been reported. Manimekalan and Das (1998) described a new species, *Glyptothorax davissinghi*(Pisces: *Sisoridae*), a new catfish from Nilambur in the Nilgiri Biosphere, South India. Manimekalan and Singh (1997) made a new record of *Schismatorhynchus (Nukta) nukta* (Sykes) (Pisces: Cyprinidae) from Moyar River. Arunachalam et al. (2005) reported a new fish species, *Neolissochilus wynaadensis*, from the Karnataka part of the Western Ghats. Arunachalam et al. (2007) reported *Psilorhynchus amplicephalus*, a new species from Balishwar River, Assam, India. Earlier Biju et al. (1996) recorded *Puntius filamentosus* (Val.) and *Puntius melanampyx* (Day) in Orukomban and Thelikal during a survey from December 1996 to May 1997. Manimekalan and Arunachalam (2002) rediscovered the critically endangered air-breathing catfish *Clarias dayi* Hora (Pisces: *Claridae)* in Mudumalai Wildlife Sanctuary. Recently Arunkumar et al. (2015) reported nearly 37 species in the Cauvery river system. Considering the above facts, the present work documents freshwater fish fauna of the southern Western Ghats, especially from the Bhavani river system, Moyar river system, Chalakudy river system, Periyar river system, Cauvery river system and Nugu river system.

2 Methodology

2.1 Collection and identification

The study area has been identified geographically (the latitude and longitude coordinates) using a GPS, and the respective topographic map has been georeferenced and digitized using ArcGIS 9.3 version software. Fish fauna were collected using cast nets, dip nets, drag nets and gill nets of different mesh sizes ranging from 8 to 22 mm. Proper care was taken to avoid damage to the specimens during collection. A total of five specimens from each species were collected from the georeferenced sampling locations, and fish fauna were photographed before being preserved in formalin so that they could be photographed in their original colour. Further, for smaller samples, specimens were preserved in 10 % formalin, and for larger samples, formalin was injected into the abdominal cavity so that the internal organs could be preserved well for further taxonomic studies. The specimens were tagged, reference numbers were given for specimen identification, and they were transported to the lab. The species were identified based on the key given by Talwar and Jhingran (1991), Menon and Devi (1992) and Jayaram (1999, 2010). The holotype and paratype species were examined at the Zoological Survey of India (Chennai and Kolkata) for species confirmation. Voucher specimens were made for each species and were deposited at the Biodiversity and Molecular Lab, Dept. of Environmental Sciences, Bharathiar University, Coimbatore, Tamil Nadu, India.

2.2 Physico-chemical analysis of the water quality at sampling sites

Water samples were collected from all the sampling stations during the post-monsoon period (February to June) at a depth of 10 cm. Water quality analyses such as pH, conductivity, turbidity, total dissolved solids (TDS), resistivity, salinity, dissolved oxygen (DO), and water temperature were done as per the regulations of APHA (1995). Field analysis of the samples was done using a portable water analyzer (X tech, Nagman Instruments Electronics, India) (Gurumurthy and Tripti, 2015; Thomas et al., 2015; Anushiya and Ramachandran, 2015).

2.3 Analysis

To quantify species diversity, a number of indices were applied. To measure the species diversity (H') the most widely used Shannon index (Shannon and Weaver, 1949), Pielou's evenness index (E) (Pielou, 1975) and Simpson's dominance index (D) (Simpson, 1949) were used. Similarity coefficients of the fish community were calculated using the Jaccard index (Southwood, 1978). The species abundance and their relative frequencies were subjected to cluster analysis; a complete linkage cluster dendrogram was drawn based on the Pearson correlation. The contribution of the variables and their influence on the species diversity has been analysed using principal component analysis (Wills, 2005). The above statistical analyses were performed using SPSS (version 21), XLSTAT and Biodiversity Pro software.

3 Results and discussion

The fish fauna were surveyed from the streams and rivers of the southern Western Ghats. The collection sites were selected based on the earlier faunal distribution published in literature. The Western Ghats is a mountain range that runs almost parallel to the western coast of the Indian peninsula.

Table 1. Study site and their habitat characteristics.

Site number	Study site	Latitude	Longitude	Altitude	Forest type in tropical region	Stream order	Stream width (m)	Stream depth (m)	Area (m^2)	Volume (m^3)	Mean velocity* ($m\,s^{-1}$)
		WGS84 (Datum)									
Moyar river system											
1	Gulithuraipatti	11°33′20.4984″ N	76°59′16.1016″ E	312	Thorn forest	4	10	6	1000	6000	4
2	Kallampalayam	11°31′25.3992″ N	77°0′16.4016″ E	300	Thorn forest	4	13	8	1300	10 400	4
3	Belemeenthurai	11°36′39.9996″ N	76°47′38.0004″ E	520	Dry deciduous	4	19	1.75	1900	3325	4
Chalakudy river system											
4	Orukomban range	10°22′53.6016″ N	76°39′21.3984″ E	450	Dry deciduous	4	6	0.5	600	300	3
5	Thenmudiparai	10°24′51.9984″ N	76°36′10.5012″ E	510	Dry deciduous	5	25	1.5	2500	3750	3
6	Baghapallam	10°24′57.6″ N	76°43′21.3996″ E	748	Dry deciduous	5	8	0.5	800	400	3
7	Thellikal	10°27′34.2″ N	76°43′48.7992″ E	840	Dry deciduous	4	4	1.0	400	400	3
8	Puliyarkutti 8th bridge	10°23′39.6996″ N	76°40′8.1984″ E	527	Dry deciduous	4	19.2	1.2	1920	2304	3
9	Puliyarkutti 3rd bridge	10°23′52.5984″ N	76°40′51.3012″ E	512	Dry deciduous	4	37	1.5	3700	5550	3
10	Thunakadavu stream	10°25′44.1012″ N	76°46′4.6992″ E	510	Dry deciduous	4	13.6	0.5	1360	680	3
11	Thunakadavu tunnel	10°20′9.3012″ N	76°34′40.6992″ E	520	Dry deciduous	5	15	10	1500	15 000	5
12	Urilikal	10°19′54.1992″ N	76°53′57.3″ E	3238	Dry deciduous	2	7	1.5	700	1050	2
13	Athirappilly	10°18′15.3598″ N	76°34′0.0012″ E	202	Semi-evergreen	4	8	3	800	2400	4
14	Pillapara	10°17′23.82″ N	76°32′21.84″ E	267	Semi-evergreen	4	5	2	500	1000	4
Bhavani river system											
15	Kovaikutralam falls	10°56′20.1516″ N	76°41′21.0084″ E	560	Semi-evergreen	2	5	1.2	500	600	4
16	Nellithurai	11°17′0.3012″ N	76°53′6.9″ E	380	Thorn forest	4	27	1.1	2700	2970	5
Periyar river system											
17	Oorpannikaham	9°28′58.1016″ N	77°16′47.7012″ E	884	Evergreen	4	12	2.1	1200	2520	2
18	Valukuparai	9°28′49.4004″ N	77°17′35.0988″ E	869	Evergreen	4	7.5	0.3	750	225	3
19	Melaparai	9°26′24.7992″ N	77°18′24.5988″ E	965	Evergreen	4	11	4.2	1100	4620	3
20	Naduthotam	9°26′5.1″ N	77°18′48.0996″ E	950	Evergreen	4	7.5	0.3	750	225	3
21	Ummikuppamthodu	9°28′20.6004″ N	77°14′57.0984″ E	943	Evergreen	4	5	3.0	500	1500	4
22	Sorrakottaodai	9°28′45.4008″ N	77°15′32.7996″ E	879	Evergreen	4	7	1.5	700	1050	3
23	Mullaithodu	9°31′58.6992″ N	77°16′15.8016″ E	869	Evergreen	4	10	0.6	1000	600	3
24	Anjurily	9°33′46.1988″ N	77°9′19.6992″ E	912	Evergreen	4	20	5	2000	10 000	2
25	Thenkasithodu	9°30′59.4″ N	77°7′5.9988″ E	872	Evergreen	4	11.3	0.5	1130	565	2
Cauvery river system											
26	Kadapilliyarthittu	12°7′18.1992″ N	77°46′28.3008″ E	1137	Dry deciduous	4	75	1.5	7500	11 250	2
27	Belikoondu	12°11′2.1012″ N	77°43′12.6012″ E	267	Dry deciduous	4	80	10	8000	80 000	5
28	Nadathittu	12°8′31.9992″ N	77°44′48.9984″ E	262	Dry deciduous	4	70	6	7000	42 000	3
29	Sinnaru	12°6′54.7992″ N	77°46′48.5004″ E	225	Dry deciduous	4	55	0.5	5500	2750	3
30	Thonanthikla	12°7′2.3988″ N	77°46′36.6996″ E	341	Dry deciduous	4	25	1	2500	2500	4
Nugu river system											
31	Noolpuzha	11°41′35.0988″ N	76°23′36.3984″ E	2810	Semi-evergreen	3	25	4.1	2500	10 250	4

Velocity ($m\,s^{-1}$): 1. very slow (< 0.05); 2. slow (0.05–0.2); 3. moderate (0.2–0.5); 4. fast (0.5–1.0); 5. very fast (> 1).

The study sites and their characteristics are recorded and presented in Table 1 and Fig. 1. In the present work, a total of 31 sites of six river systems of the southern Western Ghats were studied, of which a total of 64 species, belonging to 6 orders, 14 families and 31 genera, were recorded (Table 2). Among them, the order Cypriniformes was dominant, with 3 families, 18 genera and 49 species (76.6 %), compared to other orders (Figs. 2, 7).

3.1 Fish species diversity, abundance, and distribution

Among the 31 sampling locations, high species diversity was recorded at Sinnaru of the Cauvery river system ($H' = 1.268$) and low diversity was recorded at Thunakadavu tunnel of the Chalakudy river system ($H' = 0.357$) (Table 3, Fig. 3). The maximum species diversity was recorded at Sinnaru of the Cauvery river system ($S = 21$), and the minimum was recorded at Puliyarkutti 3rd bridge and Thunakadavu tunnel of the Chalakudy river system and Sorrakottaodai of the Periyar river system ($S = 3$) (Table 3, Fig. 4). The maximum species abundance, 152, was recorded at Naduthottam of the Periyar river system, and the lowest, 16, was recorded at Sorrakottaodai of the Periyar river system and Belikoondu of the Cauvery river system (Table 3, Fig. 5). The maximum dominance index ($D = 21.346$) was recorded at Sinnaru of the Cauvery river system, and the lowest ($D = 2.121$) was recorded at Thunakadavu tunnel of the Chalakudy river system (Table 3).

Table 2. List of freshwater fauna recorded during the present study.

S. no	Species	Distribution locations	IUCN
	Order: Cypriniformes Family: Cyprinidae Subfamily: Cyprininae		
1	*Puntius melanampyx*	18	DD
2	*Puntius carnaticus*	10	LC
3	*Puntius amphibius*	4	DD
4	*Haludaria fasciatus*	11	LC
5	*Dawkinsia filamentosa*	4	LC
6	*Puntius sarana sarana*	4	LC
7	*Puntius dorsalis*	2	LC
8	*Puntius chola*	2	LC
9	*Puntius sophore*	1	LC
10	*Eechathalakenda ophicephalus*	2	EN
11	*Puntius mahecola*	7	DD
12	*Pethia conchonius*	4	LC
13	*Sahyadria denisonii*	2	EN
14	*Sahyadria chalakkudiensis*	2	EN
15	*Puntius sarana spirulus*	1	LC
16	*Puntius bimaculatus*	3	LC
17	*Pethia ticto*	1	LC
18	*Cirrhinus cirrhosus*	2	VU
19	*Skymatorynchus nukta*	3	EN
20	*Labeo boggut*	1	LC
21	*Labeo kontius*	1	LC
22	*Labeo ariza*	3	LC
23	*Labeo calbasu*	2	LC
24	*Labeo boga*	2	LC
25	*Hypselobarbus curmuca*	4	EN
26	*Hypselobarbus periyarensis*	3	EN
27	*Hypselobarbus dubius*	6	EN
28	*Tor malabaricus*	5	EN
29	*Tor khudree*	9	EN
30	*Osteochilus longidorsalis*	2	EN
	Subfamily: Danioninae		
31	*Salmophasia acinaces*	1	LC
32	*Barilius gatensis*	16	LC
33	*Barilius bakeri*	10	LC
34	*Barilius barana*	2	LC
35	*Barilius bendelisis*	3	LC
36	*Devario aequipinnatus*	21	LC
37	*Rasbora daniconius*	13	LC
	Subfamily: Oreininae		
38	*Lepiphygopsis typus*	2	EN
	Subfamily: Garrinae		
39	*Garra mullya*	16	LC
40	*Garra surendranathi*	3	EN
41	*Garra nasuta*	1	LC
42	*Garra periyarensis*	2	VU
43	*Garra hughi*	3	EN
44	*Garra gotyla stenorhynchus*	2	LC
45	*Crossocheilus latius latius*	1	LC
	Family: Balitoridae Subfamily: Balitorinae		
46	*Travancoria jonesi*	8	EN

Table 2. Continued.

	Subfamily: Nemacheilinae		
47	*Nemacheilus denisoni*	2	LC
48	*Nemacheilus guentheri*	7	LC
	Family: Cobitidae Subfamily: Cobitinae		
49	*Lepidocephalus thermalis*	5	LC
	Order: Siluriformes Family: Bagridae Subfamily: Bagrinae		
50	*Hemibagrus punctatus*	3	CR
51	*Mystus cavasius*	4	LC
	Family: Siluridae		
52	*Ompok bimaculatus*	1	NT
	Family: Sisoridae Subfamily: Glyptosterninae		
53	*Glyptothorax housei*	1	EN
	Order: Cyprinodontiformes Family: Aplocheilidae Subfamily: Aplocheilinae		
54	*Aplocheilus lineatus*	3	LC
	Order: Synbranchiformes Suborder: Mastacembeloidei Family: Mastacembelidae Subfamily: Mastacembelidae		
55	*Macrognathus pancalus*	1	LC
56	*Mastacembelus armatus*	1	LC
	Order: Perciformes Suborder: Percoidei Family: Ambassidae		
57	*Chanda nama*	2	LC
	Family: Pristolepididae		
58	*Peristolepis marginata*	3	LC
	Suborder: Labroidei Family: Cichlidae		
59	*Oreochromis mossambica*	1	NT
60	*Etroplus suratensis*	3	LC
61	*Etroplus maculatus*	2	LC
	Suborder: Gobioidei Family: Gobiidae Subfamily: Gobiinae		
62	*Glossogobius guiris*	1	LC
	Order: Mugiliformes Suborder: Belonoidei Family: Belonidae		
63	*Xenentodon cancila*	3	LC
	Family: Hemiramphidae		
64	*Hyporhamphus limbatus*	2	LC

* EX: extinct; EW: extinct in the wild; CR: critically endangered; EN: endangered; VU: vulnerable; NT: near threatened; LC: least concern; LRlc: low risk, least concern; LRcd: low risk, conservation-dependent; DD: data-deficient.

Table 3. Indices of diversity of fish fauna and respective altitudes of six river systems.

Sampling locations	Altitude	Diversity (H')	Evenness (E)	Abundance	Richness (S)	Dominance (D)
Moyar river system						
Belemeenthurai	520	0.841	0.932	19	8	8.55
Gulithuraipatti	312	0.769	0.769	62	10	5.016
Kallampalayam	300	0.62	0.686	38	8	3.316
Chalakudy river system						
Urilikal	3238	0.734	0.869	131	7	4.598
Thellikal	840	0.805	0.843	32	9	5.701
Baghapallam	748	0.617	0.793	36	6	3.728
Puliyarkutti 8th bridge	527	0.879	0.921	39	9	7.8
Thunakadavu tunnel	520	0.357	0.748	42	3	2.121
Puliyarkutti 3rd bridge	512	0.401	0.841	17	3	2.429
Thunakadavu stream	510	0.864	0.864	68	10	6.026
Thenmudiparai	510	0.74	0.875	59	7	4.833
Orukomban range	450	0.711	0.842	49	7	4.576
Pillapara	267	0.718	0.923	25	6	5.769
Athirappilly	202	1.01	0.936	52	12	11.143
Bhavani river system						
Kovaikutralam falls	560	0.722	0.928	40	6	5
Nellithurai	380	0.757	0.896	29	7	5.639
Periyar river system						
Melaparai	965	0.798	0.944	19	7	7.773
Naduthotam	950	1.019	0.915	152	13	9.936
Ummikuppamthodu	943	0.527	0.678	41	6	2.384
Anjurily	912	0.537	0.768	19	5	3.054
Oorpannikaham	884	0.767	0.849	27	8	5.4
Sorrakottaodai	879	0.465	0.976	16	3	3.243
Thenkasithodu	872	0.638	0.668	100	9	3.327
Valukuparai	869	0.91	0.954	28	9	9.947
Mullaithodu	869	1.045	0.968	48	12	12.966
Cauvery river system						
Kadapilliyarthittu	1137	0.8	0.886	37	8	6.055
Thonanthikla	341	1.069	0.909	46	15	11.129
Belikoondu	267	0.625	0.804	16	6	3.75
Nadathittu	262	1.198	0.921	77	20	15.481
Sinnaru	225	1.268	0.959	75	21	21.346
Nugu river system						
Noolpuzha	2810	0.946	0.946	78	10	8.938

3.2 Species composition

Species similarity between the sites was very low among 31 sites of six river systems. Cluster analysis exhibited similar species composition between the sites (Table 4, Fig. 6). A total of five clusters were grouped, which clearly demonstrate the similarity of species composition among the sites. The cluster group separation is based on the following reasons: (1) rare species forms, (2) low water temperature and (3) prevalence of human disturbances.

3.3 Water quality

Water quality parameters are recorded and presented in Table 6. It is found that the parameter values lie within the IS: 10500 Permissible limits (Table 6 of BIS, 2012). The selected sites of the Western Ghats has a water pH ranging from 6.5 to 8.5. A pH of 9 was recorded at Kadapilliyarthittu of the Cauvery river system, and 7.2 was recorded at various sites like Anjurily, Athirappilly and Urilikal. A minimum conductivity value of 27.8 mS was recorded in the Chalakudy

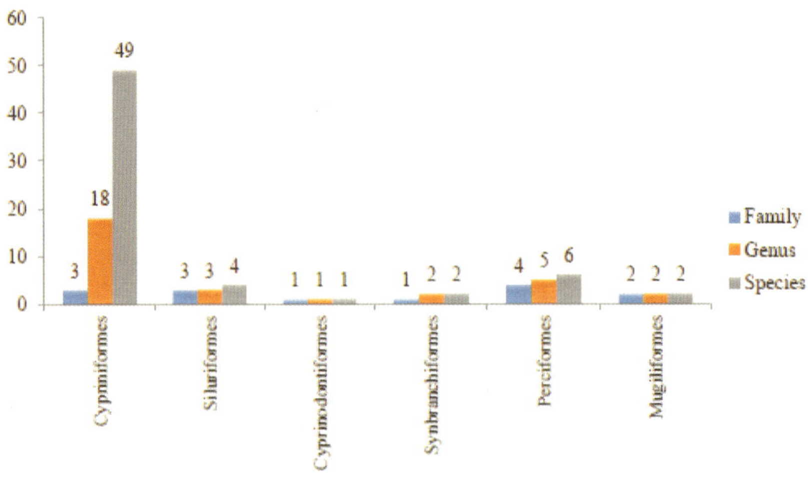

Figure 2. Representation of fish fauna orders among the six river systems.

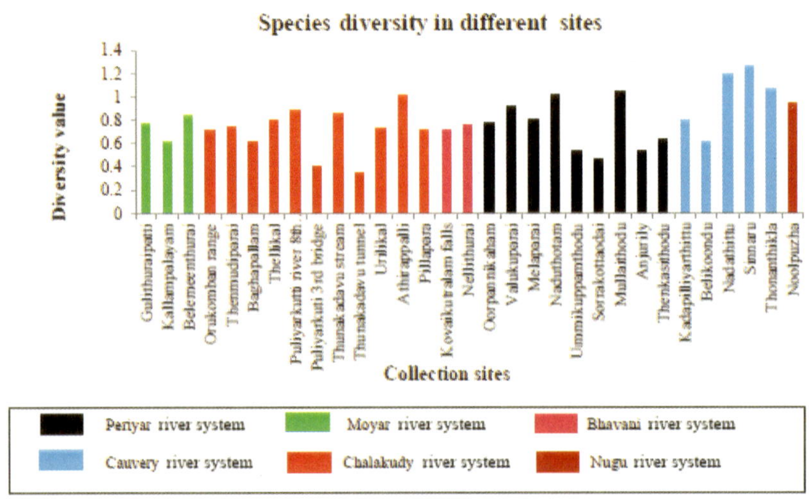

Figure 3. Species diversity among 31 sites.

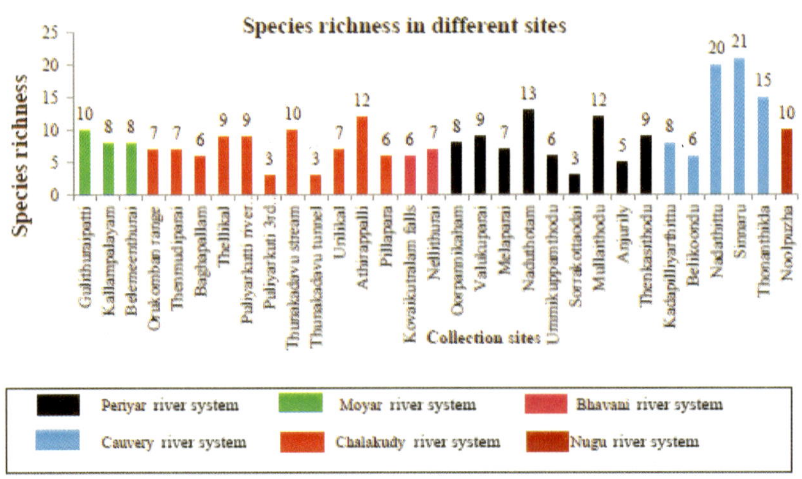

Figure 4. Species richness among 31 sites.

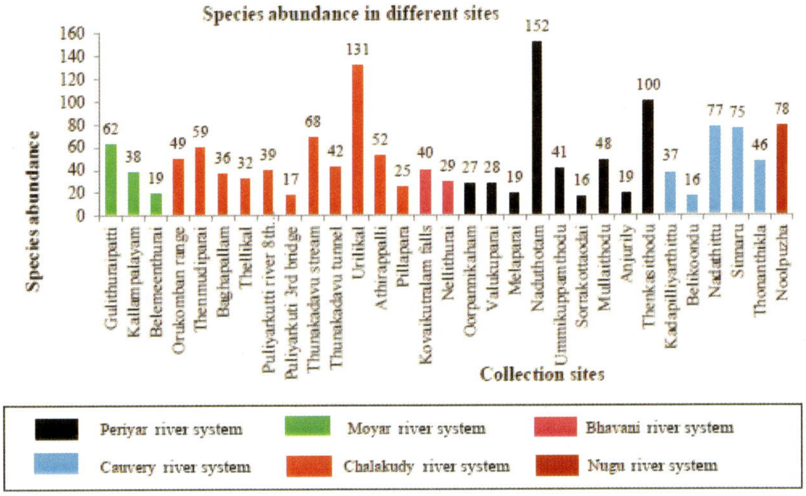

Figure 5. Species abundance among 31 sites.

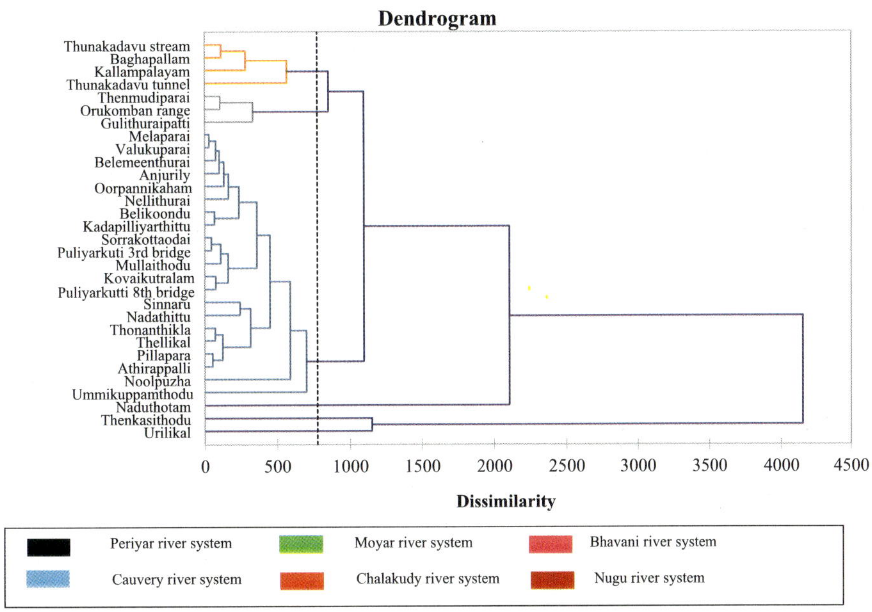

Figure 6. Cluster dendogram expressing the dissimilarity.

river system, and a maximum conductivity value of 85.2 mS was recorded at Noolpuzha of the Nugu river system. Total dissolved solids (TDS) are a measure of inorganic salts dissolved in water. These dissolved solids come from both natural and human sources. Mitchell and Stapp (1992) have suggested that changes in TDS concentrations can be harmful. If TDS concentrations are too high or too low, the population of aquatic life can be limited. Thenkasithodu of the Periyar river system witnessed a low value of TDS content of 13.7 mg L^{-1}, and at Urilikal of the Chalakudy river system, a high value of TDS of 51.9 mg L^{-1} was recorded. A minimum resistivity value of 2.58 was measured at Kadapilliyarthittu of the Cauvery river system, and a maximum of 45.6 was measured at Thenkasithodu of the Periyar river system. A high level of DO of 6.11 mg L^{-1} was recorded at Thenkasithodu of the Periyar river system, and a low DO of 0.63 mg L^{-1} was recorded at Belikoondu of the Cauvery river system. Arunkumar et al. (2015) posited that the lowest DO recorded at sampling sites is due to organic-rich domestic waste let into the river by tourists in the river system. A maximum water temperature (33.6 °C) was recorded at Pillapara of the Chalakudy river system, and a minimum water temperature (18.9 °C) was noted at Thenkasithodu of the Periyar river system.

3.4 Habitat structure

Stream habitat was measured in dimensions like length, width, depth, substratum and current. Large proportions (> 50 %) of the habitat sampled included very shallow water (< 1 cm). Typically, such areas are not habitable for fish fauna and most fish are concentrated in dispersed pools, indicating that habitat measures in upstream areas should be restricted to the pools themselves (Gorman, 1978).

3.5 Substratum types

For the present study, the fish species diversity and the habitat quality assessments of the river systems have been considered as major criteria. The results exhibit that flora and fauna flourish in the study area. This proves that the habitat provides food and shelter suitable for the fish fauna and other aquatic organisms. The habitat assessment of the study area shows that there are four habitat types (pool, riffle, run and glide) with six substratum types (fine sand, debris, silt, bedrock, gravel, rubble and boulders). The shoreline also has a sandy border and a strong rigid bottom made up of rocks, which creates a good habitat for the aquatic organisms. Moreover the water quality, substratum type and vegetation provide a good and healthy habitat and high food resource availability, which plays a key role in species diversity. The river habitat is utilized by tribal people for catching fish for their source of protein food. In the present study, the substratum types of rubble and boulders were dominant, representing 80 % in Mullaithodu of the Periyar river system. At Anjurily of the Periyar river system, gravel is the dominant substratum, representing 70 %. Moreover, substratum types like sand and silt are equally represented in all the study sites. Debris is the biological matter that occupies a major part the stream habitat and is important for providing good shelter and a feeding habitat for the fish fauna. Mostly bottom feeders like *Garra*, *Nemacheilus* and *Travanchoria* use these debris and bedrock substratum as their habitat in a total stream channel with all other substratum types. Nadathittu of Cauvery River, Naduthotam of the Periyar River, Kovaikutralam of Bhavani River and Thunakadavu of the Chalakudy river system have their base substratum as natural bedrock, which provides them with a strong rigid bottom.

Stream width and volume were high at Belikoondu of the Cauvery river system (80 m, 80 000 m^3), followed by Nadathittu (70 m, 42 000 m^3), Kadapilliyarthittu (75 m, 11 250 m^3), Kallampalayam (13 m, 10 400 m^3) and Noolpuzha (25 m, 10 250 m^3). The lowest stream width and volume were recorded at Thellikal (4 m, 400 m^3). Among the 31 sites, very fast-flowing water was noted at Nellithurai, Thunakadavu tunnel and Belikoondu. Fast-flow water and moderate-flow water was noted in most of the river systems. A slow flow of water in the channels was recorded at Thenkasithodu, Kadapilliyarthittu, Oorpannikaham and Urilikal.

3.6 Ecological structures influence characterization

Principal component analysis was used to illustrate the influence of the variables and its importance for the ecological structure of the river system and the fish species. The various habitat characteristics, like water quality, channel morphology and the substratum type, influence the species distribution. Factors like altitude (6.940), area (21.449) and volume (58.428) influence the species diversity (Table 7). All other characteristics play a supportive role in expressing the variations among the study sites. Based on the contributions, study sites like Belikoondu, Kallampalayam, Sorrakottaodai, Anjurily, Thenkasithodu, Belemeenthurai, Kovaikutralam, Naduthotam, Nadathittu, Kadapilliyarthittu and Sinnaru exhibit more variations. The results obtained conclude that altitude plays a major role in species diversity and species abundance, which supports the proposed theory that diversity changes with altitude on mountainsides, being lowest at higher elevations (Colinvaux, 1930). The present findings support the above theory as the results express that species diversity and abundance are low at high altitudes. Among the 31 sites, high species diversity was recorded at Sinnaru of the Cauvery river system ($H' = 1.268$) because of the altitude, area of the channel and the volume of flow as well. The maximum species diversity was recorded at Sinnaru of the Cauvery river system ($S = 21$), due to the channel flow, altitude and the submerged substratum types with muddy water flow. The maximum species abundance of 152 was recorded at Naduthotam of the Periyar river system due to the low area of the channel and the maximum percentage of the rocky boulder substratum. The maximum dominance of species ($D = 21.34$) was recorded at Sinnaru of the Cauvery river system influenced by the vast channel area. The rest of the sites showed low dominance due to the lower percentage of influence caused by the habitat structures.

Rajan (1955) studied the fish fauna from the Moyar river system and reported 48 species. Manimekalan (1998) reported 38 species in the Mudumalai wildlife sanctuary. Manimekalan stated that species like *Labeo dero*, *Puntius mudumaliensis*, *Schimatorhynchus nukta*, *Devario neilgherriensis*, *Crossocheilus latius latius*, *Clarias dayi* and *Gambusia affinis* were restricted to the Moyar river system. Also, *Clarias dayi*, a critically endangered species, was recorded by Manimekalan and Arunachalam (2002). *Puntius carnaticus* and *Danio aequipinnatus* were recorded as common species of the Moyar river system. Rajan (1955) and Mukerjii (1931) both studied the headwaters of Bhavani River and reported that species like *Travancoria elongata*, *Barilius canarensis*, *Rasbora caveri*, *Garra menoni* and *Silurus wynaadensis* were restricted to the river system. *Puntius filamentosus*, *Puntius melanampyx*, *Puntius carnaticus*, *Barilius gatensis*, *Danio aequipinnatus* and *Rasbora daniconius* were very common in the Bhavani river system. Arunkumkar et al. (2015) recorded nearly 37 species from the Cauvery river system. Among several fish species, only *Garra gotyla*

Table 4. Species composition among the 31 sites.

Cluster no	Cluster consists of	Study sites
1	1–4	Thunakadavu stream, Baghapallam, Kallampalayam, Thunakadavu tunnel
2	5–7	Thenmudiparai, Orukomban range, Gulithuraipatti
3	8–28	Melaparai, Valukuparai, Belemeenthurai, Anjurily, Oorpannikaham, Nellithurai, Belikoondu, Kadapilliyarthittu, Sorrakottaodai, Puliyarkutti 3rd bridge, Mullaithodu, Kovaikutralam falls, Puliyarkutti 8th bridge, Sinnaru, Nadathittu, Thonanthikla, Thellikal, Pillapara, Athirappilly, Noolpuzha, Ummikuppamthodu
4	29	Naduthotam
5	30	Thenkasithodu
6	31	Urilikal

stenorhynchus, which is locally consumed, is recorded as being an endangered species in Grand Anicut Cauvery (Murthy et al., 2015). But *Garra gotyla stenorhynchus* is still under a status of least concern of the IUCN (International Union for Conservation of Nature).

Silas (1951), in his faunal account, discussed the extension of the range of *Salmostoma acinaces* (*Chela argentea* Day), *Barbodes carnaticus* (*Barbus* (*Puntius*) *carnaticus*), *Osteochilus* (*Osteochilichthys*) *thomassi* and *Batasio travancoria* and listed two endemic species described by Herre viz. *Homoloptera Montana* and *Glyptothorax housei*. Silas further reported five species from the Cochin part of the Anamalai Hills viz. *Barilius bakeri*, *Puntius denisoni*, *Travancoria jonesi*, *Nemacheilus triangularis* and *Batasio travancoria*. *Puntius bimaculatus*, earlier considered to be a juvenile of *Puntius dorsalis*, was collected from the Anamalai Hills. Interestingly this species is found to be the most dominant *Puntius* species in the hill ranges of the Eastern Ghats, especially the Javadi Hills. *Puntius punctatus*, earlier considered as a synonym of *Puntius ticto*, was kept as a separate species and both these species were collected from the Anamalai Hills (Menon, 1999).

The Periyar River originates near Moolavaigae and reaches the Mullaperiyar reservoir, located in the premises of Periyar Tiger Reserve, which is one of the most biodiversity-rich zones in the southern Western Ghats (Silas, 1950, 1952; Kurup et al., 2004). Earliest studies on the fish fauna of the PTR date back to 1948, when Chacko (1948) listed 35 species from the Periyar Lake, including the critically endangered small-scale *Schizothoracin Lepidopygopsis typus*. Later, Menon and Devi (1995) described *Hypselobarbus kurali* from streams adjoining the Periyar River, raising the total number of fish species to 38. In the present study, 64 species were collected from 31 study sites of six river systems of the southern Western Ghats. Species like *Puntius melanampyx*, *Puntius carnaticus*, *Puntius amphibious*, *Puntius fasciatus*, *Puntius mahecola*, *Devario aequipinnatus*, *Garra mullya*, *Travancoria jonesi* and *Nemacheilus guentheri* were commonly found in all of the six river systems (Fig. 7).

Smith and Smith (1990) stated that habitat selection of the fish fauna is influenced by the body structure, by food and shelter and by physiological processes. Moreover the fish adapt and respond to the characteristics of the rivers and streams, thereby helping the survival of the fittest. Hence it is reliable information that the micro- and macrohabitat play a key role in the morphology and physiological characteristics and modifications of the species. The fish prefer habitats where the nature of the river or stream substratum type is a muddy bottom with debris; this increases high species richness of the bottom feeders. Odum (1945) stated that the flow of the water in the channel is an important factor controlling the distribution of fish fauna; species like *Barilius*, *Hypselobarbus*, *Puntius*, *Travancoria Rasbora* and *Tor* prefer fast flow. The nature of the substratum and the flow rate seem to be more or less closely interrelated in governing the distribution of the fish fauna. This induces the dominance of the Cyprinid species to flourish well in all the river systems of the Southern Western Ghats. It is clear that ecological structure plays a key role in representing river systems of the southern Western Ghats, which has rich species diversity and abundance.

Conservation of India's vast and diverse aquatic genetic resources is essential to maintain ecological as well as socio-economic equilibrium (Lakra et al., 2007). Fisheries and aquaculture have a promising role to play in social development by providing nutritional security for the Indian population and contributing to the economic advancement of farmers and fishery workers (Ambili, 2013). The concept of fish conservation is not new to India. Fishing was prohibited during the third Chatturmass (July–October) to protect the pre-spawning brood stock and juveniles. King Ashoka's prohibition period extended up to the middle of November. Renowned fisheries taxonomist Francis Day drew the attention of the government of India to the large-scale slaughter of fish and pleaded for urgent conservation measures. After persistent pressure, the Indian Fisheries Act was enacted in 1897. Destructive fishing methods, the creation of fixed engines (dams, weirs etc.) for catching fish and the use of small sized nets were banned by the law. The main threats impacting freshwater biodiversity in the Western Ghats include pollution (urban and domestic pollution ranking as the worst threats, followed by agricultural and industrial

Table 5. Distribution and abundance of fish fauna of six river systems.

Species	Collection sites																													
	1	2	3	4	5	6	7	8	9	10	11	12	13	14	15	16	17	18	19	20	21	22	23	24	25	26	27	28	29	30	31
Puntius melanampyx				4	6	12	7	10	5	11		32	5	2	5			2	2			5	4	5	2						10
Puntius carnaticus	1	1	5				1		2	6																4		9	5	2	
Puntius amphibius	1	2					1																						2		
Haludaria fasciatus								5	10	5		13			15		1	1		2		4	5								15
Dawkinsia filamentosa	1																											5	2	5	
Puntius sarana sarana	15	1														10													6		
Puntius dorsalis	1	1																													
Puntius chola	7	15																													
Puntius sophore												11																			
Eechathalakenda ophicephalus																				19	26										
Puntius mahecola											25		8	5		5										4	3	3			
Pethia conconius																										10	1	3	4		
Sahyadria denisonii													5	5																	
Sahyadria chalakkudiensis													1	1																	
Puntius sarana spirulus																3															
Puntius bimaculatus												10													2						10
Pethia ticto		1																													
Cirrhinus cirrhosus																													6	2	
Skymatorynchus nukta																2													3		5
Labeo boggut																														1	
Labeo kontius																														2	
Labeo ariza																										1			1	2	
Labeo calbasu																											1	1			
Labeo boga			2													1															
Hypselobarbus curmuca																	1						7	1							2
Hypselobarbus periyarensis																		2		10			3								
Hypselobarbus dubius			1																1	15							1	3	3		
Tor malabaricus																				17			2					2	4	3	
Tor khudree					2		2										5	3	2	10								2	3	4	
Osteochilus longidorsalis																				5											4
Salmophasia acinaces																												4			
Barilius gatensis	4		4	18	11		3	5		5			5		4	3		2		20	4		5		20						10
Barilius bakeri				2	8			2					2		5	5	2			15					21						8
Barilius barana								2		5																					
Barilius bendelisis																										2		3	3		
Devario aequipinnatus	11	15	2	5	7	14	4	6		23	15	47	5	5	6				2	10	2				47			10	2	3	
Rasbora daniconius				7	3		1					14	4					4		2	5			10	1			2	7	5	
Lepiphygopsis typus																	10			25											
Garra mullya	20	2		12	22	4	11	3		7			8	7									2	2		2		3	4	11	
Garra surendranathi													6										6		3						
Garra nasuta																	3														
Garra periyarensis																		4					2								
Garra hughi																		5	5		2										
Garra gotyla stenorhynchus																											2	4			
Crossocheilus latius latius																	1														
Travancoria jonesi						1		5							5		4	5	4		2				2						
Nemacheilus denisoni			1																3												
Nemacheilus guentheri						4	2	1		1													4		2						4

Table 5. Continued.

Species	Collection sites																													
	1	2	3	4	5	6	7	8	9	10	11	12	13	14	15	16	17	18	19	20	21	22	23	24	25	26	27	28	29	30	31
Lepidocephalus thermalis				1									2							2			3					7			
Hemibagrus punctatus			2																									1	3		
Mystus cavasius			2																										1	2	10
Ompok bimaculatus													1																		
Glyptothorax housei						1																									
Aplocheilus lineatus												4										7	5								
Macrognathus pancalus																													2		
Mastacembelus armatus																														1	
Chanda nama										3																			7		
Peristolepis marginata										2	2													1							
Oreochromis mossambica	1																														
Etroplus suratensis																										10	8	1			
Etroplus maculatus																										4		10			
Glossogobius guiris																													5		
Xenentodon cancila																												2	2	2	
Hyporhamphus limbatus																												2		1	

sources of pollution), residential and commercial development, dams and other natural system modifications, invasive species, agriculture and aquaculture and energy production and mining (IUCN, 2015). The anthropogenic perturbations to fresh water systems over the past years have escalated to enormous proportions, and it is estimated that about 3000 species will become extinct within the next 20 to 30 years (Das, 1994; Prasad, 2010).

The threat to the endangered fish species from our aquatic ecosystem can be minimized by employing both preventive and protective measures. The preventive measures may include removal of causative factors and provision of suitable legislation. The protective measures would include identification of suitable areas to declare as sanctuaries and the development of new technologies for the protection of the genetic resources of threatened and vulnerable fish species. Keeping this in view, the present investigation highlights some of the main causative factors of the decline of various fish species and some remedial measures for preserving the fish population. Degradation of aquatic systems, indiscriminate fishing of brood fish and juveniles, anthropogenic intervention, use of explosives and poisons and intrusion of exotic species are the major possible factors noticed in the present study which cause the depletion of fish population in the study area. Several authors have observed that a sharp decline in the fish population has been recorded in Indian rivers due to the indiscriminate fishing of brood stock and juveniles, fast degradation of aquatic ecosystems and the construction of dams, barrages, weirs, etc. Indiscriminate fishing of brood fish and juveniles and the use of explosives, poisons and electrocution are some of the major possible factors that cause the depletion of fish in Indian waters.

The indigenous fishermen have more preference towards fish species of large size and with medicinal properties. The use of different types of plant products by indigenous fishermen for fishing was observed, which kill all the fish fauna including juveniles. *Croton tiglium* L., *Gnidia glauca* (Fresen) Glig. *Acacia intsia*, *Acacia torta* and *Hydnocarpus pentandra* are some plants which can be used as fish poison for catching fish fauna. The parts of the plant (leaves, stem, bark, fruits and seeds) and the whole plants are used as fish poison. This method is only applicable in stagnant water, which leads to mass poisoning (Ambili, 2013). Dynamiting is another common practice seen among tribal people and it is frequently used in stagnant rock pools and deep waterbodies. In this method all the fish fauna available, from juveniles to adults, are affected on the spot. Dynamiting is also practised by tourists who visit these places illegally. The use of explosives, poisoning and electrocution and the use of small-sized nets etc. are some other fishing methods which affect the population adversely (Ambili, 2013). Use of copper sulfate is also a destructive method of fishing that leads to mass poisoning of the fish population. Irulas, Kurumbas and Mudugar are tribal settlements in the Attappadi region on the banks of Bhavani River. They use cast nets, gill nets

Table 6. Water quality of 31 study sites of six river systems.

Sampling locations	Altitude	pH	Conductivity (mS)	TDS (mg L^{-1})	Resistivity (KΩ)	DO (mg L^{-1})	Water temperature (°C)
Moyar river system							
Belemeenthurai	520	8.4	59.2	37.7	16.4	1.3	24.5
Gulithuraipatti	312	8.4	57.8	20.37	24.2	3.5	23.8
Kallampalayam	300	7.9	45.2	28.5	21.9	2.5	24.1
Chalakudy river system							
Urilikal	3238	7.2	78.7	51.9	12.9	1.4	24.1
Thellikal	840	8.8	59.2	37.7	16.4	1.3	24.5
Baghapallam	748	8	57.8	38.0	16.8	2.4	21.7
Puliyarkutti 8th bridge	527	7.79	27.8	18.0	34.8	5.4	23.5
Thunakadavu tunnel	520	5.9	38.3	28.3	22.2	5.09	21.4
Puliyarkutti 3rd bridge	512	7.79	27.8	18.0	34.8	5.4	23.5
Thunakadavu stream	510	5.9	38.3	28.3	22.2	5.09	21.4
Thenmudiparai	510	8	45.2	28.5	21.9	2.5	24.1
Orukomban range	450	7.5	33.9	26.5	22.4	3.5	23.4
Pillapara	267	7.6	34.0	19.5	29.9	0.89	33.6
Athirappilly	202	7.2	35.2	47.5	3.97	0.73	32.7
Bhavani river system							
Kovaikutralam falls	560	7.5	31.3	20.1	32.3	3.2	22.5
Nellithurai	380	7.3	30.3	20.3	31.5	2.3	25.5
Periyar river system							
Melaparai	965	9	44.7	28.8	22.5	1.3	26.1
Naduthotam	950	7.5	46.2	30.4	20.6	0.7	25.9
Ummikuppamthodu	943	7.7	64.9	43.2	17.1	1.2	24.8
Anjurily	912	7.2	21.5	13.6	47.5	4.86	19.2
Oorpannikaham	884	8.3	50.3	32.3	20.0	1.2	24.8
Sorrakottaodai	879	8	34.2	21.9	29.5	1.1	23.1
Thenkasithodu	872	5.2	22.0	13.7	45.6	6.11	18.9
Valukuparai	869	7.7	66.9	43.8	15.1	0.7	24.8
Mullaithodu	869	8.1	78.6	51.4	12.5	0.9	24.2
Cauvery river system							
Kadapilliyarthittu	1137	9.6	39.1	26.3	2.58	0.72	30.5
Thonanthikla	341	9.2	39.5	26.3	2.65	3.11	30.2
Belikoondu	267	9.4	39.8	26.3	2.63	0.63	32.7
Nadathittu	262	9.4	39.8	26.3	2.63	0.63	32.7
Sinnaru	225	9.2	39.5	26.3	2.65	3.11	30.2
Nugu river system							
Noolpuzha	2810	7.32	85.2	51.7	11.8	3.62	23.2

and bamboo traps (Kooda) for fishing. Indiscriminate fishing in the Bharathappuzha caused a large decline in the *Tor* population (Ambili, 2013). Sarkar and Srivastava (2000) noticed that because of increased anthropogenic activities, the two main species, namely *Tor putitora* and *Tor tor*, are listed under the category of endangered species and face a high risk of extinction in the wild. Due to the proximity of human settlements, aquatic ecosystems are relatively more exposed to human influences and interventions. In addition, the industrial and urban development has altered the aquatic environment (Ambili, 2013). Overfishing at various stages of the life cycle has been observed more in human settlement areas, and this causes spectacular changes in the environment that affect fishery resources. Pollution of the waterbody is also one of the main factors which cause the decline of the ichthyofaunal diversity (Ambili, 2013).

© Dr. A. Manimekalan & Dr. A. A. Arunkumar

Figure 7. Fish collected from various waterbodies of the southern Western Ghats.

The pristine riverine systems along the Western Ghats have been altered by anthropogenic activities such as dam constructions and road building, which have affected the ecology and habitat of these fish fauna. The tourist resorts starting down at Athirappilly on the bank of the Chalakudy River are altering the habitat in many ways. Sholayar Hydro Electric Project and Peringalkuthu Hydro Electric Project are the hydroelectric projects on the Chalakudy River. There are about seven dams built on the river. Peringalkuthu Dam in this river prevents the local migration of *Tor* from the lower to upper stretches of the river. There are 11 reservoirs in Bharathappuzha River, and Malampuzha Dam is the largest one. Neyyar Dam is located in the Neyyar River, and the Idukki Dam is located in the Periyar River. The construction of dams also results in less water flow and affects the migration of fish. Food availability is an important factor for the existence of fish species. MacDonald (1948) noted that Mahseer (*Tor*) is an intermittent feeder. Vegetative matter, benthic diatoms, molluscan shells, crabs, insects, small fish fauna and different types of seeds and fruits have been recorded from the stomach contents of *Tor* species (Dinesh et al., 2010). The availability of these items varies considerably during different seasons. These disruptions in the food chain also affect the species adversely. The deforestation rate all along the Western Ghats is very high, and the forest areas are being transformed for agriculture practices. This was seen

Table 7. Contribution of the variables (%) after varimax rotation for habitat characteristics.

Variables	D1	D2
Altitude	6.940	45.277
pH	0.849	0.147
Conductivity (mS)	0.424	0.002
TDS (ppm)	0.568	0.031
Resistivity (KΩ)	0.715	0.075
DO (mg L^{-1})	0.900	0.180
Salinity (ppt)	0.923	0.196
Water temperature (°C)	0.695	0.069
Rubble and boulders	0.676	0.060
Gravel	0.740	0.098
Sand	0.764	0.127
Silt	0.884	0.167
Derbies	0.828	0.120
Bedrock	0.714	0.037
Stream order	0.885	0.170
Stream width (m)	0.819	0.155
Stream depth (m)	0.909	0.196
Area (m^2)	21.449	20.245
Volume (m^3)	58.428	32.473
Mean velocity (m s^{-1})	0.891	0.177

on the banks of most of the rivers in Kerala like Chaliyar, Sholayar, Chalakudy, Kabini, Bhavani, Periyar and Kallada river systems. Cultivation of Musa, paddy, cardamom, ginger and tea plantations is observed to be the most common. The pesticides used in these areas are penetrating the river systems and severely affect aquatic organisms like insects and diatoms, vegetation such as phytoplankton and even the small fish fauna (Ambili, 2013).

The fish population is declining rapidly among the streams and rivers of the southern Western Ghats. Hence the following immediate conservation measures will help to conserve these precious species. Awareness among the tribes is very important for the conservation of fish species. Awareness can be increased about the impact of using chemicals for mass poisoning, dynamiting for catching fish, fishing during breeding seasons and the use of poisonous plant products for mass poisoning. Students, social workers, fishermen and local people should be educated about the importance of conservation of fish fauna in their area so that they can create awareness among the people in the area. More exclusive projects should be started with the co-operation of local people and students to protect the fish population. Action can be taken to change the fishing profession of those who only depend on fishing for their livelihood, which will help to reduce the fishing pressure.

In order to conserve fish genetic resources and provide adequate living space, shelter and habitat for valuable threatened fish fauna, certain areas can be declared fish sanctuaries, like national parks and wildlife sanctuaries. Menon et al. (2000) suggested that suitable segments of the rivers with fish species should be identified for the establishment of fish sanctuaries and that such sanctuaries must be heavily stocked every year with fish fingerlings. There are two fish sanctuaries protecting the *Tor* species as part of the people's religious customs, Aruvikara (Neyyar River) and Kulathupuzha (Kallada) in Kerala. The upstream part of Chalakudy River, the Karimpuzha and Manjeeri region of Chaliyar River, a part of Bhavani up to Thavalam (Attappadi region) and selected stretches of Periyar River can be declared fish sanctuaries. Ambili et al. (2014) reported the presence of three species, *Tor khudree*, *Tor malabaricus* and *Tor mussullah*, in the Chaliyar River. Long stretches of Cauvery River are a fish sanctuary of which the Karnataka Forest Department (Wildlife) has leased out 22.5 km. to the Wildlife Association of South India (WASI, 2018), Bangalore, Karnataka, which is now protecting the wildlife including fish fauna with more care.

Captive breeding is a technique widely used throughout the world for a variety of endangered animals including fish (Maitland and Evans, 1986; Keshavanath et al., 2006). It could be an important last resort measure for endangered and endemic species, which may otherwise become extinct in the wild (Reid, 1990). Fishery departments should take steps for the breeding and caring of the endangered *Tor* species. In Kerala, near Pookode Lake and Sholayar Dam, procedures for culturing the *Tor* species were attempted in hatcheries, but they did not succeed. Collection of matured brooders from the interior of the forest and maintenance of water temperature are the two major problems that were encountered. Now studies are going on to compensate the reasons for failures in *Tor* breeding. Gene banks can hold live animals or cryopreserved gametes. Gene banks can be considered as a last line of defence against species extinction. A live gene bank contributes to the delisting of threatened species by captive breeding and restocking in species-specific recovery programmes. Such gene banks can contribute to the recovery and utilization of genetic diversity and can be used in conservation programmes (e.g. NBFGR, India, and the World Fisheries Trust, Canada) and genetic enhancement (e.g. salmon in Norway and common carp in Hungary) (Lakra et al., 2007). A mini gene bank with the milt of *T. putitora* and *T. khudree* has been established by NBFGR (Ponniah et al., 1999a, b). In India *Tor* spermatozoa cryopreservation protocols have been developed by several workers (Basavaraja and Hedge, 2004, 2005; Patil and Lakra, 2005). Fish sperm cryopreservation requires the development of species-specific protocols (Lakra et al., 2006). Cryopreservation of germ plasm is a very good ex situ strategy to conserve existing allelic diversity for future use. This technique may also help to provide gametes for artificial propagation programmes in off seasons. Universities and research institutes should take care of the cryopreservation and captive breeding of *Tor* species. Reintroduction is more essential than introduction for the conservation of native species. Introduction can never neutralize the problem of the depletion of species, while reintroduction (collection and protection of wild/native fish fauna and their introduction into the rivers) can support the conservation of native species to a great extent. Introduction of *Tor* species in the rivers of Kerala from other regions or other river systems is causing more confusion and taxonomic ambiguities, and sometimes people wrongly quote this as evidence to support the Satpura Hypothesis (Kumar and Kurup, 2004).

The monitoring and documentation of fish stocks are becoming significantly important to carry out regular reviews of the distribution and status of all fish species, and this will be possible by maintaining records (Koljonen and Nyberg, 1991). The documentation of genetic resources for aquaculture is also a part of the coverage of the FishBase database (Froese and Pauly, 2013). The comprehensive listing of fish species' distribution as well as the continuous monitoring of fish species are a critical need for the protection of fish fauna.

4 Data availability

The data have been submitted in the Pangaea data bank and they can be accessed by the following link (https://doi.org/10.1594/PANGAEA.882214).

5 Conclusion

In the present study, from the fish fauna that were collected from various river systems of the southern Western Ghats, like Bhavani, Moyar, Chalakudy, Periyar, Cauvery and Kabini, it is noted that the order Cypriniformes was the most dominant order, with 3 families, 18 genera and 49 species (76.6 %), compared to other orders. Interestingly, sites like Thunakadavu stream, Gulithuraipatti, Athirappilly, Naduthotam, Nadathittu, Mullaithodu, Thonanthikla, Noolpuzha and Sinnaru revealed high species diversity. The results indicated that the species from the southern Western Ghats have an ambiguity in their taxonomy among the fishing communities. The data analyses suggested that species like *P. melanampyx*, *P. carnaticus*, *P. amphibious*, *P. fasciatus* and *P. mahecola* were found to be the dominant species in the locations considered. Among the 31 sites, maximum diversity ($H' = 1.268$) was recorded at Sinnaru (altitude = 225) of the Cauvery river system, and minimum diversity ($H' = 0.73$) was recorded at Urilikal (altitude = 3238) of the Chalakudy river system. The present findings support the Colinvaux theory, which expresses diversity changes with regards to elevation. The nature of the ecosystem and the vegetative forest which prevails along the river systems of the southern Western Ghats create suitable habitats for fish fauna. Many threats to the existence of fish fauna from the rivers of the southern Western Ghats, like the use of explosives and poison and the fishing of juveniles, are reported. Hence, urgent attention is necessary to create awareness among local communities of the importance of the stream habitat and its fish diversity, for the conservation of these important resources for future generations.

Author contributions. Both the authors have equally taken part in all aspects of the work and manuscript preparation.

Competing interests. The authors declare that they have no conflict of interest.

Acknowledgements. The authors gratefully acknowledge the University Grants Commission for providing the funding for this research, Tamil Nadu, Kerala and Karnataka forest departments for the permission and logistic support and the Department of Environmental Sciences, Biodiversity and Molecular Laboratory, Bharathiar University, Coimbatore, Tamil Nadu, India, for providing laboratory facilities for the research.

Edited by: Falk Huettmann
Reviewed by: two anonymous referees

References

Ambili, T. R.: Morphological and Molecular Characterization of Genus *Tor* from Southern Western Ghats, INDIA, PhD thesis, Dept. of Environmental Sciences, Bharathiar University, Coimbatore, September, 2013.

Ambili, T. R., Manimekalan, A., and Verma, M. S.: Genetic diversity of genus tor in river chaliyar, Southern Western Ghats, Kerala: through DNA barcoding, Journal of Science, 4, 206–214, 2014.

Anushiya, J. and Ramachandran, A.: Assessment of water availability in Chennai basin under present and future climate scenarios, in: Environmental Management of River Basin Ecosystems, edited by: Ramkumar, M., Kumaraswamy, K., and Mohanraj, R., Springer Earth System Sciences, Springer, Cham, 397–415, 2015.

APHA: Standard methods, 19th edn., American Public Health Association Washington, DC, 1995.

Armantrout, N. B.: Aquatic habitat inventory, Bureau of Land Management, Eugene, District, USA, 32 pp., 1990.

Arunkumar, A. A. and Manimekalan, A.: Freshwater fauna and habitat characteristics of rivers in the Southern Western Ghats, India, PANGAEA, https://doi.org/10.1594/PANGAEA.882214 (last access: 24 September 2018), 2017.

Arunkumar, A. A., Manimekalan, A., Manikandan, V., and Velmurugan, P.: Fish species richness and habitat quality mapping with geographical information system across Cauvery River in Tamil Nadu, India, Journal of Aridland Agriculture, 1, 43–54, 2015.

Arunachalam, M., Madhusoodanan, K., Nair, C., Vijverberg, J., and Kortmulder, K.: Food and habitat partitioning among fishes in stream pools of South Indian River, International Journal of Ecology and Environmental Sciences, 23, 271–395, 1997.

Arunachalam, M., Sankaranaraysnan, A., Johnson, J. A., and Manimekalan, A.: Fishes of Nambiyar river, Kalakkad Mundanthurai Tiger Reserve, Tamil Nadu, Journal of the Bombay Natural History Society, 97, 153–155, 2000.

Arunachalam, M., Sivakumar, P., and Muralitharan, M.: Habitat evaluation of pristine headwater streams of Western Ghat mountain ranges, Peninsular India, in: New trends in Fishery Development in India, edited by: Johal, M. S., Punjab University, India, 253–286, 2005.

Arunachalam, M., Muralidharan, M., and Sivakumar P.: *Psilorhynchus amplicephalus*, a new species from Balishwar river of Assam, India, Current Science, 92, 1352–1354, 2007.

Basavaraja, N. and Hegde, S. N.: Cryopreservation of the endangered mahseer (Tor species of Mahseers. *Tor. khudree*, *T. tor* and *T. putitora* (Cyprinidae, Pisces)), Cytobios, 85, 205–218, 2004.

Basavaraja, N. B. and Hegde, S. N.: Some characteristics and short term preservation of spermatozoa of Deccan mahseer, *Tor khudree* (Sykes), Aquacult. Res., 36, 422–430, 2005

Biju, C. R., Raju Thomas, K., and Ajithkumar, C. R.: Fishes of Parambikulam wildlife sanctuary Palakkad District, Kerala, 1996.

BIS: Indian Standard Drinking Water – Specification, 2nd edn., Bureau of Indian Standards, New Delhi, 2012.

Chacko, P. I.: A survey of the fisheries of the Tungabhandra River, P. Indian Acad. Sci. B, 28, 166–175, 1948.

Colinvaux, P. A.: Ecology congress cataloging in publication data, 1930.

Dahanukar, N., Raut, R., and Bhat, A.: Distribution, endemism and threat status of freshwater fishes in the Western Ghats of India, J. Biogeogr., 31, 123–136, 2004.

Das, P.: Conservation and Management of cold water fish genetic resources, in: Coldwater, Aquaculture and Fisheries, edited by: Singh, H. R. and Lakra, W. S., Narendra Publishing House, Delhi, 337–339 1994.

Day, F.: On the fishes of Cochin, on the Malabar Coast of India. Part II. Anacanthini, P. Zool. Soc. Lond., 1, 286–318, 1865.

Day, F.: Monograph of Indian Cyprinidae. Parts 1–3, Journal of the Asiatic Society of Bengal., 40, 95–143, 1871.

Day, F.: The fishes of India; being a natural history of fishes known to inhabit the seas and fresh waters of India, Burma and Ceylon, 4th edn., New Delhi, 778 pp., 1878.

Dinesh, K., Nandeesha, M. C., Nautiyal, P., and Aiyappa, P.: Mahseers in India: A review with focus on conservation and management, Ind. J. Ani. Sci., 80, 26–38, 2010.

Gorman, O. T. and Karr, J. R.: Habitat structure and stream fish community, Ecology, 59, 507–515, 1978.

Gurumurthy, G. P. and Tripti, M.: Geochemical Perspectives on River Water of the Tropical Basins, Southwestern India, in: Environmental Management of River Basin Ecosystems, edited by: Ramkumar, M., Kumaraswamy, K., and Mohanraj, R., Springer Earth System Sciences, Springer, Cham, 329–353, 2015.

Hamilton, F.: An Account of the Fishes of River Ganges and its Branches, George Ramsay and Co., London, vii and 405 pp., 1822.

Hora, S. L.: Notes on fishes in the Indian Museum on a new species of Nemacheilus from the Nilgiri Hills, Records of the Indian Museum, 22, 19–21, 1921.

Hora, S. L.: Notes on fishes in the Indian Museum, on three collections of fish from Mysore and Coorg, South India, Records of the Indian Museum, 39, 5–28, 1937.

Hora, S. L.: Homalopterid fishes from Peninsular India, Records of the Indian Museum, XLIII, 211–232, 1941.

Hora, S. L.: Satpura hypothesis of the distribution of Malayan fauna and flora and flora of peninsular India, Proceedings of National Institute of Science of India, 15, 309–314, 1949.

Hora, S. L. and Misra, K. S.: Fish of Deolali, Part III, Journal of Bombay Natural History Society, 40, 20–38, 1938.

IUCN: https://www.iucn.org/ (last access: 24 August 2018), 2015.

Jayaram, K. C.: The Freshwater Fishes of the Indian Region, Nerendra Publishing House, New Delhi, 551 pp., 1999.

Jayaram, K. C.: The Freshwater Fishes of Indian Region, 2nd edn., Narendra Publications, New Delhi, 119 pp., 2010.

Jerdon, T. C.: On the freshwater fishes of southern India, Madras Journal of Literature and Science, 15, 141–149, 1848.

Keshavanath, P., Gangadhar, B., Basavaraja, N., and Nandeesha, M. C.: Artificial induction of ovulation in pond-raised mahseer, Tor khudree using carp pituitary and ovaprim, Asian Fish. Sci., 19, 411–422, 2006.

Kurup, B. M., Radhakrishnan, K. V., and Manojkumar, T. G.: Biodiversity status of fishes inhabiting rivers of Kerala (S. India) with special reference to endemism, threats and conservation measures, in: Proceedings of the Second Large Rivers Symposium, Phnom Penh, Kingdom of Cambodia, 11–14 February 2003, 163–182, 2004.

Lakra, W. S., Mohindra, V., and Lal, K. K.: Fish genetics and conservation research in India: status and perspectives, Fish Physiol. Biochem., 33, 475–487, https://doi.org/10.1007/s10695-007-9168-z, 2007.

MacDonald, A.: Circumventing the Mahseer and other sporting fish of India and Burma, J. Bomb. Nat. Hist. Soc., 26, 306–307, 1948.

Maitlant, P. S. and Evans, D.: The role of captive breeding in conservation of fish species, International Zoo Yearbook, 24, 66–74, 1986.

Manimekalan, A.: Fishes of Mudumalai Wildlife Sanctuary, South India, Journal of Bombay Natural History Society, 95, 431–443, 1998.

Manimekalan, A. and Arunachalam, M.: Rediscovery of critically endangered air birthing cat fish *Clarias dayi hora* (Pisces: Claridae) from Mudumalai Wildlife Sanctuary, Tamil Nadu, Journal of Bombay Natural History Society, 99, 129–131, 2002.

Manimekalan, A. and Das, H. S.: *Glyptothorax davissinghi* (Pisces: Sisoridae) A new cat fish from Nilambur in the Nilgiri Biosphere, South India, Journal of Bombay Natural History Society, 95, 87–91, 1998.

Manimekalan, A. and Singh, D. F.: New record of *Schismatorhynchus (Nukta) nukta* (Sykes) (Pisces: Cyprinidae) from Moyar river, extends its range to Tamil Nadu, Journal of Bombay Natural History Society, 94, 170–171, 1997.

Menon, A. G. K.: Check list – Fresh water fishes of India, Records of the Zoological Survey of India Occasional Papers, 175, 366 pp., 1999.

Menon, A. G. K. and Devi, K. Rema: *Puntius sharmai*, a new cyprinid fish from Madras, Journal of Bombay Natural History Society, 89, 353–354, 1992.

Menon, A. G. K. and Devi, K. Rema: *Hypselobarbus kuali* (pisces: Cyprinidae) a new large barb from the south western rives of peninsular India, Journal of Bombay Natural History Society, 92, 389–393, 1995.

Menon, A. G. K., Singh, H. R., and Kumar, N.: Coldwater Fish and Fisheries, in: Present eco-status of cold water fish and fisheries, edited by: Singh, H. R. and Lakra, W. S., Narendra Publishing House, New Delhi, 337 pp., 2000.

Mitchell, M. K. and Stapp, W. B.: Field Manual for Water Quality Monitoring: An Environmental Education Program for Schools, 6th edn., Thomson-Shore, 240 pp., 1992.

Mukerji, D. D.: On a small collection of fish from the Bhavani river (S. India), Journal of Bombay Natural History Society, 35, 162–171, 1931.

Murthy, K. R., Dhanakumar, S., Sundararaj, P., Mohanraj, R., and Kumaraswamy, K.: GIS-based modified SINTACS model for assessing groundwater vulnerability to pollution in Vellore District (Part of Palar River Basin), Tamil Nadu, India, Environmental Management of River Basin Ecosystems, 429–453, 2015.

Myers, N., Mittermeier, R. A., Mittermeier, C. G., da Fonseca, G. A. B., and Kent, J.: Biodiversity hotspots for conservation priorities, Nature, 403, 853–858, 2000.

Odum, H. T.: Primary production in flowing water, Limnol. Oceanogr., 1, 102–117, 1945.

Pielou, E. C.: Ecological Diversity, Wiley, New York, 1975.

Ponniah, A. G., Lakra, W. S., and Ogale, S. N.: Effects of cryoprotectants and fertilization protocol on viability of Mahseer, *Tor khudree* cryopreserved spermatozoa, Journal of Aquaculture in the Tropics, 14, 153–158, 1999a.

Ponniah, A. G., Sahoo, P. K., Dayal, R., and Bharat, A.: Cryopreservation of Tor putitora spermatozoa: Effect of extender composition, activating solution, cryoprotectants and equilibrium time, P. Natl. Acad. Sci. India, 69, 1–6, 1999b.

Prasad, G.: Status and impacts of exotic fresh water fishes in Kerala, in: Conservation of Fishes in Kerala, India, edited by: Sonnenschein, L. and Benziger, A., World Aquarium and conservation for the ocean foundation, USA, 294 pp., 2010.

Rajan, S.: Notes on collection of fish from the headwaters of the Bhavani River, South IndiaJournal of Bombay Natural History Society, 53, 44–48, 1955.

Patil, R. and Lakra, W. S.: Effect of cryoprotectants, equilibration periods and freezing rates on cryopreservation of spermatozoa of mahseer, *Tor khudree* (Sykes) and *T. putitora* (Hamilton), Aquac. Res., 36, 1465–1472, 2005.

Reid, G. M.: Captive breeding for the conservation of cichlid fishes, J. Fish Biol., 37, 157–166, 1990.

Sarkar, U. K. and Srivastava, S. M.: Endangered Mahseer in North East India, in: Fish Biodiversity of North East India, edited by: Ponniah, A. G. and Sarkar, U. K., NBFGR, NATP publ, Lucknow, India, 95–96, 2000.

Shannon, C. E. and Weaver, W.: The mathematical theory of communications, Urbaba, University of Ilinois Press, Urbaba, 1949.

Silas, E. G.: On a collection of fish from Travancore, Journal of Bombay Natural History Society, 48, 792–795, 1950.

Silas, E. G.: On a collection of fish from the Annamalai and Nelliampathi hill ranges, Western Ghats, with a note on its zoogeographical significance, Journal of Bombay Natural History Society, XLIX, 470–481, 1951.

Silas, E. G.: Fishes from the high ranges of Travancore, Journal of Bombay Natural History Society, 50, 323–330, 1952.

Simpson, E. H.: Measurement of diversity, Nature, 163, 688, https://doi.org/10.1038/163688a0, 1949.

Smith, D. J. and Smith, C. R.: Summer bird species diversity and use of pastures by summer birds of the Finger Lakes National Forest, U.S. Dept. of Agric. Forest Service, Green Mountain National Forest, Final Project Report P.O. No. 40-1681-9- 0470, Middlebury, Vermont, 1990.

Southwood, T. R.: Ecological Methods, Chapmam and Hall, London, 1978.

Talwar, P. K. and Jhingran, A. G.: Inland fishes of India and adjacent countries, Oxford & IBH Publishing Co. Pvt. Ltd, New Delhi, vol. I, 541 pp., 1991.

Thomas, J., Joseph, S., and Thrivikramji, K. P.: Hydrogeochemical drivers and processes controlling solute chemistry of two Mountain River basins of contrasting climates in the Southern Western Ghats, India, Environmental Management of River Basin Ecosystems, 355–396, 2015.

WASI: http://www.wasiindia.com/, last access: 24 August 2018.

Willis, S. C., Winemiller, K. O. and Fernandez, H. L.: Habitat structural complexity and morphological diversity of fish assemblages in a Neotropical floodplain river, Oecologia. 142: 284 - 295, 2005.

8

Spatial datasets of radionuclide contamination in the Ukrainian Chernobyl Exclusion Zone

Valery Kashparov[1,3], Sviatoslav Levchuk[1], Marina Zhurba[1], Valentyn Protsak[1], Yuri Khomutinin[1], Nicholas A. Beresford[2], and Jacqueline S. Chaplow[2]

[1]Ukrainian Institute of Agricultural Radiology of National University of Life and Environmental Sciences of Ukraine, Mashinobudivnykiv str. 7, Chabany, Kiev region, 08162, Ukraine

[2]NERC Centre for Ecology and Hydrology, CEH Lancaster, Lancaster Environment Centre, Library Av., Bailrigg, Lancaster, LA1 4AP, UK

[3]CERAD CoE Environmental Radioactivity, Department of Environmental Sciences, Norwegian University of Life Sciences, 1432 Ås, Norway

Correspondence: Jacqueline S. Chaplow (jgar@ceh.ac.uk)

Abstract. The dataset "Spatial datasets of radionuclide contamination in the Ukrainian Chernobyl Exclusion Zone" was developed to enable data collected between May 1986 (immediately after Chernobyl) and 2014 by the Ukrainian Institute of Agricultural Radiology (UIAR) after the Chernobyl accident to be made publicly available. The dataset includes results from comprehensive soil sampling across the Chernobyl Exclusion Zone (CEZ). Analyses include radiocaesium (^{134}Cs and ^{134}Cs) ^{90}Sr, ^{154}Eu and soil property data; plutonium isotope activity concentrations in soil (including distribution in the soil profile); analyses of "hot" (or fuel) particles from the CEZ (data from Poland and across Europe are also included); and results of monitoring in the Ivankov district, a region adjacent to the exclusion zone.

The purpose of this paper is to describe the available data and methodology used to obtain them. The data will be valuable to those conducting studies within the CEZ in a number of ways, for instance (i) for helping to perform robust exposure estimates to wildlife, (ii) for predicting comparative activity concentrations of different key radionuclides, (iii) for providing a baseline against which future surveys in the CEZ can be compared, (iv) as a source of information on the behaviour of fuel particles (FPs), (v) for performing retrospective dose assessments and (vi) for assessing natural background dose rates in the CEZ.

The CEZ has been proposed as a "radioecological observatory" (i.e. a radioactively contaminated site that will provide a focus for long-term, radioecological collaborative international research). Key to the future success of this concept is open access to data for the CEZ. The data presented here are a first step in this process.

1 Background

The accident in reactor number four at the Chernobyl Nuclear Power Plant (ChNPP), Ukraine, on 26 April 1986 remains the worst in the history of nuclear power generation. Starting on 27 April the human population and farm animals were evacuated from an area of approximately 3500 km^2 to create what has become known as the Chernobyl Exclusion Zone (CEZ). Subsequently, the area administered as the CEZ has increased to approximately 4760 km^2. The area of the CEZ in the Ukraine is approximately 2598 km^2, with the remainder being in Belarus. The deposition of radionuclides over the CEZ is known to be highly spatially heterogeneous. Releases from the Chernobyl reactor occurred over a period of about 10 days. There is a narrow band of high radioactivity to the west of the reactor (often referred to as the western trace) which represents deposition from the initial explosion (Fig. 1). Higher levels of contamination to the north and to a lesser extent the south are a consequence of releases over the days following the accident.

The CEZ has many features making it an important radioecological study site:

- Contamination levels are such that the behaviour, transfer and mobility of a number of radionuclides can be studied (^{137}Cs, ^{90}Sr, ^{241}Am, Pu isotopes, U isotopes, ^{129}I, ^{14}C, ^{36}Cl and ^{99}Tc).
- The presence of radioactive particles means that their behaviour in the environment can be studied (Kashparov et al., 1999; Beresford et al., 2016). Fuel particles (FPs) are weathered with time such that mobilisation of ^{90}Sr has been observed to lead to increased contamination of plants (Salbu et al., 1994; Oughton et al., 1993; Kashparov et al., 1999). Fuel particles were released in two different forms (Kashparov et al., 1996): non-oxidised fuel particles of the initial release (26 April 1986) formed by the mechanical destruction of nuclear fuel – these first particles formed a 100 km long and up to 1 km wide path to the west of the plant – and oxidised fuel particles which were formed during the subsequent reactor fire (from 26 April to 5 May 1986) and deposited to the north and south of the plant.
- Dose rates remain sufficiently high that we may expect to observe effects on wildlife in some areas. Furthermore, published results on radiation effects from the CEZ are contentious with a lack of agreement on interpretation amongst scientists and a high public profile (Beresford and Copplestone, 2011).
- A wide range of terrestrial and aquatic species and habitats are present. The Ukrainian area, the focus of this paper, contains forests, abandoned farmlands, wetlands, flowing and standing waters, deserted villages and urban areas.

Recently, the CEZ was suggested as a radioecological observatory (http://www.radioecology-exchange.org/content/radioecological-observatories; Steiner et al., 2013); i.e. a radioactively contaminated site that will provide a focus for joint, long-term, radioecological research which will help address challenges identified for the field of radioecology (Hinton et al., 2013). For this to be successful, relevant data for the CEZ need to be made openly available; indeed the deficiency of open data has been highlighted as one of the causes for the lack of scientific consensus with regard to published studies from within the CEZ and more recently areas affected by the 2011 Fukushima accident (Beresford et al., 2012a, b; Barnett and Welch, 2016).

In this paper we describe a series of datasets focussed on radionuclide deposition and radioactive particles within and around the CEZ predominantly from studies conducted by the Ukrainian Institute of Agricultural Radiology (UIAR, 1998) (now part of the National University of Life and Environmental Sciences of Ukraine, NUBIP). The accompanying data are freely available (https://doi.org/10.5285/782ec845-2135-4698-8881-b38823e533bf) under the terms of the Open Government Licence. With respect to radioactive particles, data are also presented from samples collected in Poland in 1986.

The purpose of this paper is to describe the available data and methodology used to obtain them. The data will be valuable to those conducting studies within the CEZ in a number of ways, for instance (i) for helping to perform robust exposure estimates to wildlife (Beaugelin-Seiller et al., 2018), (ii) for predicting comparative activity concentrations of different key radionuclides, (iii) for providing a baseline against which future surveys in the CEZ can be compared, (iv) as a source of information on the behaviour of fuel particles, (v) for performing retrospective dose assessments and (vi) for assessing natural background dose rates in the CEZ.

Whilst many of the results of the studies described have previously been used for various purposes within refereed or other publications (e.g. Kashparov et al., 2001, 2004, 2006; Kashparov, 2003; Zhurba et al., 2009), the complete underlying datasets have not previously been published. The dataset described here results from detailed samplings in the Ukraine and add to the now openly available Chernobyl soil contamination and deposition data for wider areas of the former Soviet Union and Western Europe (Chaplow et al., 2015a, b; Evangeliou et al., 2016).

2 Data

There are six datasets available in Kashparov et al. (2017) and an overview of the contents of each is presented below and in Table 1:

1. Spatial dataset of radiocaesium, ^{90}Sr and soil chemistry parameters resulting from a soil sampling exercise conducted during the summer of 1997. The dataset com-

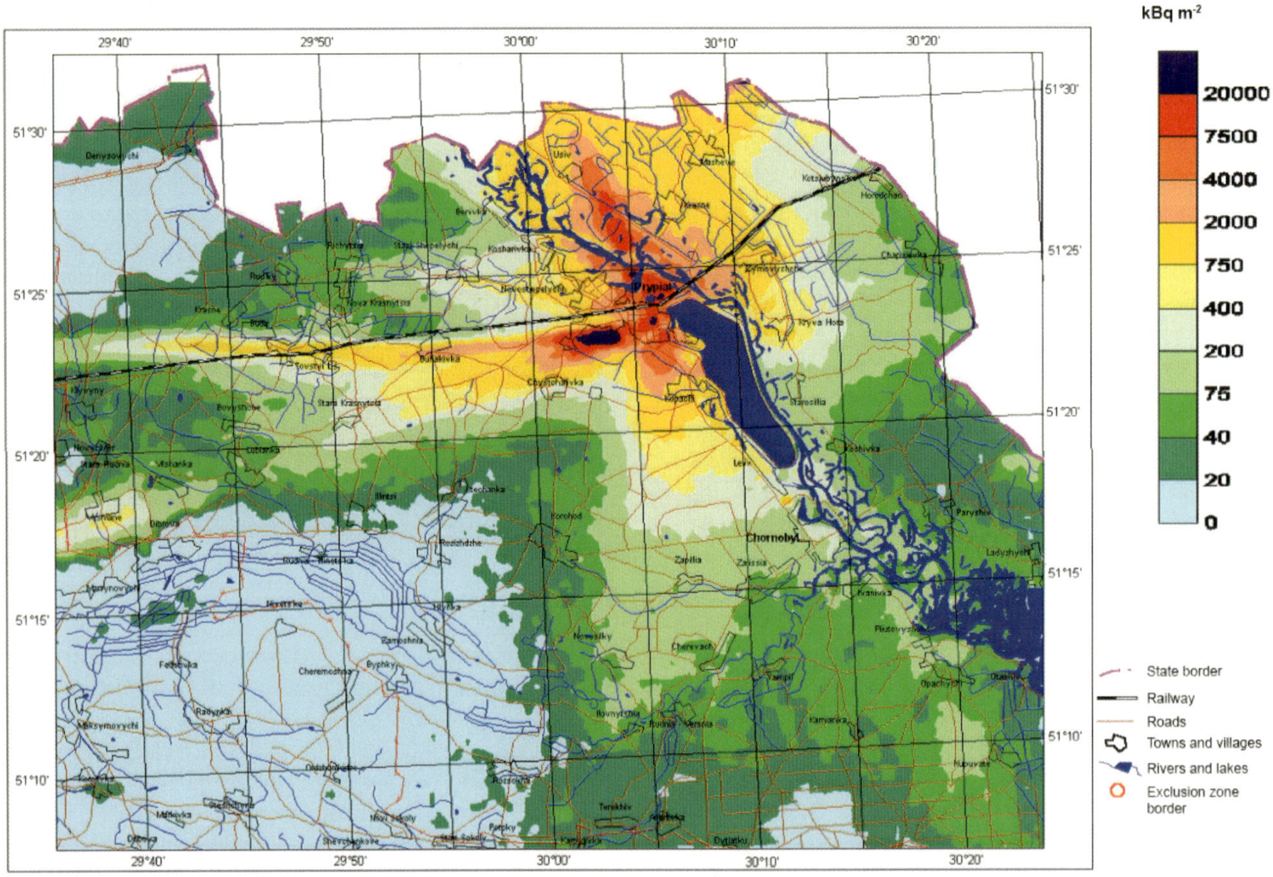

Figure 1. Spatial pattern of ^{90}Sr contamination (kBq m^{-2}) estimated for 1997 (UIAR, 1998).

Table 1. A description of the data presented in this paper and available from Kashparov et al., 2017.

Dataset ID	Date sampled	Sample description	Radionuclides reported	Other parameters reported
1	1995, 1999, 2000, 2002	Soil	^{134}Cs, ^{137}Cs, ^{90}Sr, ^{154}Eu ^{238}Pu, $^{239,240}Pu$	Soil chemistry and soil type
2	1997, 2000, 2002	Soil	^{238}Pu, $^{239,240}Pu$	None
3	2000, 2001	Soil	^{137}Cs, ^{90}Sr, ^{238}Pu, $^{239,240}Pu$	Depth and mass of soil layers
4	Gamma: 1986, 1987, 1988, 1989, 1991, 1992, 1995, 1997, 1998, 1999, 2000 Alpha: 1986, 1988, 1990, 1991 Beta: 1988, 1994, 1995	"Hot" (or fuel) particles in soil	^{95}Zr, ^{95}Nb, ^{106}Ru, ^{125}Sb, ^{134}Cs, ^{137}Cs, ^{144}Ce, ^{154}Eu, ^{155}Eu, ^{54}Mn, ^{60}Co, ^{241}Am Total alpha ^{90}Sr	Particle description (type, size, view, colour, structure), burn-up value
5	1995, 1996, 1997	Soil and hot (or fuel) particles	^{85}Sr, ^{90}Sr	None
6	2014	Soil and background dose rate	^{137}Cs, ^{90}Sr, ^{40}K, ^{226}Ra, ^{232}Th	Sample mass

prises results from 1200 soil samples reporting ^{134}Cs, ^{137}Cs, ^{90}Sr and some ^{154}Eu activities in soil (kBq m^{-2}); some chemical parameters for soils (e.g. pH, exchangeable Cs, percentage humus); soil type at the sampling site; and latitude–longitude co-ordinates. The dataset also contains information for additional samples from some of the datasets below (e.g. this dataset contains radionuclide activity in soil for all sampling though for some of these there are no corresponding soil chemistry data).

2. Pu isotope measurements in bulk soil samples reporting results for the analyses of ^{238}Pu and $^{239,240}Pu$ on a subset of 82 of the samples from the above dataset and an additional 12 samples collected from within the CEZ and outside it to a distance of 200 km between 2000 and 2001.

3. Pu isotope measurements in sectioned soil samples reporting results from cores sectioned at different depths collected at 9 of the 12 sites sampled between 2000 and 2001.

4. "Hot" or fuel particle dataset presenting radionuclide activity and some physical characteristics of (i) 1380 hot particles (predominantly fuel particles) extracted from soils collected in the CEZ (largely from in the inner 10 km zone) over the period 1987 to 2000, (ii) 206 particles collected from the northeast and Warsaw area (Poland) in September 1986 (data courtesy of Warsaw University; Dabrowska et al., 1987; Osuch et al., 1989), (iii) 294 particles collected from within the "shelter" (or sarcophagus) in 1992 and (iv) data from published literature on particles collected in different European countries between 1986 and 1987.

5. The fuel particle dissolution dataset presents results from studies on 115 soil samples, collected within the CEZ between 1995 and 1997, to determine the proportion of undissolved fuel particles.

6. The Ivankov district dataset presents unreported data of a survey of the Ivankiv Raion, a district in the Kiev Oblast immediately to the south of the CEZ, conducted in 2014. The Ivankov district borders the CEZ and Ivankov; the administrative centre is situated ~ 80 km from Kiev and ~ 50 km from the Chernobyl NPP. The dataset contains dose rates measured at approximately 3400 sites across the Ivankov district and activity concentrations of ^{90}Sr, ^{137}Cs and natural series gamma-emitting radionuclides in 547 soil samples sampled in 2014.

The above numbers are used in Table 1 and subsequently to identify the datasets.

Locations of sampling for studies 1, 2, 3 and 5 are presented in Fig. 2 and those for the Ivankov district are shown in Fig. 3. Reported activity concentrations for all datasets discussed are presented for the date of measurement which is provided in the accompanying datasets.

Overview of studies and available data

2.1 Spatial dataset of radiocaesium, ^{90}Sr and soil chemistry parameters

2.1.1 Soil sampling

Soil samples were collected at about 1200 sites in the Ukrainian CEZ in an area of 36 km radius around ChNPP between July 1995 and September 1997. Sites were selected using a grid and the approximate distance between sampling sites was about 1.2 km. For the narrow western (fuel particle) trace of the radioactive release, where contamination is characterised by a high gradient, the distance between the sampling sites was reduced to 100 to 500 m. Because of the large area of the CEZ, the majority of the sampling was conducted using a helicopter to rapidly move between sites; to sample forest areas motor vehicles were used for transport.

At each sampling point, the absorbed dose rate at a height of 1 m above ground surface was measured. Before soil sampling the homogeneity of the site (100 × 100 m) was evaluated by measurement of absorbed dose rate at several points using a DRG-01T dosimeter ("Mekhanichesky zavod", mechanical plant, Russia).

Composite soil samples were collected using the envelope method now referred to as SOU 74.14-37-425:2006. Five individual soil cores were taken to a depth of 30 cm, four from the corners and one from the centre of a plot that measured between 2 and 5 m^2. The samples were subsequently combined to form a composite. SOU refers to the standard of the Ministry of Agrarian Policy and Food of Ukraine (industry standard of Ukraine). This approach was developed to obtain a representative soil sample (Khomutinin et al., 2001). The total mass was not less than 2 kg for the five cores. The sampling depth of 30 cm was chosen because it had been shown that 10 years after the accident about 95 % of the ^{90}Sr activity is associated with the upper 10 to 20 cm layer (Ivanov et al., 1997). To test that the sampling depth was adequate the vertical distribution of ^{90}Sr in sandy soil profiles (up to 1 m depth) was determined and the $^{90}Sr : ^{154}Eu$ activity ratio in the samples was measured. Fuel particles have a characteristic $^{90}Sr : ^{154}Eu$; if this ratio deviates from that expected it is indicative of ^{90}Sr being leached down the profile. In most of the cores more that 95 % of the ^{90}Sr was located in the upper 30 cm layer. Only in a few sites (less than 0.1 % of all sites) with a low organic matter content was there a significant percentage of ^{90}Sr (> 20 %) that had migrated deeper than 30 cm (Shestopalov et al., 2003).

The co-ordinates of the sampling sites were determined by means of a GPS receiver (ScoutMaster) using the WGS-84 (world) system. In order to estimate the accuracy of the co-ordinates and their correspondence to the topographic basis,

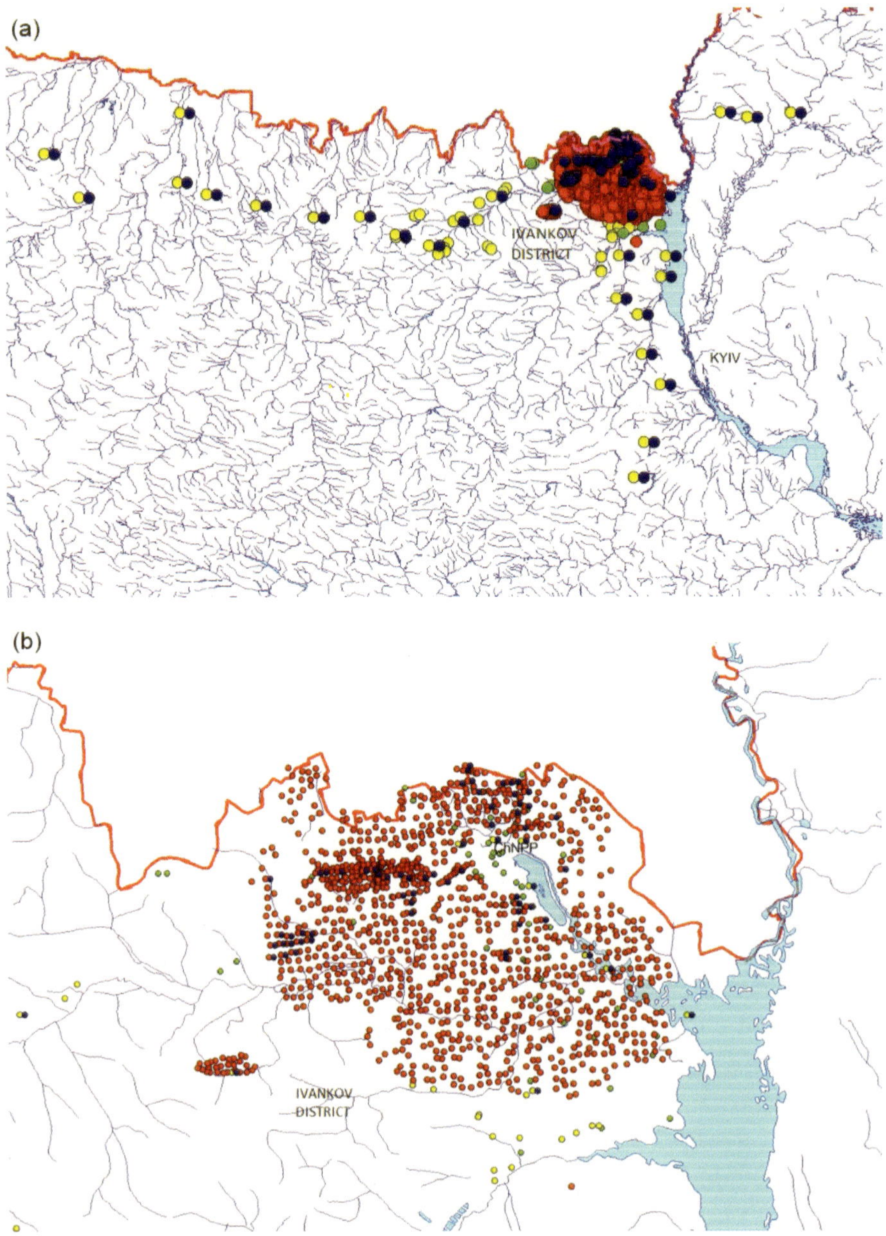

Figure 2. Sampling locations for the different studies in the CEZ (red – (1) spatial dataset of radiocaesium, ^{90}Sr and soil chemistry parameters; dark blue – (2) Pu isotope measurements in bulk soil samples; yellow – (3) Pu isotope measurements in sectioned soil samples; green – (5) fuel particle dissolution dataset; (6) Ivankov district (sampling locations see Fig. 3).

GPS measurements were performed at several points with known co-ordinates. It was shown that the accuracy of the GPS receiver was about 100 m with a systematic deviation of about 300 m.

The same methodology was used for estimating radionuclide activity concentrations in the other data described below unless otherwise indicated.

2.1.2 Soil activity measurements

The collected soil samples were dried, sieved through a 1 mm sieve and homogenised. Four subsamples were taken from each sample for the determination of the total contents of radionuclides: three subsamples of 100 cm^3 volume and one of 1000 cm^3 volume for the measurement of gamma-emitting radionuclides in a Marinelli container. Measurements of gamma-emitting radionuclide contents were performed on all four subsamples (C_i, Bq kg^{-1}) using gamma spectrometry. If the relative difference of the subsample

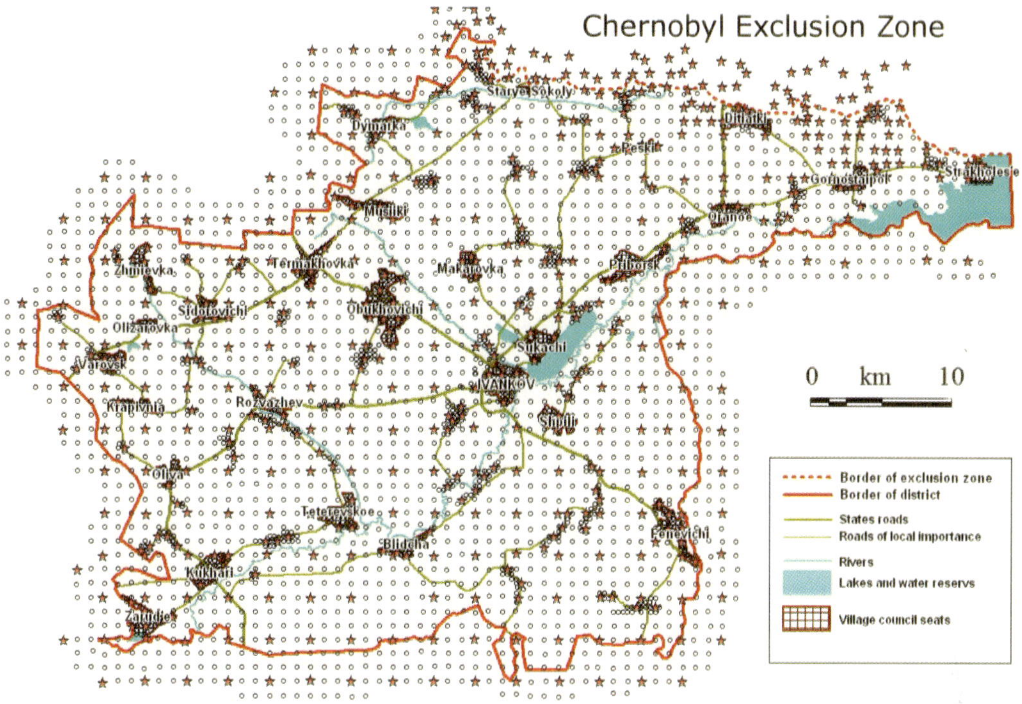

Figure 3. Map of the soil sampling locations in the Ivankov district which borders the CEZ: soil sampling and gamma survey (red stars); gamma survey only (grey circles).

measurements $(C_{max} - C_{min})/(C_{max} + C_{min})$ exceeded 0.15 for the ^{137}Cs activity concentrations, the subsamples were bulked and remixed. The resultant sample was again divided into replicates which were reanalysed. Once $(C_{max} - C_{min})/(C_{max} + C_{min})$ was < 0.15, a $100\,cm^3$ subsamples with an approximate average ^{137}Cs activity concentration was chosen for the measurement of the ^{90}Sr-specific activity. This procedure was followed to ensure that the subsample analysed for ^{90}Sr was representative of the sample as a whole.

Gamma-spectrometry measurements were performed using high-purity germanium (HPGe) detectors of 30 % relative efficiency and 1.90 keV FWHM for 1333 keV (GEM-30185, EG&G ORTEC, USA) and a multichannel analyser (ADCAM-300, ORTEC, USA) using GammaVision 32 software. The efficiency calibration was carried out for the 1 L Marinelli geometry using a spiked soil sample containing ^{241}Am, ^{243}Am, ^{152}Eu, ^{154}Eu, ^{137}Cs and ^{40}K. The average counting time was about 1 h; for samples of lower activity the counting time was extended to obtain an acceptable error on the measurement (< 30 % at 95th percentile confidence interval).

The activity of ^{90}Sr was determined using a radiochemical method (Pavlotskaya, 1997) with treatment of the samples by boiling in 6M nitric acid for 4 h. This method was used to dissolve as much of the fuel particles as possible (Kashparov, 2003); it has been estimated that for particles collected from the western trace within 5 to 10 km of the ChNPP this method may underestimate ^{90}Sr activity by up to 20 % (Kashparov et al., 2004). After filtration of the solution, hydroxides of high-valence metals (Fe, Al, Ti, Mn, Th, U) were extracted by adding ammonia into the solution. After acidification, the solution was left for 3 weeks to reach equilibrium between ^{90}Sr and ^{90}Y. Subsequently, stable Y was added to the solution and Y precipitated by addition of ammonia. The precipitate was incinerated to yield Y_2O_3 and after 1 day (to allow the decay or radium daughters) ^{90}Sr was measured using a low-background beta counter (CANBERRA-2400, USA).

The radiochemical procedure removed ^{137}Cs from the sample which increased the accuracy of the subsequent ^{154}Eu gamma-spectrometric measurements to determine the $^{90}Sr : ^{154}Eu$ ratio (gamma counting to determine the ^{154}Eu activity was conducted after extraction of high-valence metals).

2.1.3 Determination of the main agrochemical characteristics

Standard methods were used to determine soil chemistry: humus or organic matter content was determined by GOST 26213-91 (Derzhavin et al., 1991a); pH_{H_2O} by GOST 26423-85 (Derzhavin et al., 1985a); pH_{KCl} by GOST 26483-85 (Derzhavin et al., 1985b); hydrolytic acidity was determined using the Kappen method or GOST 26212-91 (Derzhavin et al., 1991c); exchangeable Ca was determined by GOST 26487-85 (Derzhavin et al., 1985c); and exchangeable K and

Table 2. Soil classifications appropriate to the CEZ.

Code used in data	Ex-USSR (including Ukraine) classification		FAO-UNESCO
1	Soddy underdeveloped		
2	Soddy slightly and mid-podzolic sandy and consolidated sandy		
3	Soddy slightly podzolic sandy loam and loamy		
4	Soddy mid-podzolic sandy loam and loamy		
6	Soddy slightly podzolic gleyic sandy and consolidated sandy	Soddy-podzolic	Podzoluvisol
7	Soddy slightly podzolic gleyic sandy loam and loamy		
8	Soddy mid- and heavy-podzolic gleyic sandy loamy		
9	Soddy slightly podzolic gleyic sandy and consolidated sandy		
10	Soddy mid- and heavy-podzolic gleyic sandy loam and loamy		
18	Grey forest	Grey forest	Greyzems
19	Dark-grey forest		
121	Meadow gleyic	Gleyic	Gleysols
122	Meadow and soddy gleyic carbonate		
124	Meadow pseudopodzolic and meadow pseudopodzolic gleyic	Meadow	Phaeozems
131	Meadow boggy		
133	Boggy		
135	Peaty boggy	Peat, boggy	Histosols
136	Peat boggy		
138	Low moor peat		
159	Soddy gleyic sandy and consolidated sandy	Soddy	Arenosols
160	Slightly soddy low-humus and non-humus sand	Alluvial	Fluvisols
161	Soddy sandy loam and loamy		
162	Soddy gleyic sandy loam and loamy	Soddy	Arenosols
167	Soddy solod		
168	Soddy pseudopodzolic gleyic		

P contents were determined using a method appropriate to the specific soil type, i.e. GOST 26207-91 (Derzhavin et al., 1991b) or GOST 26204-91 (Derzhavin et al., 1991d). GOST refers to the national standard of the USSR and Ukraine prior to the break-up of the Soviet Union in 1991.

Mechanical composition (i.e. percent clay, percent silt, sand) were determined on the soils but not reported in the data. Instead, they have been used to attribute soils to the USSR classification system. Table 2 presents soil types in the CEZ described using USSR classification and in accordance with the FAO/UNESCO system (Stolbovoi, 2000). The soil type at each sampling site was attributed to one of these codes and this is presented within the data.

Codes 116 to 143 in the data were sampled from around the village of Bober which was not within the CEZ but at which high deposition values were found. No agrochemical measurements were made at these sites except pH.

2.2 Pu isotope measurements

From the dataset described above, 82 samples from sites along the main traces of the release were selected for analysis of Pu isotopes. For determination of plutonium radioisotopes at a given site, one of the homogenised 100 cm^3 samples was used. After heating at 450 °C overnight, the samples were boiled for 4 h in 6M HNO_3. Plutonium radioisotopes were extracted by a standard radiochemical method (Pavlotskaya, 1997). Chemical yield was calculated using ^{236}Pu or ^{242}Pu as tracers. Plutonium was extracted from the resultant solution using ion-exchange resin (VP-1AP, Russia). After elution and evaporation plutonium was extracted by electrical deposition onto stainless steel plates. Activities on the plates were measured using a Soloist alpha-spectrometer equipped with a Soloist-U0300 detector (EG&G ORTEC, USA).

An additional set of 12 samples were collected in autumn 2000 and analysed to determine Pu isotopes using the same methodology. These samples were collected at distances of up to 200 km outside of the CEZ in westerly, southerly and easterly directions.

2.3 Pu isotope measurements at different depths

Apart from having total Pu isotopes determined as above for the bulked soil sample, cores from eight of these sampling sites were sectioned into layers (0–2, 2–4, 4–6, 6–8, 8–10,

10–15, 15–20, 20–25 and 25–30 cm) and analysed for Pu isotopes, ^{137}Cs and ^{90}Sr.

A further site close to Pripyat was sampled in 2001 to a depth of 110 cm and the core was sectioned into 10 cm slices (these are identified as 1_J_w in the accompanying datasets). Each of these depth samples was analysed to determine ^{238}Pu, ^{239,240}Pu, ^{137}Cs and ^{90}Sr as described above.

2.4 Hot or fuel particles

The initial explosion in the Chernobyl reactor released a large amount of uranium dioxide fuel particles with a median radius of 2–3 µm which were deposited in a narrow band up to 100 km to the west of the ChNPP (Kashparov et al., 1999). The explosion also released relatively large fuel fragments ranging from tens to hundreds of micrometres in size. These larger fragments were predominantly deposited at a distance of 2–5 km from the reactor (Kashparov et al., 1999). The deposition of particles was a distinguishing feature of radioactive contamination of the Chernobyl accident and the behaviour of radionuclides associated with such particles in the environment had not previously been considered (Loshchilov et al., 1991; Beresford et al., 2016).

In 1987, the State Committee of Hydrometeorology of the USSR and the Scientific Centre of the Defence Ministry of the USSR created a regular sampling network in the CEZ (Loshchilov et al., 1991). Fifteen points were chosen on each of 36 transects defined at 10° intervals within 60 km of the ChNPP. Soil samples were collected during 1987 and 1989 at each point, except for those located in woods, rivers and lakes; the majority of sampling sites were located within the inner 10 km of the CEZ. The samples (15 cm in diameter and 5 cm deep) were measured in Marinelli beakers using a high-purity Ge detector. From a subset of these samples, all collected within the 30 km zone, more than 1200 relatively large (size > 10 µm) hot particles (activity > 100 Bq) were identified and isolated by scanning thin soil layers with a dosimeter (Kuriny et al., 1993). In some instances more than one particle was extracted from the same soil sample (in the data the identifier code starts with the same number).

About 500 additional hot or fuel particles were collected in a similar manner from soils sampled using the same approach in the CEZ during the period 1989–1996 (Kashparov, 2003). In the inner zone of the CEZ, where these samples were collected, up to 97 % of particles comprised finely dispersed nuclear fuel particles. The other 3 % were what are known as ruthenium or condensed particles, a matrix of iron group elements with a high content of ^{103,106}Ru (Kashparov et al., 1996). Condensed particles were formed when highly volatile radionuclides released from the fuel matrix were condensed onto particles of dust, construction materials, etc. (Kashparov et al., 1996). Within the database the type of each particle is identified as fuel or condensed.

The specific activity of the fuel particles corresponded to the activity of the fuel at the moment of the accident, excluding the volatile and highly mobile radionuclides such as ^{134,137}Cs and ^{103,106}Ru (Kuriny et al., 1993; Kashparov, 2003). The presence of fissionable material in the particles was demonstrated using neutron-activation analysis, electron probe microanalysis and laser mass spectrometry. Particle enrichment with ^{235}U was found to be about 1 to 2 % (Kashparov et al., 1996); natural uranium consists of approximately 0.7 % ^{235}U (USNRC, 2017). Particle size and radionuclide activity were measured by optical microscopy and by alpha, beta and gamma spectrometry respectively. Gamma spectrometry was conducted using a high-purity germanium detector (GEM-30185, EG&G ORTEC, USA) of about 30 % efficiency and 1.85 keV energy resolution (for ^{60}Co, 1332 keV gamma rays) connected to a multichannel analyser (ADCAM™ MCA 350, EG&G ORTEC, USA). Alpha spectrometry was conducted using a gold-silicon detector (DKP-125-IA) linked to a SES-13 spectrometer. Following radiochemical separation using the methodology described above ^{90}Sr activity was determined using an alpha–beta detector (CANBERRA-2400, USA).

An additional set of 294 particles were obtained from inside of the shelter (or sarcophagus) in the reactor hall of the Chernobyl no. 4 unit early in 1992 and analysed as described above.

Using the ^{137}Cs and ^{90}Sr results for the particles, Kashparov (2003) estimated the effective duration and temperature for fuel particle annealing to be $\sim$3.5 s at 2400 K respectively for particles of > 10 µm from the western trace. This confirmed an explosion-like mechanism of fuel particle formation during the accident (Kashparov, 2003).

For particles collected in the CEZ the dataset presents "burn-up" values. Burn-up is used to describe the fraction of fuel atoms having undergone fission. The burn-up distribution of fuel particles larger than 10 µm collected from the CEZ indicated that, at the moment of the accident, particles were released from the less irradiated ("younger") part of the reactor core. Burn-up values were lower in these particles (9.8–11 MW day kg^{-1}) in comparison with the estimated most probable fuel burn-up value of 14 MW day kg^{-1} in the fourth unit of the ChNPP (Kuriny et al., 1993; Kashparov et al., 1996, 1997; Begichev et al., 1990). Table 3 presents estimated radionuclide activity concentrations at the time of the accident in fuel with different burn-up values. The lower burn-up of fuel contributing to particles within the CEZ resulted in particles with less radioactivity than if deposited particles had been largely formed from "older" fuel in the reactor.

The dataset also contains information on 206 particles collected from five areas in the northeast of Poland. Particles were collected in autumn 1986 from wasteland and forest clearings; the total area sampled was approximately 6000 m^2 (the data are supplied courtesy of Warsaw University; see Dabrowska et al., 1988, and Osuch et al., 1989, for methodology). In contrast to the CEZ, > 40 % of the particles collected in Poland were of the condensed form; the results for the

Table 3. The activity concentrations of radionuclides in the ChNPP fourth unit nuclear fuel with different burn-up, Bq g^{-1}.

Burn-up ($MW\,d\,kg^{-1}$)	^{90}Sr	^{95}Zr	^{106}Ru	^{134}Cs	^{137}Cs	^{144}Ce	^{125}Sb	^{154}Eu	^{155}Eu
15.20	1.54E+09	3.43E+10	9.66E+09	1.33E+09	1.91E+09	2.73E+10	2.64E+08	5.52E+07	6.48E+07
14.30	1.46E+09	3.46E+10	8.85E+09	1.16E+09	1.80E+09	2.69E+10	2.38E+08	4.91E+07	6.01E+07
13.40	1.38E+09	3.48E+10	8.09E+09	1.01E+09	1.69E+09	2.65E+10	2.13E+08	4.33E+07	5.54E+07
12.50	1.30E+09	3.50E+10	7.36E+09	8.74E+08	1.58E+09	2.59E+10	1.89E+08	3.81E+07	5.12E+07
11.60	1.22E+09	3.51E+10	6.67E+09	7.49E+08	1.47E+09	2.53E+10	1.68E+08	3.32E+07	4.69E+07
10.70	1.14E+09	3.53E+10	6.01E+09	6.35E+08	1.36E+09	2.46E+10	1.47E+08	2.87E+07	4.27E+07
9.84	1.05E+09	3.54E+10	5.38E+09	5.33E+08	1.24E+09	2.37E+10	1.29E+08	2.46E+07	3.86E+07
8.94	9.61E+08	3.55E+10	4.78E+09	4.41E+08	1.13E+09	2.27E+10	1.11E+08	2.09E+07	3.47E+07
8.05	8.72E+08	3.56E+10	4.21E+09	3.60E+08	1.02E+09	2.16E+10	9.48E+07	1.76E+07	3.09E+07
7.15	7.80E+08	3.55E+10	3.66E+09	2.87E+08	9.09E+08	2.03E+10	7.98E+07	1.45E+07	2.72E+07
6.26	6.88E+08	3.53E+10	3.14E+09	2.24E+08	7.96E+08	1.88E+10	6.60E+07	1.17E+07	2.36E+07
5.36	5.93E+08	3.48E+10	2.63E+09	1.69E+08	6.83E+08	1.70E+10	5.34E+07	9.22E+06	2.01E+07
4.47	4.98E+08	3.39E+10	2.15E+09	1.22E+08	5.70E+08	1.50E+10	4.18E+07	7.08E+06	1.66E+07
3.58	4.01E+08	3.23E+10	1.68E+09	8.32E+07	4.57E+08	1.27E+10	3.14E+07	5.15E+06	1.32E+07
2.68	3.03E+08	2.93E+10	1.24E+09	5.17E+07	3.43E+08	1.01E+10	2.20E+07	3.49E+06	9.86E+06
1.79	2.03E+08	2.42E+10	8.07E+08	2.74E+07	2.29E+08	7.19E+09	1.36E+07	2.09E+06	6.53E+06
0.89	1.02E+08	1.53E+10	3.95E+08	1.03E+07	1.15E+08	3.82E+09	6.30E+06	9.27E+05	3.24E+06

Polish particles were decay corrected to the day of the accident. The Polish particles were divided into three forms: A, B and C. Group A contained mostly ^{103}Ru and ^{106}Ru isotopes, group B particles were rich in other nuclides from the fission product spectrum and group C resembled group B particles in isotope contents but with a higher abundance of ruthenium.

For completeness and comparison, we have augmented the data with values from the published literature which present information on particles from Bialystok in the Masurian Lake region of northeast Poland (nine particles collected in November 1987) (Schubert and Behrend, 1987); Budapest, Hungary (15 particles collected in July 1986) (Balashazy et al., 1988); Stockholm, Gotland and Gävle in Sweden (41 particles collected in 1987) (Kerekes et al., 1991); Sofia, Bulgaria (five particles) (Mandjoukov et al., 1992). Most of the particles from these studies were also ruthenium condensation particles. Particles were selected in 1986 and activities were decay corrected to the time of the accident.

2.5 Fuel particle dissolution dataset

2.5.1 Soil sampling and preparation

Between July 1995 and May 1997, 115 soil samples were collected from sites within the CEZ at varying directions and distances from the ChNPP in order to reflect the variability of the physical and chemical characteristics of the fuel particles deposited (Kashparov et al., 1996, 1999, 2004). The sampling was carried out along the narrow western trace of fallout of the non-oxidised fuel particles of the first release (Chernobyl releases occurred over a period of 10 days, Smith and Beresford, 2005) and at the southern and northern fuel traces of fallout of oxidised fuel particles. At all sample sites, soddy podzolic sandy soils (type 2 from Table 2) were sampled. Where possible at each site, additional samples were collected from an area of peat soil within a radius of 300 m. It was assumed that within this area the deposited particles would be relatively similar, allowing the effect of soil type on particle behaviour to be studied. Density of contamination by ^{90}Sr varied at the sampling sites from 12 $kBq\,m^{-2}$ to 60 $MBq\,m^{-2}$.

Two soil cores were collected at each sampling point: one sample was collected to a depth of 5 cm with a 14 cm diameter; the second was taken to a depth of 30 cm with a 6 cm diameter. The 0 to 5 cm samples (containing the fuel particles) were used to determine soil pH_{H_2O} using the same method as above.

The ^{85}Sr and ^{90}Sr activity concentrations were determined on the 0 to 30 cm samples which were also used to estimate the fraction of remaining undissolved fuel particles. Soil samples were air dried (approximately 25 °C), passed through a 1 mm mesh sieve and homogenised prior to analysis.

2.5.2 Estimation of remaining fraction of fuel particles

The fuel particle component in the soil samples was estimated from the fraction of exchangeable ^{90}Sr determined using a 2M NH_4Ac extraction with ^{85}Sr yield monitor (Kashparov et al., 1999). Three or four subsamples of 100 cm^3 were taken from each 0 to 30 cm soil sample, and 5 mL of ^{85}Sr solution (activity of about 100 Bq) and 25 mL of water were added to each. After 6 to 48 days, the soil was subjected to extraction using 2 M NH_4Ac (solid : liquid ratio of 1 : 10). The soil-extract solution was shaken for 1 h and left for 1 day. Extractants and soil residues were separated by filtration, and the fraction of radiostrontium was determined in both the solution (A_{sol}) and the residue (A_{res}). The gamma-emitting radionuclides ^{137}Cs, ^{154}Eu, ^{241}Am and ^{85}Sr were measured using gamma spectrometry and ^{90}Sr by beta counting after radiochemical separation using the approach outlined above.

The percentages of ^{85}Sr and ^{90}Sr leached from soil (D^{90}Sr and D^{85}Sr respectively) were calculated as follows:

$$\Delta^{85}\mathrm{Sr} = \left\{ \frac{A_{sol}\left(^{85}\mathrm{Sr}\right)}{A_{sol}\left(^{85}\mathrm{Sr}\right) + A_{res}\left(^{85}\mathrm{Sr}\right)} \right\} \times 100\,\%,$$

$$\Delta^{90}\mathrm{Sr} = \left\{ \frac{A_{sol}\left(^{90}\mathrm{Sr}\right)}{A_{sol}\left(^{90}\mathrm{Sr}\right) + A_{res}\left(^{90}\mathrm{Sr}\right)} \right\} \times 100\,\%.$$

These percentages were used to calculate the fraction of undissolved fuel particles, estimated as

$$\Delta \mathrm{FP} = \left[1 - \frac{\Delta^{90}\mathrm{Sr}}{\Delta^{85}\mathrm{Sr}} \right].$$

For samples with a low ^{90}Sr : ^{154}Eu ratio, it had previously been shown that leached ^{90}Sr had been removed from the 0 to 30 cm layer (Kashparov et al., 1999). This migration was particularly significant for sandy soils with very low humus content. Here, for such samples, the total estimated ^{90}Sr content was used to assess the fraction of undissolved fuel particles rather than A_{res} (^{90}Sr). The total ^{90}Sr content was derived from the vertical distribution of ^{90}Sr and ^{154}Eu activity. From estimates of initial activity concentrations in the fuel the ^{90}Sr : ^{154}Eu ratio in 1996 of 56 would be expected (see discussion of predicted activity levels in fuel presented below).

The data contain a limited number of negative values which are the consequence of a low percentage of ^{90}Sr in the exchangeable form and a high associated error. Users of the data should assume these values to be zero.

2.6 Survey of the Ivankov district

A total of 3389 sampling points were identified for survey in 2014, using a 1 km grid placed over the Ivankov district and surrounding 3 km (Fig. 3) (Kashparov et al., 2014). The coordinates of the sampling sites were determined by means of a GPS receiver (Garmin, GPSMap 78s) using the WGS-84 (world) system with the accuracy of 10 m or less.

Measurements of ambient equivalent dose rate (hereafter referred to as dose rate) were carried out at heights of 1 and 0.1 m above ground surface at 1 km intervals using this grid. Equivalent dose was measured using certified gamma–beta-irradiation dosimeters (RKS-01, Stora-TU, Ukraine). The measurement range of the dosimeters was approximately 0.1 to 1000 μSv h^{-1} and the time taken to obtain a statistically reliable estimate was approximately 20 s. Seven of the sites selected for survey were inaccessible; however, the locations have been left in the data for completeness.

An estimation of "zero background" for the dosimeters was made by taking measurements over a frozen water body in the centre of the Ivankov district. The contribution of gamma irradiation from natural and anthropogenic radionuclides was seen to be negligible and hence the readings were the sum of cosmic radiation and the detectors' own background (Kashparov et al., 2014). The background dose rates of the devices themselves were determined in the laboratory by taking measurements within a lead-shielded (10 cm thick) area. The background dose rate of the detectors was estimated to be 0.05 μSv h^{-1} and the zero background determined over the frozen lake was 0.08 μSv h^{-1}. Therefore, the contribution of cosmic radiation in the study area can be estimated to be approximately 0.03 μSv h^{-1}. Contributions from cosmic and instrument background were subtracted from measured values reported in the dataset.

A total of 547 soil samples were collected. Sampling locations were every third dose rate measurement point apart from within settlements where soils were sampled at every measurement point. Sample sites were selected that were undisturbed, relatively flat and with consistent vegetation cover, as well as where dose rate estimates did not vary by more than 30 %. Samples were taken at a distance from the nearest buildings or trees which equated to 2 times the height of the buildings or trees. The exception was for samples collected from forest or scrubland where the sample site was positioned equidistantly from the nearest trees or shrubs. Sampling points were located no closer than 20 m to roads or places where accumulation or wash-off of radioactive contamination was possible.

The envelope method was used to collect soil samples using a 37 mm corer down to a depth of not less than 20 cm. The mass was not less than 3 kg for the composite sample.

Soil samples were oven dried at 105 °C to constant weight, homogenised and sieved through a 1 mm sieve. Subsamples of 100 to 150 g were taken for ^{90}Sr analysis and for gamma analysis a 1 L Marinelli container was filled.

The approach used to determine ^{90}Sr activity concentration in soil samples was based on the standard ISO 18589-5:2009 (International Organization for Standardization, 2009). Strontium-90 activity was estimated through measurement of its daughter product, ^{90}Y. The radiochemical preparation involved digestion of the ashed sample in 8 M HNO_3 followed by oxalate precipitation. Strontium was purified using ammonia and saturated sodium carbonate solution. The resultant strontium carbonate was dissolved in 2.5 M HNO_3; Y carrier was added and stored for 2 weeks such that ^{90}Y reached equilibrium. Yttrium was then precipitated as oxalate and ^{90}Y was measured using a beta spectrometer (SEB-70, AKP, Ukraine). The radiochemical separation yield was calculated using carriers such as stable Sr and Y measured using atomic absorption spectroscopy (Varian). Gamma analyses were conducted as described above. Quality assurance of radioanalytical procedures was based on ISO/IEC 17025 standards. The laboratory regularly participated in international and national proficiency tests (e.g. International Atomic Energy Agency).

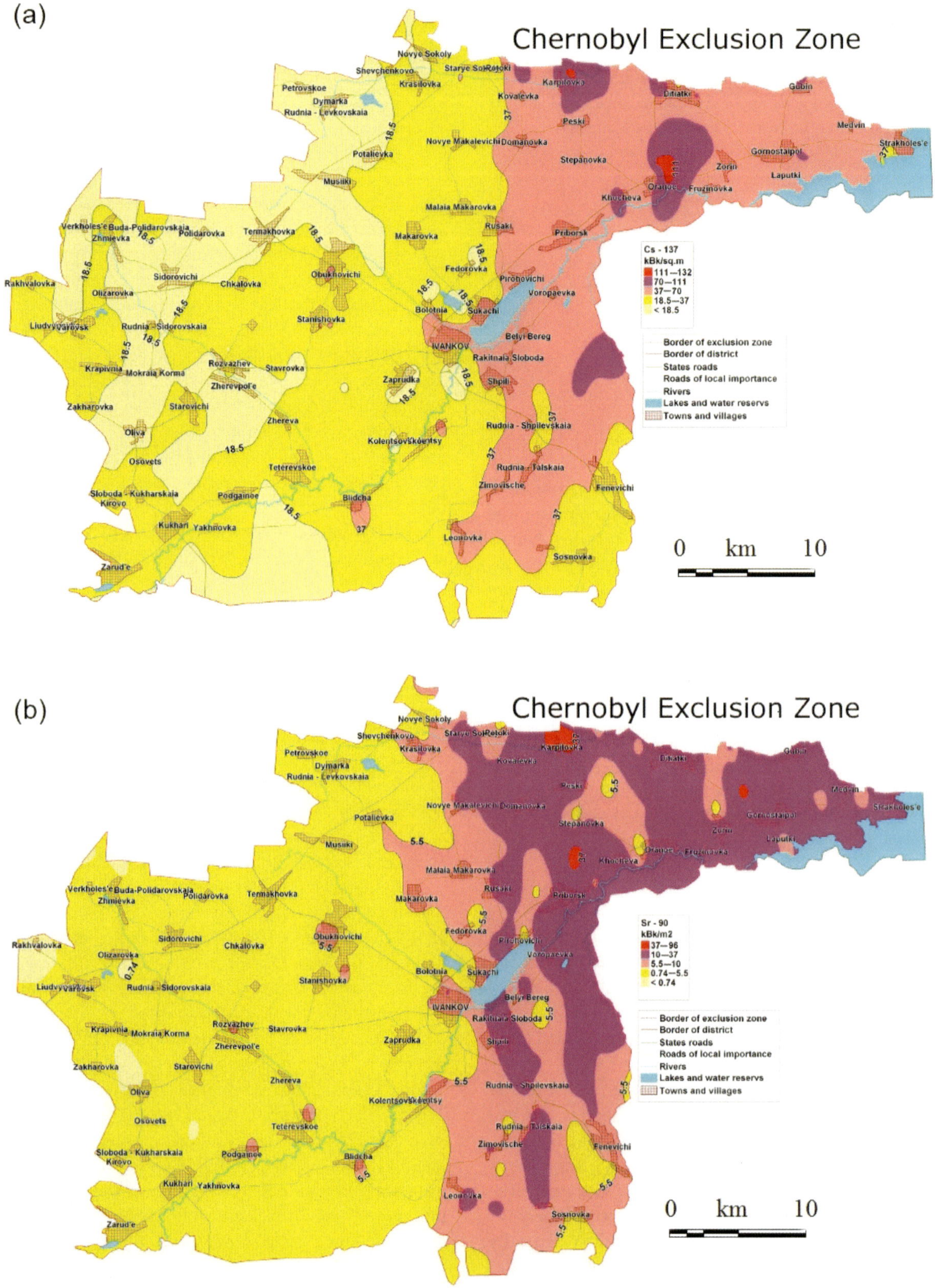

Figure 4. Spatial variation in ^{137}Cs **(a)** and ^{90}Sr **(b)** soil concentrations ($kBq\,m^{-2}$) in the Ivankov district in 2014.

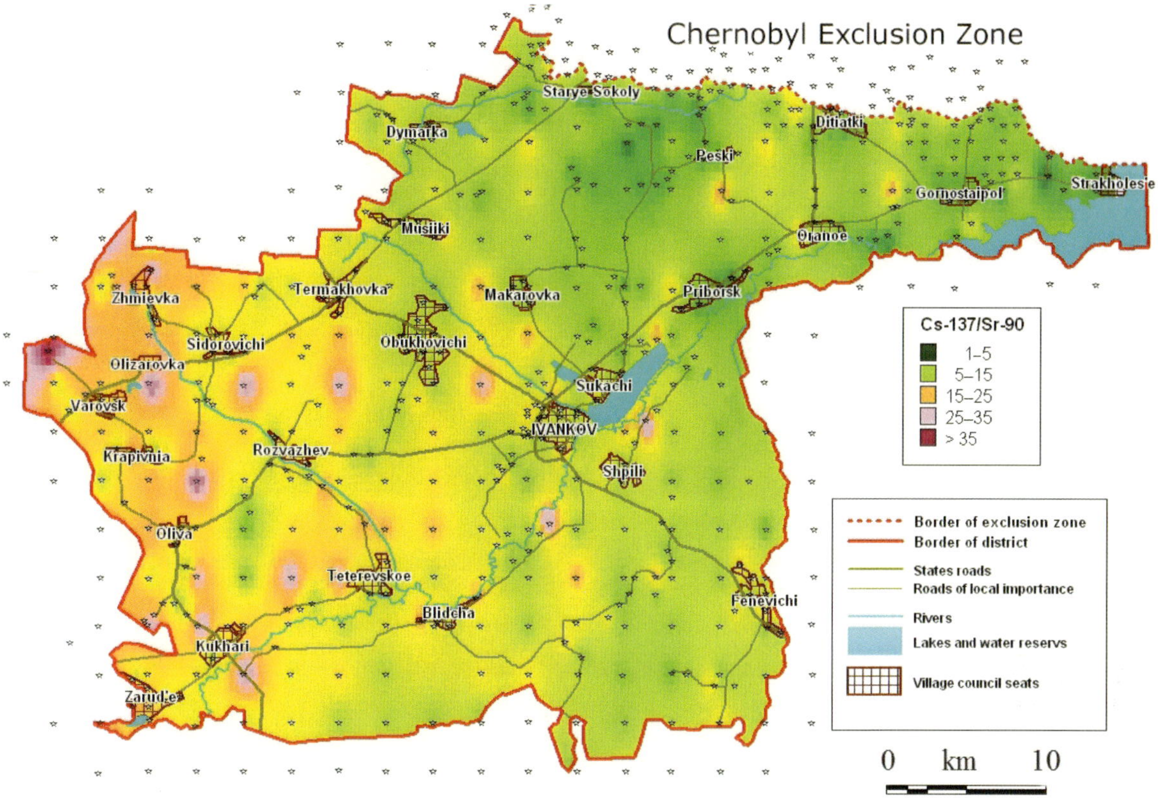

Figure 5. Spatial variation in $^{137}Cs : ^{90}Sr$ ratio in the Ivankov district (sampling points are shown as black stars).

The dataset reports activity concentrations for the natural radionuclides ^{40}K, ^{226}Ra and ^{232}Th in addition to ^{137}Cs and ^{90}Sr. The mean activity concentrations for these radionuclides were 140 Bq kg^{-1} ^{40}K, 12 Bq kg^{-1} ^{226}Ra and 10 Bq kg^{-1} ^{232}Th. Natural background radionuclides are sometimes cited as being relatively low in the CEZ (Møller and Mousseau, 2011) though there are few data in the international literature. On the basis of soil types we could expect the CEZ to have similar natural radionuclide activity concentrations in soils to the Ivankov district. These activity concentrations do appear to be relatively low compared to average values for e.g. the UK (Beresford et al., 2008).

Caesium-137 activity concentrations in the Ivankov district soils range from 6 to 390 Bq kg^{-1} dry matter. Strontium-90 concentrations range from 1 to 160 Bq kg^{-1} dry matter. These compare to anticipated global fallout values of approximately 4 Bq kg^{-1} ^{137}Cs and 1 Bq kg^{-1} ^{90}Sr. Activity concentrations of both ^{137}Cs and ^{90}Sr were highest in the east of the Ivankov district (Fig. 4). On an aerial basis ^{137}Cs ranged from 2 to 140 kBq m^{-2} and ^{90}Sr < 1 to 60 kBq m^{-2}. The $^{137}Cs : ^{90}Sr$ ratio increased to the west and south of the Ivankov district with distance from the ChNPP (Fig. 5).

Dose rates at 1 m across the Ivankov district ranged from 0.1 to 0.24 µSv h^{-1}. From measurements of natural radionuclides in soil and using conversion coefficients by UNSCEAR (1977, 1988), a dose rate of 0.01 to 0.07 µSv h^{-1} was estimated. There was a tendency for dose rates to increase with increasing ^{137}Cs activity concentration in soils although correlation was poor ($R^2 = 0.11$ for measurements at 1 m).

3 Data availability

All samples after gamma-spectrometry measurements are stored at NUBIP and can be made available on request.

The data described here have a digital object identifier (doi:10.5285/782ec845-2135-4698-8881-b38823e533bf) and are freely available for registered users from the NERC Environmental Information Data Centre (http://eidc.ceh.ac.uk/) under the terms of the Open Government Licence.

The data must be fully referenced for every use as follows: Kashparov, V., Levchuk, S., Zhurba, M., Protsak, V., Khomutinin, Y., Beresford, N. A., and Chaplow, J. S. (2017): Spatial datasets of radionuclide contamination in the Ukrainian Chernobyl Exclusion Zone. NERC-Environmental Information Data Centre, doi:10.5285/782ec845-2135-4698-8881-b38823e533bf. Supporting documentation to aid in the reuse of these data are also available from the EIDC.

Competing interests. The authors declare that they have no conflict of interest.

Acknowledgements. Funding for preparing this paper and the accompanying dataset (Kashparov et al., 2017) was provided by the EU COMET project (http://www.radioecology-exchange.org/content/comet) and TREE (http://www.ceh.ac.uk/tree) project funded by NERC, the Environment Agency and Radioactive Waste Management Ltd under the RATE programme.

Edited by: David Carlson
Reviewed by: Andrey Zaitsev and one anonymous referee

References

Aronov, V. I.: Methods of creation of maps of geological and geophysical characteristics and application of geometry for mapping of oil and gas deposits using computers, Book Company "Nedra", Moscow, Russia, 1990 (in Russian).

Balashazy, I., Hofmann, W., and Martonen, T. B.: A model of particle deposition at airway bifurcations, J. Aerosol Med., 1, 190–191, 1988.

Barnett, C. and Welch, S.: Thirty years after the Chernobyl accident: what do we know about the effects of radiation on the environment?, COMET project, European Commission, available at: http://www.radioecology-exchange.org/sites/www.radioecology-exchange.org/files/files/Deliverable_56_COMET_workshop_4_final.pdf (last access 13 June 2017), 2016.

Beaugelin-Seiller, K., Garnier-Laplace, J., and Beresford, N. A.: Estimating radiological exposure of wildlife in the field, J. Environ. Radioactiv., submitted, 2018.

Begichev, S. N., Borovoy, A. A., Burlakov, E. V., Havrilov, S. L., Dovbenko, A. A., Levina, L. A., Markushev, D. M., Marchenko, A. E., Stroganov, A. A., and Tataurov A. K.: Preprint IAE-5268/3: Reactor Fuel of Unit 4 of the Chernobyl NPP (a brief handbook), Moscow, Publisher Kurchatov Institute of Atomic Energy, 21 pp., 1990 (in Russian).

Beresford, N. A. and Copplestone, D.: Effects of ionizing radiation on wildlife: what knowledge have we gained between the Chernobyl and Fukushima accidents?, Integr. Environ. Asses., 7, 371–373, soi:10.1002/ieam.238, 2011.

Beresford, N. A., Barnett, C. L., Jones, D. G., Wood, M. D., Appleton, J. D., Breward, N., and Copplestone, D.: Background exposure rates of terrestrial wildlife in England and Wales, J. Environ. Radioactiv., 99, 1430–1439, https://doi.org/10.1016/j.jenvrad.2008.03.003, 2008.

Beresford, N. A., Barnett, C. L., Howard, B. J., Howard, D. C., Wells, C., Tyler, A. N., Bradley, S., and Copplestone, D.: Observations of Fukushima fallout in Great Britain, J. Environ. Radioactiv., 114, 48–53, https://doi.org/10.1016/j.jenvrad.2011.12.008, 2012a.

Beresford, N. A., Adam-Guillermin, C., Bonzom, J.-M., Garnier-Laplace, J., Hinton, T., Lecomte, C., Copplestone, D., Della Vedova, C., and Ritz, C.: Response to authors' reply regarding "Abundance of birds in Fukushima as judged from Chernobyl" by Møller et al., 2012, Environ. Pollut., 169, 139–140, https://doi.org/10.1016/j.envpol.2012.05.013, 2012b.

Beresford, N. A., Fesenko, S., Konoplev, A., Smith, J. T., Skuterud, L., and Voigt, G.: Thirty years after the Chernobyl accident, J. Environ. Radioactiv., 157, 38–40, https://doi.org/10.1016/j.jenvrad.2016.01.023, 2016.

Chaplow, J. S., Beresford, N. A., and Barnett, C. L.: Post Chernobyl surveys of radiocaesium in soil, vegetation, wildlife and fungi in Great Britain, NERC Environmental Information Data Centre, https://doi.org/10.5285/d0a6a8bf-68f0-4935-8b43-4e597c3bf251, 2015a.

Chaplow, J. S., Beresford, N. A., and Barnett, C. L.: Post-Chernobyl surveys of radiocaesium in soil, vegetation, wildlife and fungi in Great Britain, Earth Syst. Sci. Data, 7, 215–221, https://doi.org/10.5194/essd-7-215-2015, 2015b.

Dabrowska, M., Jaracz, P., Jastrzebski, J., Kaczanowski, J., Mirowski, S., Osuch, S., Piasecki, E., Pienkowski, L., Szeflinska, G., Szeflinski, Z., Tropilo, J., and Wilhlemi, Z.: Isotopic composition of hot particles collected in North-East Poland, International Workshop on Hot Particles in the Chernobyl Fallout, Theuern, Germany, 28–29 October, 1987.

Evangeliou, N., Hamburger, T., Talerko, N., Zibtsev, S., Bondar, Y., Stohl, A., Balkanski, Y., Mousseau, T. A., and Møller A. P.: Reconstructing the Chernobyl Nuclear Power Plant (CNPP) accident 30 years after, A unique database of air concentration and deposition measurements over Europe, Environ. Pollut., 216, 408–418, 2016.

Derzhavin, L. M., Samohvalov, S. G., Sokolov, N. V., Prizhukov, V. G., Shaymukhametova, A. A., Yerinov, A. L., Luchkina, L. E., Konkina, V. L., Pankova, N. G., Vasilevskaya, N. V., Pleshkova, A.P., Yakovleva, A. A., Gruzdeva, T. S., Chebotareva, N. A., Molkanova, L. I., Soboleva, O. V., and Kondratieva, G. K.: GOST 26423-85, Soil, Methods for determination of specific electric conductivity, pH and dense residue of aqueous extract, Ministry of Agriculture of the USSR, available at: http://docs.cntd.ru/document/gost-26423-85 (last access: 19 February 2018), 1985a.

Derzhavin, L. M., Samohvalov, S. G., Sokolov, N. V, Prizhukova, V. G., Shaymukhametova, A. A., Molkanova, L. I., Chukova, T. V., Belyanina, E. N., Mayortz, G. A., Mezhova, N. P., Erinov, A. E., Arsenyeva, M. N., Gruzdeva, T. S., Vasilevskaya, N. V., Sukhareva, V. N., Pleshkova, A. P., Yakovleva, A. A., and Orlov, A. N.: GOST 26483-85, Soil, Preparation of salt extract and determination of its pH by the CINAO method, Ministry of Agriculture of the USSR, available at: http://docs.cntd.ru/document/gost-26483-85 (last access: 19 February 2018), 1985b.

Derzhavin, L. M., Samohvalov, S. G., Sokolov, N. V, Prizhukova, V. G., Shaymukhametova, A. A., Molkanova, L. I., Chukova, T. V., Belyanina, E. N., Mayortz, G. A., Mezhova, N. P., Erinov, A. E., Arsenyeva, M. N., Gruzdeva, T. S., Vasilevskaya, N. V., Sukhareva, V. N., Pleshkova, A. P., Yakovleva, A. A., and Orlov, A. N.: GOST 26487-85, Soils, Determination of exchangeable calcium and exchangeable (mobile) magnesium by CINAO methods, Ministry of Agriculture of the USSR, available at: http://docs.cntd.ru/document/gost-26487-85 (last access: 19 February 2018), 1985c.

Derzhavin, L. M., Samohvalov, S. G., Sokolov, N. V., Orlov, A. N., Khabarov, K. A., Prizhukov, V. G., and Privalenkov S. Y.: GOST 26213-91, Soil, Methods for the determination of organic matter, Committee for Standardization and Metrology of the

USSR, available at: http://docs.cntd.ru/document/gost-26213-91 (last access: 19 February 2018), 1991a.

Derzhavin, L. M., Samohvalov, S. G., Sokolov, N. V., Orlov, A. N., Khabarov, K. A., Sukharev, V. N., Fedotov, M. I., and Sokolov, Y. V.: GOST 26207-91, Soils, Determination of mobile compounds of phosphorus and potassium by Kirsanov method modified by CINAO, Committee for Standardization and Metrology of the USSR, available at: http://docs.cntd.ru/document/gost-26207-91 (last access: 19 February 2018), 1991b.

Derzhavin, L. M., Samohvalov, S. G., Sokolov, N. V., Orlov, A. N., Khabarov, K. A., Vasilevskaya, N. V., and Erinov, A. E.: GOST 26212-91, Soils, Determination of hydrolytic acidity by Kappen method modified by CINAO, Committee for Standardization and Metrology of the USSR, available at: http://docs.cntd.ru/document/gost-26212-91 (last access: 19 February 2018), 1991c.

Derzhavin, L. M., Samohvalov, S. G., Sokolov, N. V., Orlov, A. N., , Sukharev, V. N., Fedotov, M. I., and Sokolov, V. V.: GOST 26204-91, Soils, Determination of mobile compounds of phosphorus and potassium by Chiricov method modified by CINAO, Committee for Standardization and Metrology of the USSR, available at: http://docs.cntd.ru/document/gost-26204-91 (last access: 21 February 2018), 1991d.

Hinton, T. G., Garnier-Laplace, J., Vandenhove, H., Dowdall, M., Adam-Guillermin, C., Alonzo, F., Barnett, C., Beaugelin-Seiller, K., Beresford, N. A., Bradshaw, C., Brown, J., Eyrolle, F., Fevrier, L., Gariel, J.-C., Gilbin, R., Hertel-Aas, T., Horemans, N., Howard, B. J., Ikäheimonen, T., Mora, J. C., Oughton, D., Real, A., Salbu, B., Simon-Cornu, M., Steiner, M., Sweeck, L., and Vives i Batlle, J.: An invitation to contribute to a strategic research agenda in radioecology, Environ. Radioactivity, 115, 73–82, https://doi.org/10.1016/j.jenvrad.2012.07.011, 2013.

International Organization for Standardization: Measurement of radioactivity in the environment – Soil – Part 5: Measurement of strontium 90, ISO 18589-5:2009, available at: https://www.iso.org/standard/40877.html (last access 17 May 2017), 2013.

Ivanov, Y. A., Lewyckyj, N., Levchuk, S. E., Prister, B. S., Firsakova, S. K., Arkhipov, N. P., Arkhipov, A. N., Kruglov, S. V., Alexakhin, R. M., Sandalls, J., and Askbrant, S.: Migration of ^{137}Cs and ^{90}Sr from Chernobyl fallout in Ukrainian, Belarussian and Russian soils, J. Environ. Radioactiv., 35, 1–21, 1997.

Kashparov, V. A.: Hot Particles at Chernobyl, Environ. Sci. Pollut. R., 10, 21–30, 2003.

Kashparov, V. A., Ivanov, Y. A., Zvarich, S. I., Protsak, V. P., Khomutinin, Y. V., Kurepin, A. D., and Pazukhin, E. M.: Formation of Hot Particles During the Chernobyl Nuclear Power Plant Accident, Nucl. Technol., 114, 246–253, 1996.

Kashparov, V. A., Ivanov, Y. A., Protsak, V. P., Khomucinin, Y. V., Yoschenko, V. I., and Pazukhin, E. M.: Assessment of the maximal temperature and of the duration of annealing of Chernobyl fuel particles during the accident, Radiokhimiya (Radiochemistry) 39, 66–70, 1997 (in Russian).

Kashparov, V. A., Oughton, D. H., Zvarich, S. I., Protsak, V. P., and Levchuk, S. E.: Kinetics of fuel particle weathering and ^{90}Sr mobility in the Chernobyl 30-km exclusion zone, Health Phys., 76, 251–259, 1999.

Kashparov, V. A., Lundin, S. M., Khomutinin, Y. V., Kaminsky, S. P., Levtchuk, S. E., Protsak, V. P., Kadygrib, A. M., Zvarich, S. I., Yoschenko, V. I., and Tschiersch, J.: Soil contamination with ^{90}Sr in the near zone of the Chernobyl accident, J. Environ. Radioactiv., 56, 285–298, 2001.

Kashparov, V. A., Ahamdach, N., Zvarich, S. I., Yoschenko, V. I., Maloshtan, I. M., and Dewiere, L.: Kinetics of dissolution of Chernobyl fuel particles in soil in natural conditions, J. Environ. Radioact., 72, 335–353, https://doi.org/10.1016/j.jenvrad.2003.08.002, 2004.

Kashparov, V. A.: Assessment of ecological risk caused by the long-living radionuclides in the environment, in: Ecotoxicology, Ecological Risk Assessment and Multiple Stressors, edited by: Arapis, G., Goncharova, N., and Baveye, P., Nato Sci. Peace Secur., 6, 155–164, 2006.

Kashparov, V. A., Levchuk, S., Protsak, V., Khomutinin, Yu, Maloshtan, I., Yoshenko, L., Otreshko, L., Pavlushenko, N., Kadygryb, A., Glukhovsky, A., Landshyn, V., Korol, V., and Lyubatchevskii, S. V.: Mapping the contamination of Ivankiv district territory with radionuclides Final report November 2013–September 2014 on the studies carried out by UIAR within the frameworks of contract No. 2013-04 from 19 November 2013, 2014.

Kashparov, V., Levchuk, S., Zhurba, M., Protsak, V., Khomutinin, Y., Beresford, N. A., and Chaplow, J. S.: Spatial datasets of radionuclide contamination in the Ukrainian Chernobyl Exclusion Zone, NERC-Environmental Information Data Centre, https://doi.org/10.5285/782ec845-2135-4698-8881-b38823e533bf, 2017.

Kerekes, A., Falk, R., and Suomela, J.: Analysis of hot particles collected in Sweden after the Chernobyl accident, Statens Stralskyddinstitut, SSI-rapport 91-02, 1991.

Khomutinin, Y. V., Kashparov, V. A., and Zhebrovska, K. I.: Sampling optimisation when radioecological monitoring, Book publisher "VIPOL" Kiev, Ukraine, available at: http://www.uiar.org.ua/Eng/7.pdf (last access: 19 February 2018), 2004.

Kuriny, V. D., Ivanov, Y. A., Kashparov, V. A., Loschilov, N. A., Protsak, V. P., Yudin, E. B., Zhurba, M. A., and Parshakov, A. E.: Particle associated Chernobyl fall-out in the local and intermediate zones, Ann. Nucl. Energy, 20, 415–420, 1993.

Loshchilov, N. A., Kashparov, V. A., Yudin, Y. B., Protsak, V. P., Zhurba, M. A., and Parshakov, A. E.: Experimental assessment of radioactive fallout from the Chernobyl accident, Sicurezza e Protezione, 25–26, 46–49, 1991.

Mandjoukov, I. G., Burin, K., Mandjoukova, B., Vapirev, E. I., and Tsacheva T. S.: Spectrometry and visualization of "standard" hot particles from the Chernobyl accident, Radiat. Prot. Dosim., 40, 235—244, 1992.

Møller, A. P. and Mousseau, T. A.: Efficiency of bio-indicators for low-level radiation under field conditions, Ecol. Indicators, 11, 424–430, 2011.

Osuch, S., Dabrovska, M., Jaracz, P., Kaczanowski, J., Le Van Khoi., Mirowski, S., Piasecki, E., Szeflinska, G., Szeflinski, Z., Tropilo, J., Wilheimi, Z., Jastrzebski, J., and Pienkowski, L.: Isotopic composition of high-activity particles released in the Chernobyl accident, Health Phys., 57, 707–716, 1989.

Oughton, D. H., Salbu, B., Brand, T. L., Day, J. P., and Aarkrog, A.: Under-determination of strontium-90 in soils containing particles of irradiated uranium oxide fuel, Analyst, 118, 1101–1105, 1993.

Pavlotskaya, F. I.: Main principles of radiochemical analysis of environmental objects and methods of measurements of strontium and transuranium elements radionuclides, J. Anal. Chem., 52, 126–143, 1997 (in Russian).

Salbu, B., Krekling, T., Oughton, D. H., Ostby, G., Kashparov, V. A., Brand, T. L., and Day, J. P.: Hot particles in accidental releases from Chernobyl and Windscale nuclear installations, Analyst, 119, 125–130, 1994.

Schubert, P. and Behrend, U.: Investigations of radioactive particles from the Chernobyl fall-out, Radiochim. Acta, 41, 149–155, 1987.

Shestopalov, V. M., Kashparov, V. A., and Ivanov, Y. A.: Radionuclide Migration into the Geological Environment and Biota Accident, Environ. Sci. Pollut. R., 10, 39–47, 2003.

Smith, J. T. and Beresford, N. A.: Radioactive fallout and environmental transfers, in: Chernobyl – catastrophe and consequences, edited by: Smith, J. T. and Beresford, N. A., Chichester, Praxis Publishing/Springer, 35–80, 2005.

SOU 74.14-37-424:2006. The quality of the soil, Determine the density of contamination of agricultural land anthropogenic radionuclides, Ministry of Agrarian Policy and Food of Ukraine, 2006 (in Ukrainian).

Steiner, M., Willrodt, C., Wichterey, K., Ikäheimonen, T., Ioshchenko, V., Hutri, K. L., Muikku, M., Outola, I., Beresford, N. A., Bradshaw, C., Dowdall, M., Eyrolle-Boyer, F., Guillevic, J., Hinton, T., Howard, B. J., Liland, A., Michalik, B., Mora, J. C., Oughton, D., Real, A., Robles, B., Salbu, B., Stark, K., and Sweeck, L.: Observatories for Radioecological Research – Description, STAR project, European Commission, available at: http://www.radioecology-exchange.org/sites/www.radioecology-exchange.org/files/STAR_Deliverable-2.3.pdf (last access: 3 June 2017), 2013.

Stolbovoi, V.: Soils of Russia: Correlated with the revised legend of the FAO soil map of the world and world reference base for soil resources, IIASA Research Report, IIASA, Laxenburg, Austria: RR-00-013, available at: http://pure.iiasa.ac.at/6111 (last access: 10 July 2017), 2000.

UIAR: The map of the 30-km Chernobyl zone terrestrial density of contamination with strontium-90 (in 1997), UIAR, Kyiv, Ukraine, 1998.

United Nations Scientific Committee on the Effects of Atomic Radiation (UNSCEAR): UNSCEAR 1977 Report, Sources and effects of ionizing radiation, available at: http://www.unscear.org/unscear/en/publications/1977.html (last access: 7 June 2017), 1977.

United Nations Scientific Committee on the Effects of Atomic Radiation (UNSCEAR): UNSCEAR 1988 Report, Sources, effects and risks of ionizing radiation, available at: http://www.unscear.org/unscear/en/publications/1988.html (last access: June 2017), 1988.

United States Nuclear Regulatory Commmission (USNRC): Uranium Enrichment, available at: https://www.nrc.gov/materials/fuel-cycle-fac/ur-enrichment.html, last access: 12 June 2017.

Zhurba, M., Kashparov, V., Ahamdach, N., Salbu, B., Yoschenko, V., and Levchuk, S.: The "hot particles" data base. Radioactive Particles in the Environment, NATO Science for Peace and Security Series C: Environmental Security, edited by: Oughton, D. and Kashparov, V., Springer, the Netherlands 187–195 https://doi.org/10.1007/978-90-481-2949-2_12, 2009.

9

Environmental conditions of a salt-marsh biodiversity experiment on the island of Spiekeroog

Oliver Zielinski[1], Daniela Meier[1], Kertu Lõhmus[2], Thorsten Balke[3], Michael Kleyer[2], and Helmut Hillebrand[1,4]

[1]Institute for Chemistry and Biology of the Marine Environment, University of Oldenburg, Wilhelmshaven, 26382, Germany
[2]Institute of Biology and Environmental Sciences, University of Oldenburg, Oldenburg, 26129, Germany
[3]School of Geographical and Earth Sciences, University of Glasgow, Glasgow, G128QQ, UK
[4]Helmholtz-Institute for Functional Marine Biodiversity, University of Oldenburg, Oldenburg, 26129, Germany

Correspondence: Oliver Zielinski (oliver.zielinski@uol.de)

Abstract. Field experiments investigating biodiversity and ecosystem functioning require the observation of abiotic parameters, especially when carried out in the intertidal zone. An experiment for biodiversity–ecosystem functioning was set up in the intertidal zone of the back-barrier salt marsh of Spiekeroog Island in the German Bight. Here, we report the accompanying instrumentation, maintenance, data acquisition, data handling and data quality control as well as monitoring results observed over a continuous period from September 2014 to April 2017. Time series of abiotic conditions were measured at several sites in the vicinity of newly built experimental salt-marsh islands on the tidal flat. Meteorological measurements were conducted from a weather station (WS, https://doi.org/10.1594/PANGAEA.870988), oceanographic conditions were sampled through a bottom-mounted recording current meter (RCM, https://doi.org/10.1594/PANGAEA.877265) and a bottom-mounted tide and wave recorder (TWR, https://doi.org/10.1594/PANGAEA.877258). Tide data are essential in calculating flooding duration and flooding frequency with respect to different salt-marsh elevation zones. Data loggers (DL) for measuring the water level (DL-W, https://doi.org/10.1594/PANGAEA.877267), temperature (DL-T, https://doi.org/10.1594/PANGAEA.877257), light intensity (DL-L, https://doi.org/10.1594/PANGAEA.877256) and conductivity (DL-C, https://doi.org/10.1594/PANGAEA.877266) were deployed at different elevational zones on the experimental islands and the investigated salt-marsh plots. A data availability of 80 % for 17 out of 23 sensors was achieved. Results showed the influence of seasonal and tidal dynamics on the experimental islands. Nearby salt-marsh plots exhibited some differences, e.g., in temperature dynamics. Thus, a consistent, multi-parameter, long-term dataset is available as a basis for further biodiversity and ecosystem functioning studies.

1 Introduction

Biodiversity is changing at an unprecedentedly high rate (Mace et al., 2005), reflecting the anthropogenic alteration of Earth's ecosystems (Rockström et al., 2009). As a consequence, research on biodiversity–ecosystem function (BEF) relationships has become a major facet of ecology and evolutionary biology (Balvanera et al., 2006; Cardinale et al., 2006; Hillebrand and Matthiessen, 2009; Cadotte et al., 2008; Gravel et al., 2011). Research on BEF has been dominated by experimental studies manipulating the number of species in a trophic group (Hillebrand and Matthiessen, 2009). However, the change in the number of species is not the predominant pattern in biodiversity change, which more frequently comprises altered dominances and species turnover (Hillebrand et al., 2017). Moreover, the mechanisms involved in altering species composition, i.e., immigration and extinction, are usually experimentally prohibited in most

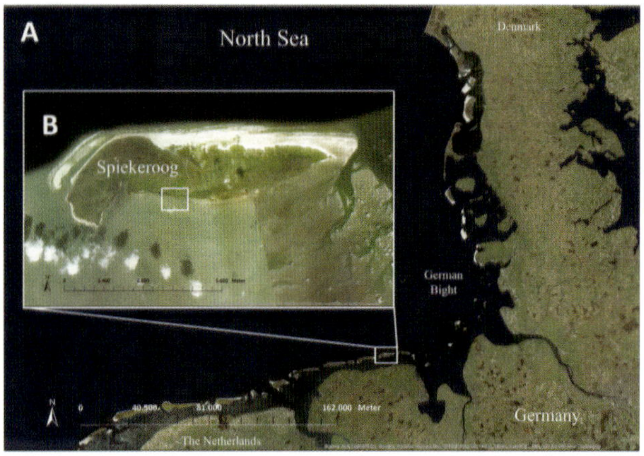

Figure 1. **(a)** Island of Spiekeroog located in the German Bight (North Sea). **(b)** The study site is located south of Spiekeroog (white box).

BEF studies, e.g., via weeding. On the other hand, there have been recent advances in understanding the interaction between regional community assembly, local dynamics and the biodiversity-functioning aspects (Hodapp et al., 2016; Leibold et al., 2017) that should be incorporated in BEF experiments.

The intertidal zone represents an interface between terrestrial and marine processes and biodiversity. This area is sensitive to climate change and is heavily impacted by anthropogenic activities (Cooley and Doney, 2009; Ekstrom et al., 2015; Haigh et al., 2015; Mathis et al., 2015). Within the intertidal environment, salt marshes have increasingly gained attention in times of sea-level rise (Kirwan and Megonigal, 2013; Balke et al., 2016), but studies on meta-community dynamics remain scarce.

Based on this backdrop the project BEFmate "Biodiversity–Ecosystem Functioning across marine and terrestrial ecosystems" was conducted in March 2014–December 2017, aiming to quantify the dynamics of biodiversity and the associated functions of salt marsh and tidal flat ecosystems. For this purpose, a series of experimental islands were set up in September 2014 on the back-barrier tidal flat of Spiekeroog island in the German East Frisian Wadden Sea (Balke et al., 2017) and were mirrored by salt-marsh enclosed plots located within the nearby salt marsh. Here we report abiotic parameters observed from 23 sensors installed either near the experimental islands, within the island structures themselves or within the nearby salt marsh as well as meteorological data from a locally installed weather station. We describe the instrumentation, data handling and results observed over a period of 32 months starting from the middle of September 2014. We further discuss the data and results with respect to validity and potential limitations of the observational setup.

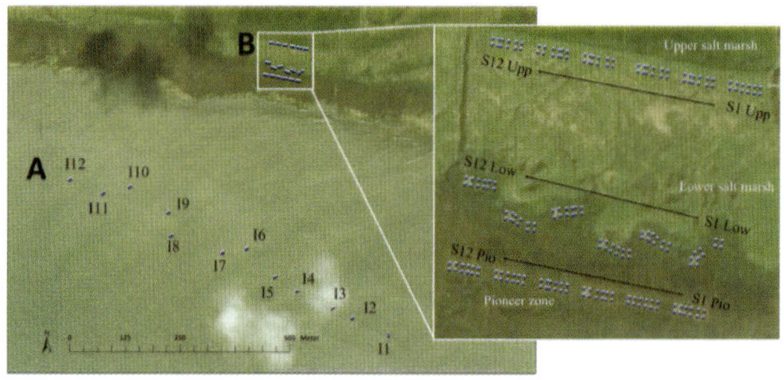

Figure 2. Experiment setup with **(a)** experimental islands in the back-barrier tidal flat and **(b)** salt-marsh control plots at different elevational zones in the salt marsh. Both, island and salt-marsh experimental plots are numbered from 1 to 12 from east to west.

2 Materials and methods

2.1 Research area

The island of Spiekeroog is located in the southern North Sea (Fig. 1) and is part of the Wadden Sea, which has been renowned as an UNESCO world natural heritage site since 2012. Twelve experimental islands (I1–I12) were built in the back-barrier tidal flats at distances of 240 m (I12) to 460 m (I1) from the southern salt marsh of the island of Spiekeroog. The experimental islands were distributed unevenly over 810 m from east to west at an elevation of 0.8 m NHN (standard height zero), with a mean tidal range of 2.7 m (Fig. 2, numbered 1–12 from east to west). The islands were built in a northeast–southwest direction, with the lowest elevation at the northeastern end of the island. The actual sensor position on the islands was determined by the local bathymetry, since the experimental islands encompass three different elevation levels (Fig. 3) that consist of the reflecting pioneer zone (Pio; 1.5 m NHN), lower salt-marsh zone (Low; 1.8 m NHN) and upper salt-marsh zone (Upp; 2.1 m NHN) of natural salt marshes. Half of the experimental islands were filled with mudflat sediment and left bare, whereas half of them were additionally transplanted with sods from the lower salt-marsh zone of the natural adjacent salt marsh. Experimental islands represent a treatment of dispersal limitation, constraining community assembly on the islands. Additionally 12 equally treated salt-marsh enclosed plots (S 1–12) were created (Fig. 2) that reflect unlimited dispersal. In addition to experimental plots, six control plots (C) per salt-marsh elevation zone were marked but left natural to compare established communities with the community assembly on dispersedly limited island plots and unlimited salt-marsh plots. The experimental design and setup of the experimental islands are not subject of this work and are described in detail in Balke et al. (2017). Abiotic conditions were measured at several sites due to the involvement of wide selection of parameters. Details concerning the individual sensors, their location, data provided and the associated methods of data

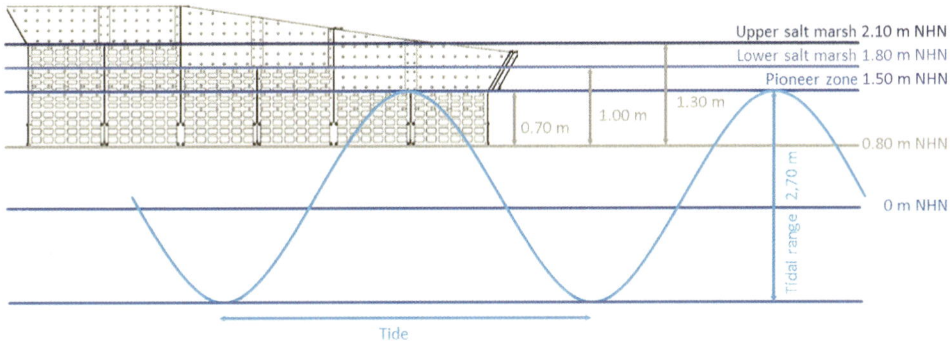

Figure 3. Scheme of an experimental island showing three elevational levels representing natural salt-marsh zones. The islands are built on an average of 0.8 m NHN (normal height null). The experimental islands' outer hull (galvanized steel) is patterned in the upper area to allow for a lateral water exchange. Lower area pattern is only meant to reduce weight and ease construction.

handling are provided in the following subsections. All positions, coordinates and elevations of sensors are indicated and provided in Table 1.

2.2 Instrumentation and data processing

2.2.1 Weather station (WS)

Meteorological data were collected near the experimental setup (see Table 1), with a locally installed weather station located approximately 500 m north of the southern shoreline (53°45′57.10″ N, 007°43′34.11″ E). The system was installed at the end of a glass fibre pole at a height of 10 m. The weather station system used here was a ClimaSensor US 4.920x.00.00x that was pre-calibrated by the manufacturer (Adolf Thies GmbH & Co. KG, D-Göttingen). Data were recorded and saved within the Meteo-Online (V4.5.0.20253) software in a sampling interval of 1 min, with an averaging time of 10 s, and the date and time were given in UTC. The position, solar azimuth angle and solar elevation were derived from the internal GPS system. Ultrasonic propagation time measurements were used for the determination of wind speed and direction. Two sensors were integrated for measuring air temperature and relative humidity (precision combination sensor) as well as atmospheric pressure (micro-electro-mechanical system technology). For recording and calculating precipitation, the back-reflected signal of a Doppler radar was used. An additional four photo sensors were used for the identification of light direction and light intensity. The post-processing of collected data was done using MATLAB (R2012b). Further quality control was performed by (a) erasing negative readings and data covering maintenance activities and (b) removing the outliers, defined as data exhibiting changes of more than two standard deviations within one time step. As the weather station was not oriented directly towards north, a manual correction to the north had to be done afterwards for accurate wind direction values (+20° from November 2014 to March 2016 and +10.1° since October 2016).

2.2.2 Recording current meter (RCM)

A RCM9 LW recording current meter (AADI, Aanderaa, Bergen/Norway; RCM DCS 4220) with additional temperature (3621), conductivity (3919) and pressure (4017) probes was deployed for deriving hydrographic conditions (see Table 1). The device was bottom mounted through a buried H-anchor between islands 6 (I6) and 7 (I7) (53°45′29.34″ N, 007°43′16.50″ E), approximately 35 m southeast of island 7 and 50 m southwest of island 6 in a shallow tidal creek (0.71 m NHN). The position was derived from a portable DGPS system (Leica Differential-GPS, SR530). The acoustic sensor head was placed 0.4 m above the sediment at 1.2 m above the mean low-water height. As a consequence, the sensor head became dry during low tide, and data had to be examined and eliminated accordingly. The sensor was pre-calibrated by the manufacturer. Recorded data were internally logged on a memory card (DSU 2990 E), with a sampling interval of 10 min until the readout with the Data Reading Program DRP 5059 software. The sampling interval was chosen to resolve the tidal changes expected (dominant M2 lunar tide with 12.4 h period) while at the same time providing a minimum deployment time of 2 months between services to account for the environmental protection regulations in this area. The same sampling interval was applied to the other sensors in this setup, except for the weather station. Date and time was given in UTC. Post-processing was performed using MATLAB (R2012b) to remove low-tide data. Conductivity values were used as an indicator of low-tide periods. Data were erased when conductivity values fell below 25 mS cm^{-1}. Further quality control was applied as described in Sect. 2.2.1 for the weather station (WS).

2.2.3 Tide and wave recorder (TWR)

Local tide and wave conditions were recorded with a RBRduo TD|wave sensor (RBR Ltd., Ontario/Canada). The sensor was bottom mounted in a shallow tidal creek (0.71 m NHN) through a steel girder (buried 0.3 m deep in

Table 1. Overview of all installed loggers at the experimental islands (I) and the salt-marsh enclosed plots (S) as well as the nearby installed sensors. WS: weather station, RCM: recording current meter, TWR: tide and wave recorder, DL: data logger.

Logger	Name	Location	Elevation zone	Latitude	Longitude	Elevation (m NHN)	Parameter
WS	WS	Shore		53°45′57.10″ N	7°43′34.11″ E	10 m	Meteorology
RCM	RCM	Seafloor		53°45′29.34″ N	7°43′16.50″ E	0.8 m	Current
TWR	RBR	Seafloor		53°45′29.34″ N	7°43′16.50″ E	0.8 m	Tide/wave
DL-W0	Hobo-P	I 3	Pole	53°45′26.33″ N	7°43′29.16″ E	3.5 m	Atm pressure
DL-W1	Hobo-P	I 3	Upp	53°45′26.33″ N	7°43′29.16″ E	1.0 m	Depth
DL-W2	Hobo-P	I 3	Low	53°45′26.38″ N	7°43′29.24″ E	0.9 m	Depth
DL-W3	Hobo-P	I 4	Pio	53°45′27.68″ N	7°43′24.85″ E	1.0 m	Depth
DL-W4	Hobo-P	I 1	Pio	53°45′24.61″ N	7°43′36.26″ E	1.1 m	Depth
DL-W5	Hobo-P	S 3	Low	53°45′43.62″ N	7°43′23.91″ E	0.9 m	Depth
DL-W6	Defi-D	I 4	Low	53°45′27.64″ N	7°43′24.75″ E	1.0 m	Depth
DL-T1	Defi-T	I 3	Pio	53°45′26.42″ N	7°43′29.34″ E	1.4 m	Temperature
DL-T2	Defi-T	I 3	Low	53°45′26.38″ N	7°43′29.24″ E	1.7 m	Temperature
DL-T3	Defi-T	I 3	Upp	53°45′26.33″ N	7°43′29.16″ E	2.0 m	Temperature
DL-T4	Defi-T	S 3	Low	53°45′43.62″ N	7°43′23.91″ E	1.7 m	Temperature
DL-T5	Defi-T	S 2	Upp	53°45′45.17″ N	7°43′24.80″ E	2.0 m	Temperature
DL-T6	Defi-T	S 2	Low	53°45′43.54″ N	7°43′24.27″ E	1.7 m	Temperature
DL-L1	Defi-L	I 3	Pio	53°45′26.42″ N	7°43′29.34″ E	1.4 m	Light
DL-L2	Defi-L	I 3	Low	53°45′26.38″ N	7°43′29.24″ E	1.7 m	Light
DL-L3	Defi-L	I 3	Upp	53°45′26.33″ N	7°43′29.16″ E	2.0 m	Light
DL-L4	Defi-L	S 3	Low	53°45′43.62″ N	7°43′23.91″ E	3.5 m	Light
DL-L5	Defi-L	I 3	Pole	53°45′26.33″ N	7°43′29.16″ E	2.5 m	Light
DL-L6	Defi-L	Seafloor		53°45′29.34″ N	7°43′16.50″ E	0.8 m	Light
DL-C1	Hobo-C	I 3	Pio	53°45′26.42″ N	7°43′29.34″ E	0.9 m	Salinity
DL-C2	Hobo-C	S 3	Pio	53°45′42.88″ N	7°43′23.55″ E	0.9 m	Salinity

the sediment) next to the RCM 9 LW recording current meter between islands 6 and 7 (53°45′29.34″ N, 007°43′16.50″ E) (see Table 1). The sensor was positioned 10 cm above sediment surface, as was determined by using a portable differential GPS (Leica Differential-GPS, SR530). This resulted in the sensor falling dry during low tide and data had to be flagged accordingly. The sensor was pre-calibrated by the manufacturer and the sampling rate was 3 Hz with 1024 samples per burst at a sample interval of 10 min. Recorded data were internally logged until the readout with the Ruskin (V1.13.10) software and post-processed using MATLAB (R2012b). Date and time was given in UTC. For accurate depth calculations, raw pressure data were manually corrected for atmospheric pressure derived from the locally installed weather station. Low-tide data were not removed, but were easily identified through the manually calculated water depth data, where all depths < 0.05 m represented low-tide data. Again, quality control was applied as described in Sect. 2.2.1 for the WS.

2.2.4 Data logger (DL)

Several data loggers were installed on the experimental islands as well as in the salt-marsh enclosed plots for the observation of groundwater level, temperature, light and salinity (see Table 1). In all cases the date and time is given in UTC and all post-processing was performed using MATLAB (R2012b). Quality control was applied as described in Sect. 2.2 for the weather station (WS).

To continuously observe the flooding and the groundwater levels on the experimental islands as well as in the salt marsh, pressure loggers were deployed in dip wells within the experimental setup at different elevational levels. Six HOBO® U20L Water Level Logger (onset® HOBO® Data Loggers, Bourne, MA, USA; S/N 10685287, 10685288, 10685289, 10685290, 10685291, 10685292; Hobo-P) as well as a DEFI-D Miniature Pressure Recorder (S/N OA5K008; DEFI-D) were deployed. All water level loggers were pre-calibrated by the manufacturer. Recorded data were internal logged until the readout afield with the Hobo Underwater Shuttle (U-DTW-1) and the HOBOware Pro (V3.7.4) software respectively with the DEFI Series software (V1.02). For depth calculations, pressure data were manually corrected by atmospheric pressure. Accordingly, a HOBO® U20L Water Level Logger was installed outside the dip wells at a higher elevation, attached on a steel pole at the upper zone of island 3. All loggers were initially calibrated to get the exact height inside the dip well. Data of one of the HOBO loggers had to be corrected with −0.082 m due to a wrong initial measurement. Further corrections were applied to the DEFI-D Log-

Table 2. Structure of the weather station (WS) datasets at PANGAEA.

#	Name	Short name	Unit
1	DATE/TIME	Date/Time	
2	ALTITUDE	Altitude	m
3	Wind speed	ff	$m\,s^{-1}$
4	Wind direction	dd	deg
5	Temperature, air	TTT	degC
6	Humidity, relative	RH	%
7	Brightness, North	Brightness	lux
8	Brightness, East	Brightness	lux
9	Brightness, South	Brightness	lux
10	Brightness, West	Brightness	lux
11	Brightness, Max	Brightness	lux
12	Brightness direction	Brightness dir	deg
13	Indicator for inclusion or omission of precipitation data	Indicator precipitation	code
14	Precipitation, daily total	Precip day total	mm
15	Precipitation	Precip	$mm\,h^{-1}$
16	Precipitation description	Precip descry	
17	Solar elevation	SE	deg
18	Solar azimuth angle	SAA	deg

Table 3. Structure of the recording current meter (RCM) datasets at PANGAEA.

#	Name	Short name	Unit
1	DATE/TIME	Date/Time	
2	Depth, water	Depth water	m
3	Current speed	V	$cm\,s^{-1}$
4	Current direction	DIR	deg
5	Temperature, water	Temp	degC
6	Conductivity	Cond	$mS\,cm^{-1}$
7	Pressure, raw	Press raw	dbar

Table 4. Structure of the tide and wave recorder (TWR) datasets PANGAEA.

#	Name	Short name	Unit
1	DATE/TIME	Date/Time	
2	Depth, water	Depth water	m
3	Tidal slope	Tidal slope	
4	Wave height, significant	$H(1/3)$	m
5	Wave period, significant	$T(1/3)$	s
6	Wave height, tenth	$H(1/10)$	m
7	Wave period, tenth	$T(1/10)$	s
8	Wave height, maximum	H(max)	m
9	Wave period, maximum	T(max)	s
10	Wave height, average	H(avg)	m
11	Wave period, average	T(avg)	s
12	Wave energy	Wave energy	$J\,m^{-2}$
13	Temperature, water	Temp	degC
14	Pressure, raw	Press raw	dbar

ger, since a manual correction with atmospheric pressure was not possible. As the local maxima of other water level loggers were very similar, a mean value of other loggers was calculated to correct the DEFI-D logger values with +0.25 m.

Temperature in the sediment surface layer (in approximately 0.05 m depth) was measured with six DEFI-T miniature temperature recorders (JFE Advantech Co., Ltd., Tokyo; DEFI-T). The manufacturer pre-calibrated temperature recorders and were installed on the experimental islands and in salt-marsh enclosed plots at different elevation levels. Recorded data were internally logged until the readout with the DEFI Series software (V1.02).

Light availability was measured with six locally installed light intensity loggers on the experimental islands as well as in the salt-marsh plots at different elevation levels. The DEFI-L Miniature Light Intensity Recorder (JFE Advantech Co., Ltd., Tokyo; DEFI-L) used here were pre-calibrated by the manufacturer. Recorded data were internally logged until the readout with the DEFI Series software (V1.02). Due to different calibrations for underwater or above-water applications, raw data were processed differently and merged depending on water levels. The processed pressure and depth data of the RBRduo TD|wave sensor were used to identify the flooding times and durations of the different elevations.

Two HOBO conductivity loggers (Onset Computer Corporation, Bourne, MA, USA) were installed inside of dip wells on the experimental islands as well as in the salt-marsh enclosed plots at the pioneer zone. The conductivity logger used here was the HOBO U24 Conductivity Logger U24-002-C (S/N 10570000, 10599255). The conductivity loggers

Table 5. Structure of the water level logger (DL-W) datasets at PANGAEA.

#	Name	Short name	Unit
1	DATE/TIME	Date/Time	
2	Water level	Water level	m

Table 6. Structure of the temperature logger (DL-T) datasets at PANGAEA.

#	Name	Short name	Unit
1	DATE/TIME	Date/Time	
2	Depth, sediment/rock	Depth	m
3	Temperature, in rock/sediment	*t*	degC

Table 7. Structure of light logger (DL-L) datasets at PANGAEA.

#	Name	Short name	Unit
1	DATE/TIME	Date/Time	
2	Light intensity	Io	$\mu mol\, m^{-2}\, s^{-1}$

Table 8. Structure of conductivity logger (DL-C) datasets at PANGAEA.

#	Name	Short name	Unit
1	DATE/TIME	Date/Time	
2	Depth, sediment/rock	Depth	m
3	Salinity	Sal	

were pre-calibrated by the manufacturer. Recorded data were internally logged until the readout afield with the Hobo Underwater Shuttle (U-DTW-1) and in the following with the HOBOware Pro (V3.7.4) software. An automatic calculation of salinity was conducted within the software according to PSS-78 using the measured conductivity and temperature. Due to fluctuations in the groundwater level, conductivity loggers periodically became dry, especially in the beginning of the deployment. Data until October 2015 are therefore very scattered. The depth of conductivity loggers was thereupon adjusted to the bottom of the dip wells, assuring a constant coverage with water. Data from dry sensors was removed, using a salinity of 20 as a threshold value. As a reference, soil samples of all plots on the experimental islands and salt-marsh enclosed plots were sampled to analyze pore water salinity in laboratory (data not shown here). Comparative data of meteorological and hydrographic conditions for validation processes were taken from the nearby Time Series Station – Spiekeroog (TSS) at Otzumer Balje (Holinde et al., 2015; Baschek et al., 2017).

2.3 Data provenance, structure and availability

All datasets described herein are available at the World Data Center PANGAEA (https://www.pangaea.de/) when searching for "BEFmate" as identifier (Zielinski et al., 2017a–g). Due to extensive datasets as well as individual questions of potential users, a collection (parent) for each sensor (type) was created. Each parent serves as a basis where data are divided into monthly sections enabling an easy selection of required time periods. The raw data and MATLAB routines applied in the curation process are available on request.

2.3.1 Weather station (WS)

Data of meteorological observations are available at PANGAEA from November 2014 to April 2017 (https://doi.org/10.1594/PANGAEA.870988). Gaps in the dataset resulted from a malfunction of the pressure sensor and an adjacent maintenance and re-calibration by the manufacturer. The structure of dataset is shown in Table 2.

2.3.2 Recording current meter (RCM)

Continuous current data are available on PANGAEA for September 2014 to October 2015 (https://doi.org/10.1594/PANGAEA.877265). There are additional datasets before the main sampling period of November and December 2013, March 2014 and June 2014. These data were used as a basis for deciding on the final placement and orientation of the experimental islands. Gaps in the dataset are the result of local readouts and maintenance. Low-tide data were removed from the dataset. Data after October 2015 is missing due to sensor malfunctions. The structure of the datasets is shown in Table 3.

2.3.3 Tide and wave recorder (TWR)

Tide and wave data are available on PANGAEA from October 2014 to April 2017 (https://doi.org/10.1594/PANGAEA.877258). Gaps in the dataset are the result of local readouts and maintenance. Low-tide data were not removed from the dataset. The depth was calculated from raw pressure, which was corrected beforehand with atmospheric pressure. The structure of the datasets is described in Table 4.

2.3.4 Water level loggers (DL-W)

Data for groundwater level on the experimental islands and salt-marsh enclosed plots are available on PANGAEA from June 2015 to April 2017 (https://doi.org/10.1594/PANGAEA.877267). Gaps in the dataset are the result of local readouts and maintenance. For depth calculations, pressure data were manually corrected by atmospheric pressure. Each dataset includes date/time and water level (water level measured in m; Table 5).

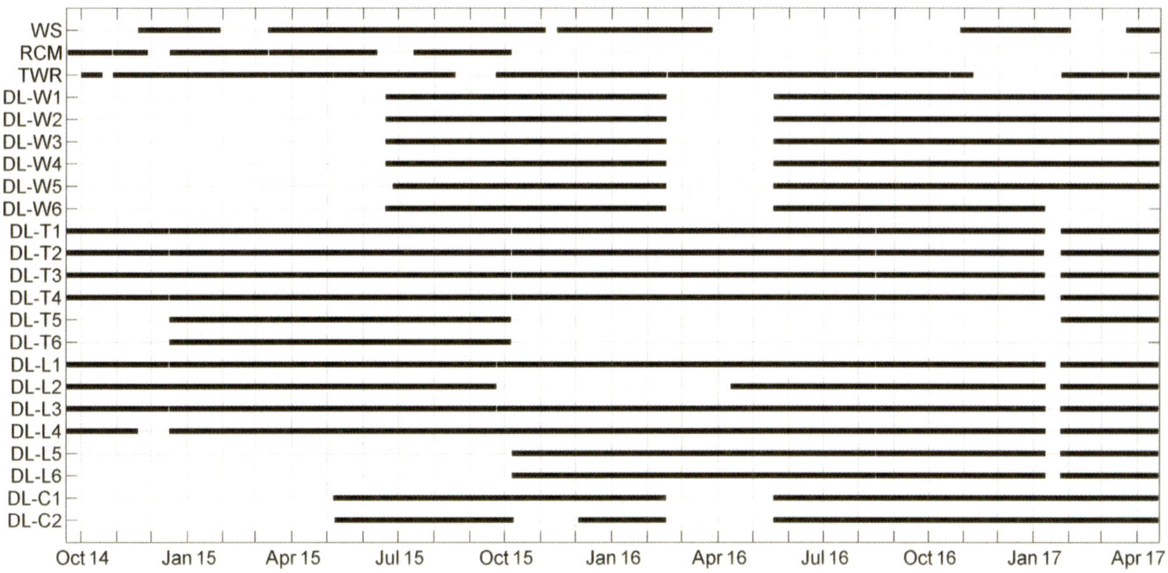

Figure 4. Available data shown in black bars for each sensor or logger type over the sampling period from 18 September 2014 to 18 April 2017.

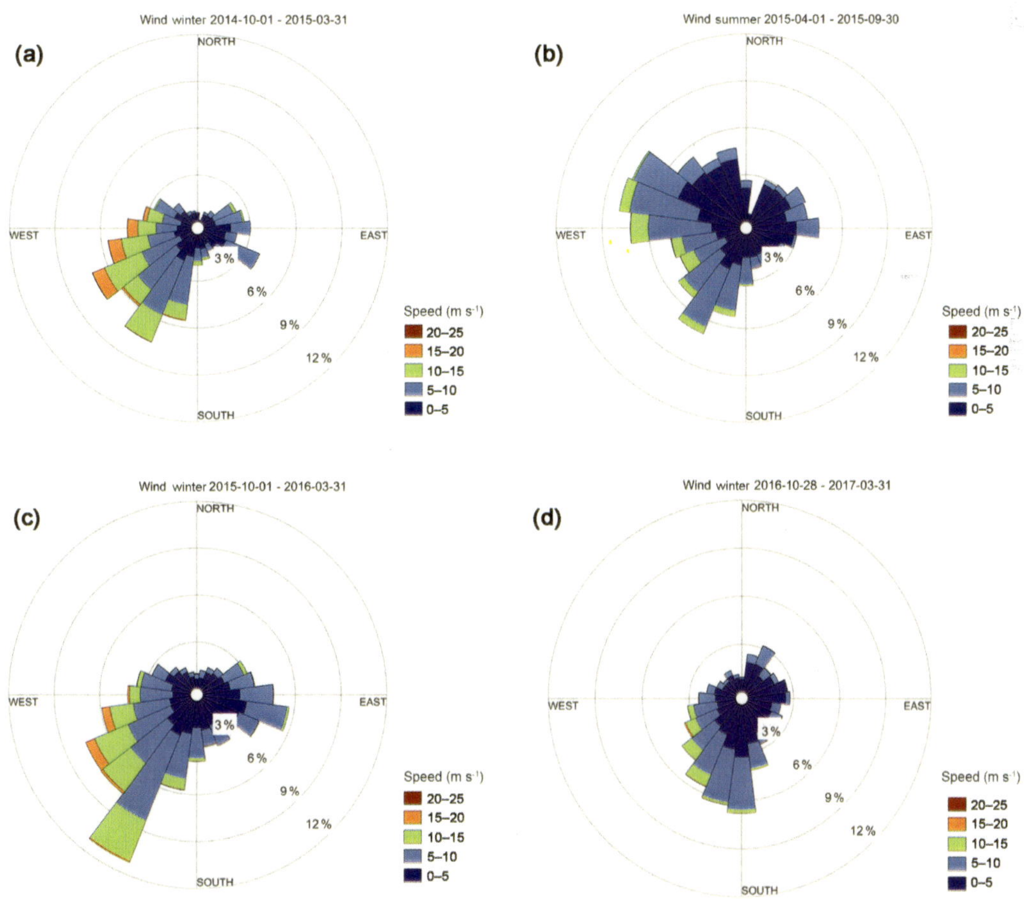

Figure 5. Local wind roses (wind speed and wind direction) for storm seasons (1 October–31 March) in 2014 and 2015 (**a**), 2015 and 2016 (**c**) and 2016 and 2017 (**d**) as well as for one summer season (1 April–30 September) in 2015 (**b**).

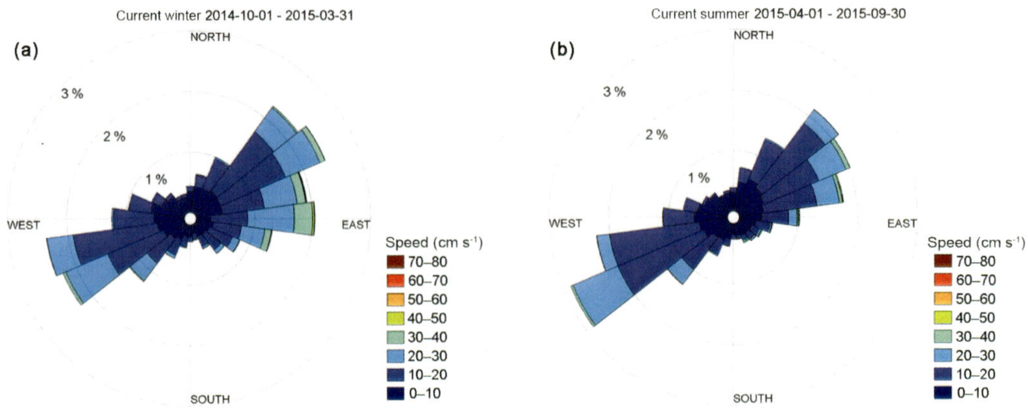

Figure 6. Current conditions in the storm season **(a)** from October 2014 to March 2015 and the summer season **(a)** from April 2015 to September 2015.

2.3.5 Temperature loggers (DL-T)

Temperature data for the surface sediment layer of the experimental islands and salt-marsh enclosed plots are available on PANGAEA for September 2014 to April 2017 (https://doi.org/10.1594/PANGAEA.877257). Gaps in the dataset are the result of local readouts and maintenance. Each dataset includes date/time, depth in sediment (depth measured in m) and temperature in sediment (represented by t, expressed in degC; Table 6).

2.3.6 Light loggers (DL-L)

Measured light availability on the experimental islands and salt-marsh enclosed plots are available on PANGAEA from September 2014 to April 2017 (https://doi.org/10.1594/PANGAEA.877256). Gaps in the dataset result from local readouts and maintenance. Due to different calibrations for underwater or above-water application, raw data were processed differently and were merged depending on low- or high-water times. Each dataset includes the date/time and light intensity (Io, expressed in $\mu mol\, m^{-2}\, s^{-1}$; Table 7).

2.3.7 Conductivity loggers (DL-C)

Data of conductivity measurements inside dip wells on the experimental islands and salt-marsh enclosed plots are available on PANGAEA for May 2015 to April 2017 (https://doi.org/10.1594/PANGAEA.877266). Gaps in the dataset result from local readouts and maintenance. Low-tide data were removed from the dataset. An automatic calculation of salinity were performed within the HOBOware Pro (V3.7.4) software according to PSS-78. Each dataset includes date/time, depth in sediment (depth measured in m) and salinity (Sal; Table 8).

3 Data availability

Sensor operation encompasses the time span of 32 months, from 18 September 2014 to 18 April 2017 (944 days). Figure 4 illustrates the data availability over the whole period, and Table 9 provides total availability in days as well as in percent.

3.1 Weather station (WS)

Within this time frame meteorological data were available from 19 November 2014 to 18 April 2017 on 571 days (corresponding to 60.49 % availability). Absent days resulted from malfunctions and maintenances from 29 January 2015 to 11 March 2015, 4 November 2015 to 14 November 2015, 25 March 2016 to 28 October 2016 and 1 February 2017 to 21 March 2017.

3.2 Recording current meter (RCM)

Current meter operation was possible from 18 September 2014 until 6 October 2015 on 335 days (35.49 %). Further operations occurred in November and December 2013, March 2014, and June 2014 for preliminary investigations. Missing days were resulted from local maintenance and readouts. Due to a malfunction, the sensor had not been operating since October 2015.

3.3 Tide and wave recorder (TWR)

Continuous observations of wave and tide data were feasible from 1 October 2014 to 18 April 2017 on 812 days (corresponding to 86.02 % availability). Absent days were the result of local maintenance and readouts as well as some extended services from 19 October 2014 to 28 October 2014, 17 August 2015 to 23 September 2015 and 8 November 2016 to 24 January 2017.

Table 9. An overview of sensor and logger data availability over the 944-day sampling period. D 1 – sum of days with data available; D 0 – sum of days with data absent; % 1 – proportion of days with data available; % 0 – proportion of days with data absent; D x – total available days specified for each sensor application, i.e., from first deployment to last measurement; % x1 and % x0 are the corresponding percent availabilities. Sum 1 – number of measurements available; Sum 0 – number of measurements missing; % S 1 – proportion of measurements available; % S 0 – proportion of measurements missing; Sum x – total available measurements specified of each sensor application, i.e., from first data point to last measurement; %S x1 and %S x0 are the corresponding percent availabilities. WS: weather station, RCM: recording current meter, TWR: tide and wave recorder, DL: data logger.

	WS	RCM	TWR	DL-W1	DL-W2	DL-W3	DL-W4	DL-W5	DL-W6	DL-C1	DL-C2
		Seafloor	Seafloor	I4 Pio	I3 Upp	I3 Low	I1 Pio	S3 Low	I4 Low	I3 Pio	S3 Pio
D 1	571	335	812	578	578	578	578	572	480	624	567
D 0	373	50	132	366	366	366	366	372	464	320	377
% 1	60.49	35.49	86.02	61.23	61.23	61.23	61.23	60.59	50.85	66.10	60.06
% 0	39.51	5.30	13.98	38.77	38.77	38.77	38.77	39.41	49.15	33.90	39.94
D x	882	384	931	669	669	669	669	663	571	714	713
% x1	64.74	87.24	87.22	86.40	86.40	86.40	86.40	86.27	84.06	87.39	79.52
% x0	35.26	12.76	12.78	13.60	13.60	13.60	13.60	13.73	15.94	12.61	20.48
Sum 1	816 050	47 490	115 535	82 997	82 999	83 000	83 002	82 112	68 850	89 481	81 297
Sum 0	542 410	88 356	20 311	52 849	52 847	52 846	52 844	53 734	66 996	46 365	54 549
%S 1	60.07	34.96	85.05	61.10	61.10	61.10	61.10	60.44	50.68	65.87	59.84
%S 0	39.93	65.04	14.95	38.90	38.90	38.90	38.90	39.56	49.32	34.13	40.16
Sum x	1 269 390	55 093	134 064	96 288	96 288	96 288	96 288	95 407	82 118	102 745	102 613
%S x1	64.29	86.20	86.18	86.20	86.20	86.20	86.20	86.06	83.84	87.09	79.23
%S x0	35.71	13.80	13.82	13.80	13.80	13.80	13.80	13.94	16.16	12.91	20,77

	DL-T1	DL-T2	DL-T3	DL-T4	DL-T5	DL-T6	DL-L1	DL-L2	DL-L3	DL-L4	DL-L5	DL-L6
	I3 Pio	I3 Low	I3 Upp	S3 Low	S2 Low	S2 Upp	I3 Pio	I3 Low	I3 Upp	S3 Low	I3 Pole	Seafloor
D 1	931	930	931	931	380	295	931	731	931	905	547	547
D 0	13	14	13	13	564	649	13	213	13	39	397	397
% 1	98.62	98.52	98.62	98.62	40.25	31.25	98.62	77.44	98.62	95.87	57.94	57.94
% 0	1.38	1.48	1.38	1.38	59.75	68.75	1.38	22.56	1.38	4.13	42.06	42.06
D x	944	944	944	944	854	295	944	944	944	944	560	560
% x1	98.62	98.52	98.62	98.62	44.50	100.0	98.62	77.44	98.62	95.87	97.68	97.68
% x0	1.38	1.48	1.38	1.38	55.50	0.00	1.38	22.56	1.38	4.13	2.32	2.32
Sum 1	133 311	133 208	133 386	133 331	54 371	42 241	133 577	104 887	133 577	129 900	78 534	78 532
Sum 0	2535	2638	2460	2515	81 475	93 605	2269	30 959	2269	5946	57 312	57314
%S 1	98.13	98.06	98.19	98.15	40.02	31.09	98.33	77.21	98.33	95.62	57.81	57.81
%S 0	1.87	1.94	1.81	1.85	59.98	68.91	1.67	22.79	1.67	4.38	42.19	42.19
Sum x	135 846	135 846	135 846	135 846	123 018	42 282	135 846	135 846	135 846	135 846	80 556	80 556
%S x1	98.13	98.06	98.19	98.15	44.20	99.90	98.33	77.21	98.33	95.62	97.49	97.49
%S x0	1.87	1.94	1.81	1.85	55.80	0.10	1.67	22.79	1.67	4.38	2.51	2.51

3.4 Data logger (DL)

3.4.1 Water level loggers (DL-W)

Within the total time period water level measurements were performed from 20 June 2015 to 18 April 2017 on 578 days (572 days within the salt-marsh enclosed plot), resulting in a 61.23 % (60.59 %) availability. Absent days resulted from malfunctions and maintenances from 16 February 2016 to 18 May 2016 as well as some local readouts. Groundwater level data of the DEFI-D logger ends at 10 January 2017 due to a missing readout in April 2017. The coverage of water level measurements is therefore 50.85 % (480 days).

3.4.2 Temperature loggers (DL-T)

Recordings of the surface layer temperature were available on 931 days, excluding the period from 10 January 2017 to 24 January 2017 due to maintenance (98.62 % availability). However, temperature data of two loggers within the salt-marsh enclosed plots (S2 Upp and S2 Low) exhibit a shorter time of deployment. An application was possible at 295/380 days (31.25 %/40.25 % availability) from 16 December 2014 to 6 October 2015 (S2 Upp and S2 Low) and further from 24 January 2017 to 18 April 2017 (S2 Low).

3.4.3 Light Loggers (DL-L)

Light intensity and availability was recorded over the whole period from 18 September 2014 to 18 April 2017 for

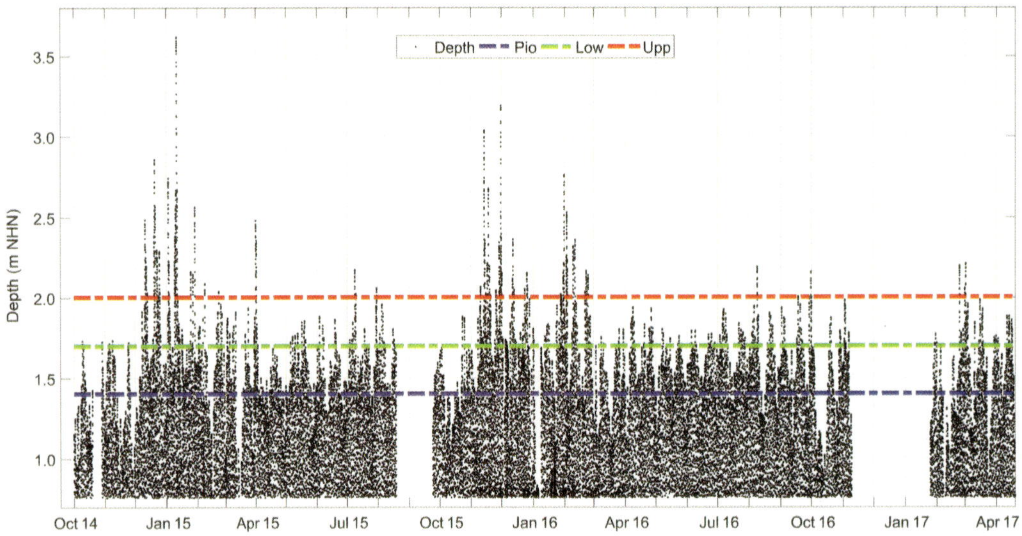

Figure 7. Calculated water depth (black dots) at 0.71 m NHN, next to islands 6 and 7. The blue line represents the elevation of the pioneer zone (Pio), the green line shows the height of the lower salt marsh (Low) and the red line describes the upper salt-marsh zone (Upp). Thus, water level data can give information about flooding periods at the three elevation levels. Low-tide data (all data below 0.05 m) were removed before plotting the data. To provide m NHN as a basis, 0.71 m were added to the data points.

Table 10. Minima, maxima and mean values as well as median and standard deviation of the weather station (WS) for the time frame of 19 November 2014 to 18 April 2017 on 571 days. See Table 2 for parameter list.

WS	Min.	Max.	Mean	Median	SD
ff (m s^{-1})	0.00	24.68	5.70	5.04	3.21
dd (°)	0.00	379.99	197.52	210.02	87.20
TTT (°C)	−8.10	32.20	8.23	7.50	5,57
RH (%)	2.00	100.00	87.75	90.00	10.37
PPPP (hPa)	977.58	1042.07	1018.12	1018.68	10.37
Brightness North (lux)	0.00	89 502.54	4785.86	6.23	9441.66
Brightness East (lux)	0.00	149 582.40	8520.22	6.59	19 157.07
Brightness South (lux)	0.00	150 000.00	9685.35	6.41	21 869.61
Brightness West(lux)	0.00	143 925.20	6701.04	6.90	15 295.02
Brightness Max. (lux)	0.00	150 000.00	12 376.23	8.40	25 729.80
Brightness dir (°)	0.00	360.00	71.71	0.00	101.66
Precip day total (mm)	0.00	6.00	0.00	0.00	0.01
Precip (mm h^{-1})	0.00	68.00	0.14	0.00	1.00
SE (°)	−60.40	60.40	−4.75	−4.06	29.43
SAA (°)	0.00	359.83	180.00	180.07	100.02

931 days, except for the period from 10 January 2017 to 24 January 2017 due to maintenance (98.62 % availability). Two loggers (I3 Pole, Seafloor) were first installed on 7 October 2015 with a total application time of 547 days, corresponding to 57.94 % availability. A light logger within the salt-marsh enclosed plot (S3 Low) had to be removed from 19 November 2014 to 16 December 2014 due to maintenance. A light logger on the experimental island (I3 Low) had to be temporarily removed between 23 September 2015 and 11 April 2016.

3.4.4 Conductivity Loggers (DL-C)

Data of the conductivity logger on the experimental island (I3 Pio) are available from 6 May 2015 to 16 February 2016 and from 18 May 2016 to 18 April 2017 on 624 days (66.10 % availability). The logger located in the salt-marsh enclosed plot (S3 Pio) recorded water conductivity from 7 May 2015 to 8 October 2015, 3 December 2015 to 16 February 2016 and from 18 May 2016 to 18 April 2017, 567 days in total (60.06 % availability). Gaps in the dataset resulted from a malfunction and maintenance.

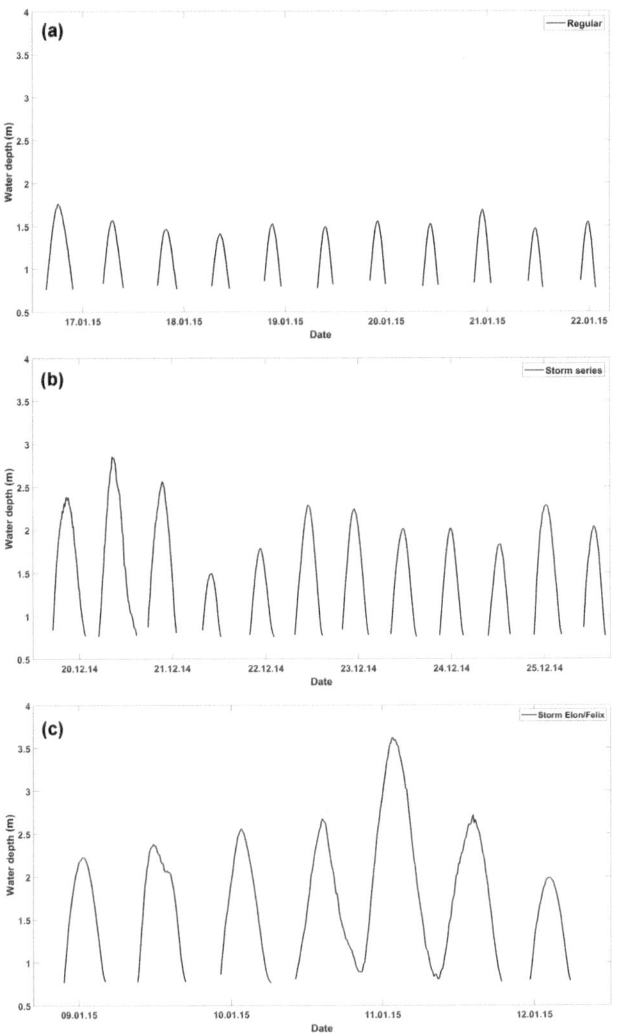

Figure 8. Calculated water depth (black line) during different weather conditions. **(a)** shows regular weather and wave conditions from 16 to 22 January 2015. **(b)** represents a series of storms with higher water levels from 19 until 25 December 2015. At this point the highest wave energy values were observed as well as highest significant wave heights. **(c)** shows the two big storms Elon and Felix, the latter occurring directly after the first storm. Here the highest water levels during application time were measured as well as the maximum wave height.

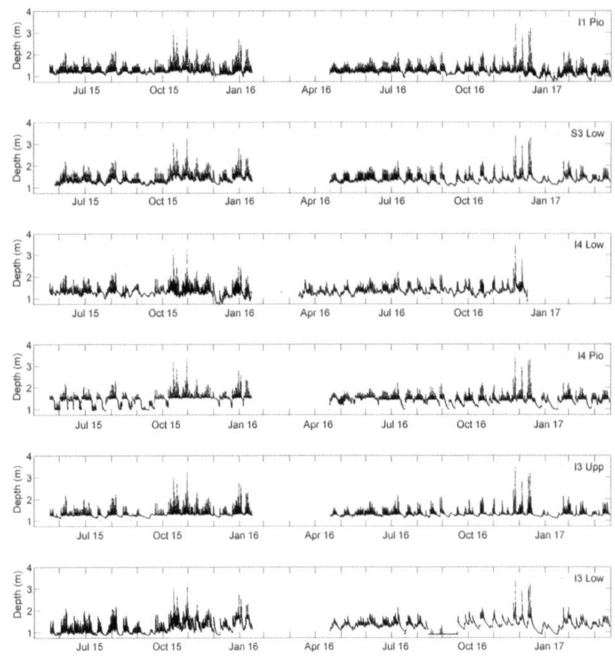

Figure 9. Groundwater level on the experimental islands and the salt-marsh enclosed plot.

Table 11. Minima, maxima and mean values as well as median and standard deviation of the recording current meter (RCM) for its application time from 18 September 2014 until 6 October 2015 on 335 days.

RCM	Min.	Max.	Mean	Median	SD
Speed ($cm\,s^{-1}$)	0.29	107.05	14.09	12.90	8.62
Direction (°)	0.00	359.70	157.41	134.66	98.54
Temperature (°C)	−0.68	27.30	10.69	10.25	5.43
Conductivity ($mS\,cm^{-1}$)	24.39	50.60	33.82	32.08	4.97
Pressure (kPa)	93.74	136.71	103.78	101.60	4.03

4 Results and discussion

4.1 Weather station (WS)

Local wind diagrams for three winter storm seasons and one summer season are shown in Fig. 5. Furthermore, minima, maxima and mean values as well as the median and standard deviation of the weather station for the time frame of 19 November 2014 to 18 April 2017 on 571 days are listed in Table 10. Within the application time, wind speeds of less than $25\,m\,s^{-1}$ were present. Mean wind speed was $5.7\,m\,s^{-1} \pm 3.21$, but it strongly differs between the single seasons.

4.2 Recording current meter (RCM)

Current conditions for one storm season and one summer season are shown in Fig. 6. A main current direction from the southwest to the northeast was clearly identified, showing the good orientation of the experimental islands. Furthermore, minima, maxima and mean values as well as the median and standard deviation of the RCM data from 18 September 2014 until 6 October 2015 on 335 days are listed in Table 11. A maximum current speed of $107.05\,cm\,s^{-1}$ was observed in December 2013 during storm Xaver. Mean current speed represent $12.9\,cm\,s^{-1} \pm 8.62$, but with variances between the single seasons.

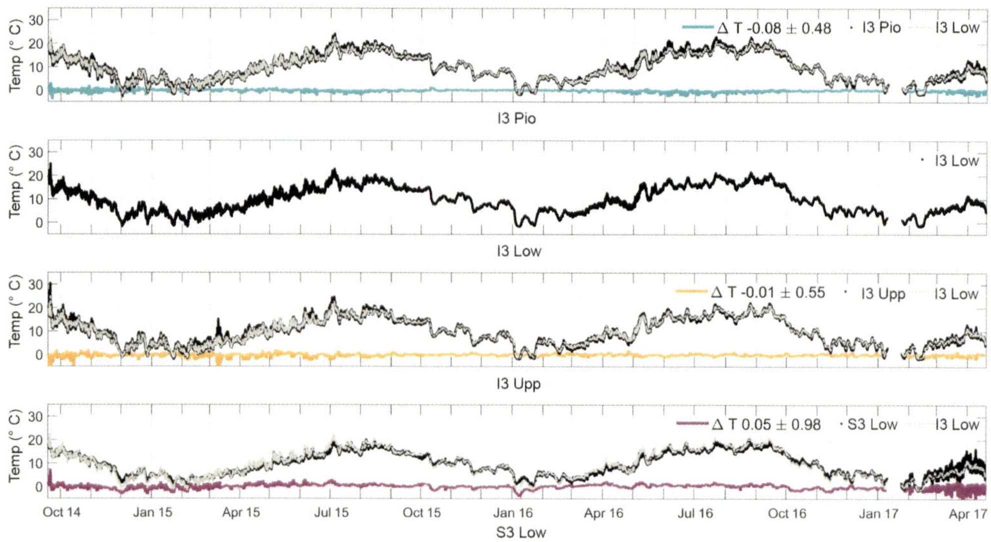

Figure 10. Surface layer temperature and temperature differences over the whole time frame with the DL-T2 (I3 Low).

Table 12. Minima, maxima and mean values as well as median and standard deviation of the tide and wave recorder (TWR) from 1 October 2014 to 18 April 2017 on 812 days.

TWR	Min.	Max.	Mean	Median	SD
Depth (m)	0.69	3.62	1.23	1.19	0.35
Tdslope	−1.48	1.08	0.00	0.00	0.21
$H(1/3)$ (m)	0.00	0.73	0.03	0.00	0.06
$T(1/3)$ (s)	0.00	271.00	3.58	1.66	9.60
$H(1/10)$ (m)	0.00	0.96	0.04	0.00	0.08
$T(1/10)$ (s)	0.00	272.67	5.47	1.68	14.12
$H(\max)$ (m)	0.00	2.14	0.05	0.00	0.10
$T(\max)$ (s)	0.00	288.33	7.97	1.67	20.29
$H(\mathrm{avg})$ (m)	0.00	0.51	0.02	0.00	0.04
$T(\mathrm{avg})$ (s)	0.00	288.33	2.01	1.46	4.84
Energy ($\mathrm{J\,m^{-2}}$)	0.00	400.90	3.67	0.00	14.02
Temp (°C)	−6.10	31.87	10.00	9.29	6.20
Pressure raw (dbar)	9.71	12.98	10.36	10.25	0.33

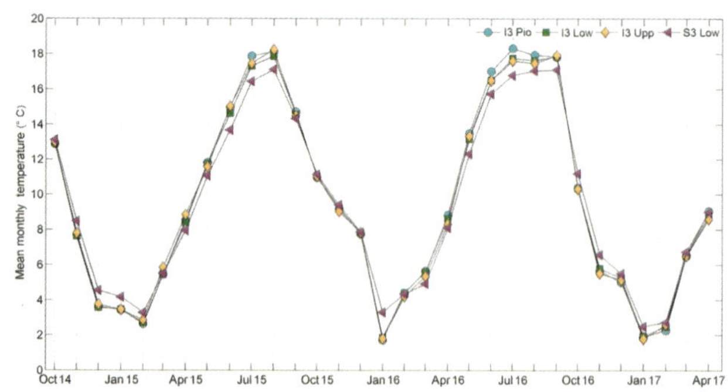

Figure 11. Mean monthly temperature over the whole application time. Since the temperatures on the experimental islands are very similar, a clear offset of the salt-marsh enclosed plot temperature could be identified in winter and summer months.

4.3 Tide and wave recorder (TWR)

The water depth calculated from pressure data of the tide and wave recorder (TWR) can be seen in Fig. 7. A seasonal pattern of water depth is exhibited and the highest water depths were reached during storm season. However, water level reached the upper salt-marsh zone (2.0 m NHN) several times (e.g., April 2015, July 2015 and August 2016). During the storms Elon and Felix in January 2015, the highest water level was observed at 3.62 m as well as the highest wave at 2.14 m (Fig. 8). The highest wave energy (400.90 $\mathrm{J\,m^{-2}}$) and significant wave height $H(1/3)$, which was measured at 0.73 m, were detected during several shortly sequenced storm events in the weeks before in December 2014 (Fig. 8). Mean water depths for the whole application are 1.23 m ± 0.35 m. Statistic values of the TWR from 1 October 2014 to 18 April 2017 on 812 days are listed in Table 12.

4.4 Data logger (DL)

4.4.1 Water level loggers (DL-W)

The depth of the groundwater level achieved from pressure logger on the experimental islands and in the salt-marsh enclosed plots can be seen in Fig. 9. Differences in the water level were observed, especially during low water. This could be a result of various factors, ranging from diverse water consumption of plants to less flooding on higher elevations or leaks in the plastic bags inside an experimental island. Statistics of groundwater level data (DL-W 1-6) for each logger application time from 20 June 2015 to 18 April 2017 on

Table 13. Minima, maxima and mean values as well as median and standard deviation of the groundwater level data (DL-W 1-6) for each logger application time from 20 June 2015 to 18 April 2017 on 578 days (572 days within the salt-marsh enclosed plot).

DL-W	Min.	Max.	Mean	Median	SD
DL-W1 Water level I3 Upp (m)	1.03	3.44	1.33	1.30	0.15
DL-W2 Water level I3 Low (m)	0.88	3.34	1.28	1.27	0.27
DL-W3 Water level I4 Pio (m)	0.93	3.56	1.45	1.48	0.20
DL-W4 Water level I1 Pio (m)	0.78	3.39	1.25	1.23	0.17
DL-W5 Water level S3 Low (m)	0.87	3.38	1.39	1.35	0.19
DL-W6 Water level I4 Low (m)	0.64	3.43	1.35	1.33	0.20

Table 14. Minima, maxima and mean values as well as median and standard deviation of the temperature data (DL-T 1-6) from 18 September 2014 until 18 April 2017 on 931 days.

DL-T	Min.	Max.	Mean	Median	SD
DL-T1 Temperature I3 Pio (°C)	−2.60	24.14	9.78	9.12	5.74
DL-T2 Temperature I3 Low (°C)	−1.75	25.09	9.70	9.05	5.56
DL-T3 Temperature I3 Upp (°C)	−1.87	30.43	9.70	9.07	5.66
DL-T4 Temperature S3 Low (°C)	−1.52	21.12	9.64	9.16	5.02

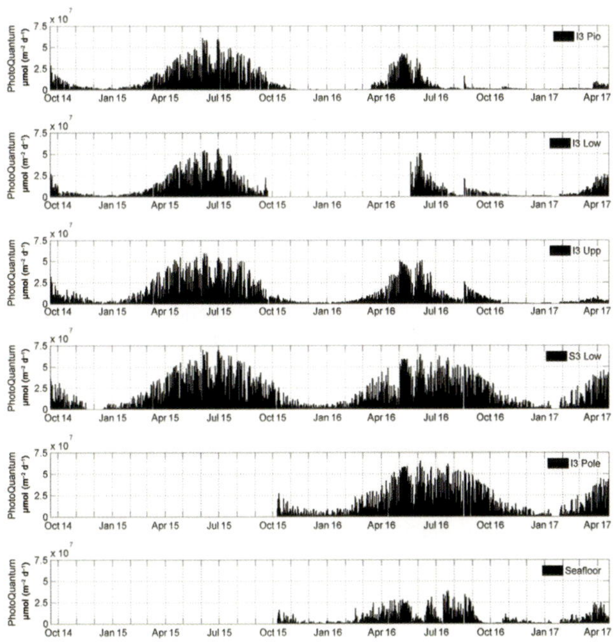

Figure 12. Calculated sum of light intensity per day.

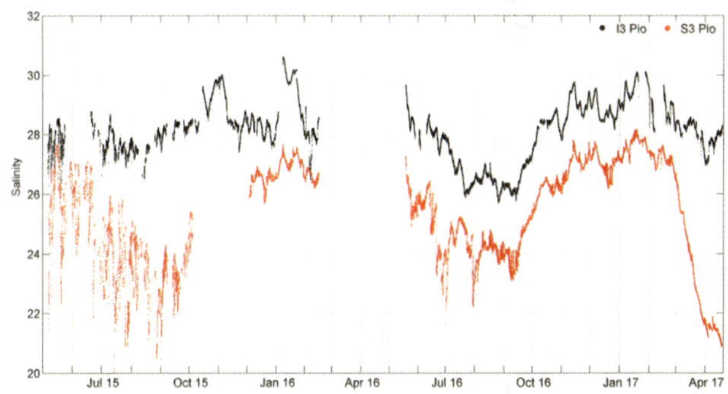

Figure 13. Salinity values within dip wells on the experimental island and salt-marsh enclosed plot. Due to fluctuations in the groundwater level, conductivity loggers periodically became dry, especially in the beginning. Thus, data until October 2015 are scattered. The depth of conductivity loggers was thereupon adjusted deeper in the dip wells, assuring a constant covering of water.

578 days (572 days within the salt-marsh enclosed plot) are listed in Table 13.

4.4.2 Temperature loggers (DL-T)

Surface layer temperatures for one experimental island and its three elevational zones as well as one elevation zone within the salt-marsh enclosed plots are shown in Fig. 10. To assess temperature differences the DL-T2 Logger (I3 Low) was taken to compare both differences within one island's elevational zone (DL-T1 I3 Pio, DL-T3 I3 Upp) and between the same elevational level and those of the salt-marsh control plots (DL-T4, S3 Low). All temperatures are very similar and showing only less differences, especially I3 Upp and I3 Low, with $\Delta T 0.01 \pm 0.55\,°C$. Since the temperatures on the experimental islands are very similar considering the mean monthly temperatures of the four plots, a clear offset of the salt-marsh enclosed plot temperature could be identified in winter and summer months (Fig. 11). The lowest temperature was observed at I3 Pio with −2.6 °C, whereas the highest value of 30.43 °C was measured at the upper zone of the experimental island. Maximum temperatures of S3 Low and I3 Upp differ by 9.31 °C. Mean values are very similar and ranging

Table 15. Minima, maxima and mean values as well as median and standard deviation of the light intensity data (DL-L 1-6) for each logger application time.

DL-L	Min.	Max.	Mean	Median	SD
DL-L1 Light I3 Pio (μmol ($m^{-2}\,s^{-1}$))	0.00	2774.60	105.16	0.50	291.63
DL-L2 Light I3 Low (μmol ($m^{-2}\,s^{-1}$))	−0.40	2603.30	120.22	1.10	301.43
DL-L3 Light I3 Upp (μmol ($m^{-2}\,s^{-1}$))	0.00	2609.70	141.10	0.80	343.06
DL-L4 Light S3 Low (μmol ($m^{-2}\,s^{-1}$))	0.20	2657.80	244.58	4.20	435.98
DL-L5 Light I3 Upp pole (μmol ($m^{-2}\,s^{-1}$))	−0.30	2706.30	222.82	1.10	412.30
DL-L6 Light seafloor (μmol ($m^{-2}\,s^{-1}$))	0.00	2058.10	85.15	0.50	196.23

Table 16. Minima, maxima and mean values as well as median and standard deviation of the salinity data (DL-C 1-2) for each logger application time.

DL-C	Min.	Max.	Mean	Median	SD
DL-C1 Sal I3 Pio (ppt)	20.00	30.61	27.95	28.17	1.22
DL-C2 Sal S3 Pio (ppt)	20.04	28.17	25.45	25.91	1.74

from 9.64 °C ± 5.02 (S3 Low) to 9.78 ± 5.74 (I3 Pio). More statistics of temperature data (DL-L 1-6) for each logger application time from 18 September 2014 to 18 April 2017 on 931 days are listed in Table 14.

4.4.3 Light loggers (DL-L)

To calculate the light sum per day, measured light intensity (Quantum (μmol $m^{-2}\,s^{-1}$)) was extrapolated from seconds to the measuring interval of 10 min, and afterwards values were added up for 1 day. Polycarbonate covers were installed on the experimental islands to prevent the scouring during the storm season (Balke et al., 2017). In consequence, lower light availabilities were detected under the covers (Fig. 12) at island 3 in I3 Pio, I3 Low and I3 Upp other than at the seafloor, I3 Pole and S3 Low (all uncovered). Minima, maxima and mean values as well as the median and standard deviation of the light intensity data (DL-L 1-6) for each logger application over the whole period from 18 September 2014 to 18 April 2017 for 931 days are listed in Table 15.

4.4.4 Conductivity loggers (DL-C)

Salinity values achieved from both conductivity logger on the experimental island and the salt-marsh enclosed plot can be seen in Fig. 13. Due to fluctuations in the groundwater level, conductivity loggers periodically became dry, especially in the beginning. Thus, data until October 2015 are scattered. The conductivity loggers were thereupon adjusted deeper in the dip wells assuring a constant covering of water. The mean salinity for the experimental island is 27.95 ± 1.22 and 25.45 ± 1.74 for the salt-marsh enclosed plot. Both loggers were installed at the pioneer zone. Statistics of salinities (DL-C 1-2) for each logger application time from 6 May 2015 to 16 February 2016 and from 18 May 2016 to 18 April 2017 on 624 days are shown in Table 16.

Sample availability. The dataset described within this document is based on sensor information that has been uploaded to the PANGAEA database (see Sect. 2.3).

5 Conclusions and future directions

The BEFmate project included a variety of experiments dedicated to investigate biodiversity–ecosystem functioning across marine and terrestrial ecosystems, utilizing 12 experimental islands in the back-barrier tidal flat of Spiekeroog Island. Abiotic conditions were recorded from a suite of 23 different sensors installed at different locations in the vicinity of the experimental islands, on the islands themselves and in the nearby salt marsh. Data described here covers the period from September 2014 to April 2017 and has been published in seven datasets in the World Data Center PANGAEA. Data coverage within the period ranged from 35 % for the recording current meter (RCM) that failed in October 2015 to 99 % for six data loggers. With 17 sensors covering 80 % or more of the period of interest, a very good coverage was achieved. Additional data from pre-experiment investigations with the RCM between November 2013 and June 2017 was added to the dataset. Seasonal and tidal dynamics as well as storms were covered, and these data are available for interpretation in further contexts.

For future operations, data availability can be further increased if a rotational system for maintenance is applied, provided that spare sensors are available. Furthermore, the online data transfer of central information would not only increase the data availability, but meteorological and hydrographic conditions could also indicate the need to attend the experiment. For example, indicating a need to take action to prevent damage during extreme weather events. Non-invasive remote sensing sensor techniques can provide complementary data, to avoid fouling issues, as demonstrated successfully at the nearby Time Series Station – Spiekeroog (Garaba et al., 2014; Schulz et al., 2016). Additionally camera systems should be applied, providing a visual impression of the overall scene and detailed information, e.g., the process of flooding for different levels. Recently the RCM failure within the BEFmate project has inspired the development of a machine-learning environment that creates a virtual sensor enabling the compensation for single sensor dropouts (Oehmcke et al., 2017a, b). Finally, data quality assurance and quality measures should be further developed to reduce the workload of manual data curation while improving data availability in near-real time.

Author contributions. OZ was responsible for the scientific approach and performed the interpretation of results. He drafted and prepared the manuscript. DM was responsible for technical maintenance and laboratory analysis and performed fieldwork, data sampling and analysis, created figures, and contributed to the writing process of the manuscript. HH was the principle investigator of the BEFmate project and contributed to the writing process. MK was the principle investigator of one of the BEFmate subprojects and was responsible for the conception of the experimental islands. TB and KL were responsible for maintenance of the experimental islands as well as fieldwork coordination. All authors contributed to proofreading of the manuscript.

Competing interests. The authors declare that they have no conflict of interest.

Acknowledgements. The authors are very grateful to Kai Schwalfenberg and Claudia Thölen for their tremendous support in the field and laboratory. Sincere thanks to Daniela Voß, Ursel Gerken, Kathrin Dietrich, Nick Rüssmeier, Rohan Henkel, Jule Beßler, Franziska Wöhrmann and Hauke Haake for technical, laboratory and fieldwork support. Special thanks to Helmo Nicolai and Gerrit Behrens for their technical and logistical support. The support and cooperation with Nationalparkverwaltung Niedersächsisches Wattenmeer and the Umweltzentrum Wittbülten is acknowledged as well as the support of Rainer Sieger at PANGAEA. Thanks to the BEFmate colleagues and all other helping hands in the field during sampling campaigns, especially Regine Redelstein. The BEFmate project (Biodiversity–Ecosystem Functioning across marine and terrestrial ecosystems) was funded by the Ministry for Science and Culture of Lower Saxony, Germany under project number ZN2930. The feedback from two reviewers is gratefully acknowledged.

Edited by: Giuseppe M. R. Manzella
Reviewed by: Jaume Piera and one anonymous referee

References

Balke, T., Stock, M., Jensen, K., Bouma, T. J., and Kleyer, M.: A global analysis of the seaward salt marsh extent: The importance of tidal range, Water Resour. Res., 52, 3775–3786, https://doi.org/10.1002/2015WR018318, 2016.

Balke, T., Lõhmus, K., Hillebrand, H., Zielinski, O., Haynert, K., Meier, D., Hodapp, D., and Kleyer, M.: Experimental salt marsh islands: A model system for novel metacommunity experiments, Estuarine, Coastal Shelf Sci., 198, 288–298, https://doi.org/10.1016/j.ecss.2017.09.021, 2017.

Balvanera, P., Pfisterer, A. B., Buchmann, N., He, J. S., Nakashizuka, T., Raffaelli, D., and Schmid, B.: Quantifying the evidence for biodiversity effects on ecosystem functioning and services, Ecol. Lett., 9, 1146–1156, https://doi.org/10.1111/j.1461-0248.2006.00963.x, 2006.

Baschek, B., Schroeder, F., Brix, H., Riethmüller, R., Badewien, T.H., Breitbach, G., Brügge, B., Colijn, F., Doerffer, R., Eschenbach, C., Friedrich, J., Fischer, P., Garthe, S., Horstmann, J., Krasemann, H., Metfies, K., Merckelbach, L., Ohle, N., Petersen, W., Pröfrock, D., Röttgers, R., Schlüter, M., Schulz, J., Schulz-Stellenfleth, J., Stanev, E., Staneva, J., Winter, C., Wirtz, K., Wollschläger, J., Zielinski, O., and Ziemer, F.: The Coastal Observing System for Northern and Arctic Seas (COSYNA), Ocean Sci. 13, 379–410, https://doi.org/10.5194/os-13-379-2107, 2017.

Cadotte, M. W., Cardinale, B. J., and Oakley, T. H.: Evolutionary history and the effect of biodiversity on plant productivity, P. Natl. Acad. Sci. USA, 105, 17012–17017, https://doi.org/10.1073/pnas.0805962105, 2008.

Cardinale, B. J., Srivastava, D. S., Duffy, J. E., Wright, J. P., Downing, A. L., Sankaran, M., and Jouseau, C.: Effects of biodiversity on the functioning of trophic groups and ecosystems, Nature, 443, 989–992, https://doi.org/10.1038/nature05202, 2006.

Cooley, S. R. and Doney, S. C.: Anticipating ocean acidification's economic consequences for commercial fisheries, Environ. Res. Lett., 4, 024007, https://doi.org/10.1088/1748-9326/4/2/024007, 2009.

Ekstrom, J. A., Suatoni, L., Cooley, S. R., Pendleton, L. H., Waldbusser, G. G., Cinner, J. E., Ritter, J., Langdon, C., van Hooidonk, R., Gledhill, D., Wellman, K., Beck, M. W., Brander, L. M., Rittschof, D., Doherty, C., and Edwards, P. E. T.: Vulnerability and adaption of US shellfisheries to ocean acidification, Nat. Clim. Change, 5, 207–214, https://doi.org/10.1038/nclimate2508, 2015.

Garaba, S. P., Badewien, T. H., Braun, A., Schulz, A. C., and Zielinski, O.: Using ocean colour remote sensing products to estimate turbidity at the Wadden Sea time series station Spiekeroog, J. Europ. Opt. Soc., 9, 14020-1–14020-6, https://doi.org/10.2971/jeos.2014.14020, 2014.

Gravel, D., Bell, T., Barbera, C., Bouvier, T., Pommier, T., Venail, P., and Mouquet, N.: Experimental niche evolution alters the strength of the diversity-productivity relationship, Nature, 469, 89-U1601, https://doi.org/10.1038/nature09592, 2011.

Haigh, R., Ianson, D., Holt, C. A., Neate, H. E., and Edwards, A. M.: Effects of Ocean Acidification on Temperate Coastal Marine Ecosystems and Fisheries in the Northeast Pacific, PLoS ONE 10, e0117533, https://doi.org/10.1371/journal.pone.0117533, 2015.

Hillebrand, H. and Matthiessen, B.: Biodiversity in a complex world: consolidation and progress in functional biodiversity research, Ecol. Lett., 12, 1405–1419, https://doi.org/10.1111/j.1461-0248.2009.01388.x, 2009.

Hillebrand, H., Blasius, B., Borer, E. T., Chase, J. M., Downing, J. A., Eriksson, B. K., Filstrup, C. T., Harpole, W. S., Hodapp, D., Larsen, S., Lewandowska, A. M., Seabloom, E. W., Van de Waal, D. B., and Ryabov, A. B.: Biodiversity change is uncoupled from species richness trends: consequences for conservation and monitoring, J. Appl. Ecol., 55, 169–184, https://doi.org/10.1111/1365-2664.12959, 2017.

Hodapp, D., Hillebrand, H., Blasius, B., and Ryabov, A. B.: Environmental and trait variability constrain community structure and the biodiversity-productivity relationship, Ecology, 97, 1463–1474, https://doi.org/10.1890/15-0730.1, 2016.

Holinde, L., Badewien, T. H., Freund, J. A., Stanev, E. V., and Zielinski, O.: Processing of water level derived from water pressure data at the Time Series Station Spiekeroog, Earth Syst. Sci. Data, 7, 289–297, https://doi.org/10.5194/essd-7-289-2015, 2015.

Kirwan, M. L. and Megonigal, J. P.: Tidal wetland stability in the face of human impacts and sea-level rise, Nature, 504, 53–60, https://doi.org/10.1038/nature12856, 2013.

Leibold, M. A., Chase, J. M., and Ernest, S. K. M.: Community assembly and the functioning of ecosystems: how metacommunity processes alter ecosystems attributes, Ecology, 98, 909–919, https://doi.org/10.1002/ecy.1697, 2017.

Mace, G. M., Masundire, H., Baillie, J., Ricketts, T., Brooks, T., Hoffmann, M., Stuart, S., Balmford, A., Purvis, A., Reyers, B., Wang, J., Revenga, C., Kennedy, E., Naeem, S., Alkemade, R., Allnutt, T., Bakarr, M., Bond, W., Chanson, J., Cox, N., Fonseca, G., Hilton-Taylor, C., Loucks, C., Rodrigues, A., Sechrest, W., Stattersfield, A. J., Van Rensburg, B. J., and Whiteman, C.: Biodiversity, in: Biodiversity in Ecosystems and Human Wellbeing: Current State and Trends, edited by: Hassan, H., Scholes, R., Ash, N., https://doi.org/10.1111/acv.12222, 2005.

Mathis, J. T., Cooley, S. R., Lucey, N., Colt, S., Ekstrom, J., Hurst, T., Hauri, C., Evans, W., Cross, J. N., and Feely, R. A.: Ocean acidification risk assessment for Alaska's fishery sector, Prog. Oceanogr., 136, 71–91, https://doi.org/10.1016/j.pocean.2014.07.001, 2015.

Oehmcke, S., Zielinski, O., and Kramer, O.: Recurrent neural networks and exponential PAA for virtual marine sensors, IEEE International Joint Conference on Neural Networks (IJCNN), 4459–4466, https://doi.org/10.1109/IJCNN.2017.7966421, 2017a.

Oehmcke, S., Zielinski, O., and Kramer, O.: Input quality aware convolutional lSTM networks for virtual marine sensors, Neurocomputing, 275, 2603–2615, https://doi.org/10.1016/j.neucom.2017.11.027, 2017b.

Rockström, J., Steffen, W., Noone, K., Persson, Å, Chapin III, F. S, Lambin, E. F., Lenton, T. M., Scheffer, M., Folke, C., Schellnhuber, H. J., Nykvist, B., de Wit, C. A., Hughes, T., van der Leeuw, S., Rodhe, H., Sörlin, S., Snyder, P. K., Costanza, R., Svedin, U., Falkenmark, M., Karlberg, L., Corell, R. W., Fabry, V. J., Hansen, J., Walker, B., Liverman, D., Richardson, K., Crutzen, P., and Foley, J. A.: A safe operating space for humanity, Nature, 461, 472–475, 2009.

Schulz, A. C., Badewien, T. H., Garaba, S. P., and Zielinski, O.: Acoustic and optical methods to infer water transparency at Time Series Station Spiekeroog, Wadden Sea, Ocean Sci., 12, 1155–1163, https://doi.org/10.5194/os-12-1155-2016, 2016.

Zielinski, O., Meier, D., and Hillebrand, H.: Continuous meteorological observations at BEFmate automatic weather station, Spiekeroog, Germany, 2014-11 to 2017-04, Institute for Chemistry and Biology of the Marine Environment, Carl-von-Ossietzky University of Oldenburg, Germany, PANGAEA, https://doi.org/10.1594/PANGAEA.870988, 2017a.

Zielinski, O., Meier, D., Kleyer, M., and Hillebrand, H.: Continuous temperature observations in surface sediment within BEFmate experimental islands and saltmarsh control plots at different elevation levels, Spiekeroog, Germany, 2014-09 to 2017-04, Institute for Chemistry and Biology of the Marine Environment, Carl-von-Ossietzky University of Oldenburg, Germany, PANGAEA, https://doi.org/10.1594/PANGAEA.877257, 2017b.

Zielinski, O., Meier, D., Kleyer, M., and Hillebrand, H.: Continuous wave and tide observations at BEFmate experimental islands in the back-barrier tidal flat, Spiekeroog, Germany, 2014-10 to 2017-04, Institute for Chemistry and Biology of the Marine Environment, Carl-von-Ossietzky University of Oldenburg, Germany, PANGAEA, https://doi.org/10.1594/PANGAEA.877258, 2017c.

Zielinski, O., Meier, D., Kleyer, M., and Hillebrand, H.: Continuous current observations at BEFmate experimental islands in the back-barrier tidal flat, Spiekeroog, Germany, 2013-11 to 2015-10, Institute for Chemistry and Biology of the Marine Environment, Carl-von-Ossietzky University of Oldenburg, Germany, PANGAEA, https://doi.org/10.1594/PANGAEA.877265, 2017d.

Zielinski, O., Meier, D., Kleyer, M., and Hillebrand, H.: Continuous porewater salinity observations within BEFmate experimental islands and saltmarsh control plots, Spiekeroog, Germany, 2015-05 to 2017-04, Institute for Chemistry and Biology of the Marine Environment, Carl-von-Ossietzky University of Oldenburg, Germany, PANGAEA, https://doi.org/10.1594/PANGAEA.877266, 2017e.

Zielinski, O., Meier, D., Kleyer, M., and Hillebrand, H.: Continuous light intensity observations within BEFmate experimental islands and saltmarsh control plots at different elevation levels, Spiekeroog, Germany, 2014-09 to 2017-04, Institute for Chemistry and Biology of the Marine Environment, Carl-von-Ossietzky University of Oldenburg, Germany, PANGAEA, https://doi.org/10.1594/PANGAEA.877256, 2017f.

Zielinski, O., Meier, D., Kleyer, M., and Hillebrand, H.: Continuous water level observations within BEFmate experimental islands and saltmarsh control plots at different elevation levels, Spiekeroog, Germany, 2014-09 to 2017-04, Institute for Chemistry and Biology of the Marine Environment, Carl-von-Ossietzky University of Oldenburg, Germany, PANGAEA, https://doi.org/10.1594/PANGAEA.877267, 2017g.

10

Gross and net land cover changes in the main plant functional types derived from the annual ESA CCI land cover maps (1992–2015)

Wei Li[1], Natasha MacBean[2], Philippe Ciais[1], Pierre Defourny[3], Céline Lamarche[3], Sophie Bontemps[3], Richard A. Houghton[4], and Shushi Peng[5]

[1]Laboratoire des Sciences du Climat et de l'Environnement, LSCE/IPSL, CEA-CNRS-UVSQ, Université Paris-Saclay, 91191 Gif-sur-Yvette, France

[2]School of Natural Resources and the Environment, The University of Arizona, Tucson, AZ 85721, USA

[3]Earth and Life Institute, Environmental Sciences, Université catholique de Louvain, Louvain-la-Neuve, Belgium

[4]Woods Hole Research Center, Falmouth, MA, USA

[5]Sino-French Institute for Earth System Science, College of Urban and Environmental Sciences, Peking University, Beijing 100871, China

Correspondence: Wei Li (wei.li@lsce.ipsl.fr)

Abstract. Land-use and land-cover change (LULCC) impacts local energy and water balance and contributes on global scale to a net carbon emission to the atmosphere. The newly released annual ESA CCI (climate change initiative) land cover maps provide continuous land cover changes at 300 m resolution from 1992 to 2015, and can be used in land surface models (LSMs) to simulate LULCC effects on carbon stocks and on surface energy budgets. Here we investigate the absolute areas and gross and net changes in different plant functional types (PFTs) derived from ESA CCI products. The results are compared with other datasets. Global areas of forest, cropland and grassland PFTs from ESA are 30.4, 19.3 and 35.7 million km^2 in the year 2000. The global forest area is lower than that from LUH2v2h (Hurtt et al., 2011), Hansen et al. (2013) or Houghton and Nassikas (2017) while cropland area is higher than LUH2v2h (Hurtt et al., 2011), in which cropland area is from HYDE 3.2 (Klein Goldewijk et al., 2016). Gross forest loss and gain during 1992–2015 are 1.5 and 0.9 million km^2 respectively, resulting in a net forest loss of 0.6 million km^2, mainly occurring in South and Central America. The magnitudes of gross changes in forest, cropland and grassland PFTs in the ESA CCI are smaller than those in other datasets. The magnitude of global net cropland gain for the whole period is consistent with HYDE 3.2 (Klein Goldewijk et al., 2016), but most of the increases happened before 2004 in ESA and after 2007 in HYDE 3.2. Brazil, Bolivia and Indonesia are the countries with the largest net forest loss from 1992 to 2015, and the decreased areas are generally consistent with those from Hansen et al. (2013) based on Landsat 30 m resolution images. Despite discrepancies compared to other datasets, and uncertainties in converting into PFTs, the new ESA CCI products provide the first detailed long-term time series of land-cover change and can be implemented in LSMs to characterize recent carbon dynamics, and in climate models to simulate land-cover change feedbacks on climate. The annual ESA CCI land cover products can be downloaded from http://maps.elie.ucl.ac.be/CCI/viewer/download.php (Land Cover Maps – v2.0.7; see details in Sect. 5).

1 Introduction

LULCC (Land-use and land-cover change) is the essential human perturbation on natural ecosystems (Klein Goldewijk et al., 2016) and one of the main drivers of climate change (Alkama and Cescatti, 2016; Bonan, 2008) through biophysical (e.g. albedo and transpiration change; Peng et al., 2014; Zhao and Jackson, 2014) and biogeochemical effects (e.g. carbon emissions from gross deforestation and carbon sinks in secondary forest regrowth; Houghton and Nassikas, 2017). Forest loss from 2003 to 2012 was found to have caused a local increase in air temperature of about 1 °C in temperate and tropical regions, despite less solar energy being absorbed by non-forest secondary vegetation with a higher albedo (Alkama and Cescatti, 2016). Global net LULCC carbon emissions (E_{LUC}) are estimated to be $1.1 \pm 0.4\,\mathrm{Pg\,C\,yr^{-1}}$ during the past decade (2006–2015) by the bookkeeping model of Houghton and Nassikas (2017) based on the national land cover data from the Food and Agriculture Organization of the United Nations (FAO, 2015; Keenan et al., 2015). The E_{LUC} diagnosed from an ensemble of land surface models (LSMs) is $1.3 \pm 0.3\,\mathrm{Pg\,C\,yr^{-1}}$ during 2006–2015 (Le Quere et al., 2016) based on different (successive) versions of expanding cropland and pasture area from the HYDE dataset (Klein Goldewijk et al., 2016).

Accurate, well defined and spatially explicit gridded LULCC data are a prerequisite for calculating E_{LUC} in models, either under the form of annual area change in bookkeeping models or converted to changes in plant functional type (PFT) areas in LSMs. In fact, uncertain historical LULCC data are one of the largest contributors to the uncertainties in E_{LUC} estimation (Bayer et al., 2017; Houghton and Nassikas, 2017). In addition to the inventory data (e.g. FAO data reported by individual countries), satellite observations in the recent three decades offer the possibility to characterize the vegetation distributions as well as their temporal changes due to both natural and anthropogenic activity. Global satellite data include the Global Land Cover 2000 (GLC2000) map based on SPOT VEGETATION (SPOT-VGT; 1 km resolution; Bartholomé and Belward, 2005), the MODIS Collection 5 Land Cover Product (500 m resolution; Friedl et al., 2010), forest cover maps based on Landsat (30 m resolution; Hansen et al., 2013), the GlobCover 2005 and 2009 products (300 m resolution; Bontemps et al., 2011; Defourny et al., 2012) and the European Space Agency Climate Change Initiative (ESA CCI) epoch maps based on MERIS (300 m resolution; Bontemps et al., 2013). These satellite land cover products, however, differ in terms of land cover type, spatial resolution, time span, stability and accuracy due to the different sensor designs, classification procedures and validation methods (Bontemps et al., 2012). In order to use satellite land cover (LC) products in LSMs, these maps of LC classes are usually translated into maps of PFTs to drive the carbon dynamics in vegetation and soils (Poulter et al., 2015); however, the cross-walking table between LC classes and PFTs is complicated by subjective decisions related to the interpretation of LC class descriptions, and therefore is a source of uncertainty in model simulations (Hartley et al., 2017). Because LC transitions of opposite directions can happen simultaneously in a $0.5° \times 0.5°$ grid cell, which is a typical spatial resolution of LSMs, gross transitions instead of net transitions are gradually implemented in LSMs to more accurately simulate E_{LUC} (Bayer et al., 2017; Shevliakova et al., 2009; Stocker et al., 2014; Wilkenskjeld et al., 2014; Yue et al., 2017). Thus, high-resolution and successive long-term data on LC change are needed to generate the gross transition matrix used in LSMs. Although the products from Hansen et al. (2013) have a resolution of 30 m, they only provide forest area change rather than changes between all LC types. Further, the gross forest gain is only available for the whole period of 2000–2012 rather than at annual time steps (Hansen et al., 2013). The previous ESA CCI epoch maps contain all LC types (Bontemps et al., 2013) but the LC transitions are not appropriate to be used in LSMs because these epoch products represent 5-year composite maps and thus do not allow us to assess annual LC change dynamics; furthermore, only transitions to or from forest cover were considered at that time (Li et al., 2016).

The newly released annual ESA CCI land cover maps from 1992 to 2015 partly overcome these challenges with 300 m resolution and long and successive annual time series for all major land cover transitions (i.e. the maps now include transitions between non-forest classes, including grasses, crops and urban areas; ESA, 2017) and thus can be potentially translated into PFT maps used in the LSMs. The objectives of this study are to document the major gross and net changes and transitions in PFT maps derived from annual ESA CCI LC products and to evaluate whether they can be used in LSMs. Geographical distributions and temporal trends of the translated PFT maps from ESA CCI products are characterized and compared with those from other datasets. It should be noted that our analyses are based on the PFT maps that have been translated from the ESA CCI LC maps, rather than the original LC classes, because we aim to demonstrate the differences between different datasets and provide suggestions to modellers for implementing them in LSMs.

2 Methods

2.1 ESA CCI land cover products

The annual ESA CCI LC maps cover a period of 24 years from 1992 to 2015 at a spatial resolution of 300 m (ESA, 2017). These maps describe the Earth's terrestrial surface in 37 original LC classes based on the United Nations Land Cover Classification System (UN-LCCS; Di Gregorio, 2005).

This unique long-term land cover time series was achieved by combining the global daily surface reflectance of five different observation systems while aiming to maintain a good consistency over time. This was identified as a key requirement from the modeling community (Bontemps et al., 2012). Each of these global daily measurements of multispectral radiance recorded from 1992 to 2015 have been pre-processed to complete radiometric calibration, geometric and atmospheric correction, and clouds and cloud shadow screening. The full archive of MERIS (2003–2012) providing 15 spectral bands at 300 m resolution was classified to establish a baseline by fusing the outputs of machine learning and unsupervised algorithms (ESA, 2017). The 1 km time series recorded respectively by AVHRR from 1992 to 1999, SPOT-VGT from 1999 to 2013 and PROBA-V from 2014 and 2015 were used to detect and confirm the change which was eventually delineated more precisely at the 300 m spatial resolution whenever possible, i.e. later than 2004. This last step results in both back- and forward-dating the 10 year baseline LC map to produce the 24 annual LC maps from 1992 to 2015. In order to avoid false change detections due to the interannual variability in classifications, each change has to persist over more than 2 successive years in the classification time series to be confirmed (for more information see Sect. 3.1.2 of the ESA CCI LC Product User Guide; ESA, 2017). The resulting series of consistent 300 m annual LC maps from 1992 to 2015 is delivered with a pixel-based uncertainty value indicating the confidence at which a LC class was assigned for each pixel.

The accuracy of ESA CCI LC products was evaluated on a global scale according to international standards, using an independent validation dataset to produce a confusion matrix and derive overall an accuracy figure. An object-based validation database of 2600 primary sampling units was built by a panel of international experts to specifically assess the accuracy of both the LC classes and changes (ESA, 2017). Research is currently ongoing to find how to address the new challenges underlying this database, i.e. following a per-object approach and interpreting not just a unique land cover class but also a distribution of land cover classes within a primary sampling unit. The uniqueness of these two concepts in the framework of global land cover validation requires more time to derive reliable figures about LC classes and LC change accuracy. It will also prevent any comparison with previous validation figures.

In this respect, for the sake of comparison, the accuracy of the ESA CCI LC product from 2010 was assessed using the GlobCover 2009 validation database (Bontemps et al., 2011). Using all the points interpreted as "certain" by the experts, whether "homogeneous" (i.e. made of a single LC class) or "heterogeneous" (i.e. made of several or mosaic LC classes), the overall accuracy was found to be 71.5 %. Accounting for only the "homogeneous" and "certain" points, the overall accuracy raised to 75.4 % (ESA, 2017). The highest user accuracy values were found for the classes of rainfed cropland, irrigated cropland, broadleaved evergreen forest, urban areas, bare areas, water bodies and permanent snow and ice. Conversely, mosaic classes of natural vegetation were associated with the lowest user accuracy values, as well as the three classes of lichens and mosses, sparse vegetation and flooded forest with fresh water.

The overall accuracy of the ESA CCI LC products was also assessed by independent studies over specific regions (e.g. Tsendbazar et al., 2015, over Africa and Yang et al., 2017, over China), which can give valuable insights for specific applications.

2.2 PFT area and net changes

The original 37 ESA CCI LC classes were first aggregated into $0.5^\circ \times 0.5^\circ$ resolution and then translated into 14 different PFTs based on the cross-walking table (Table S1 in the Supplement) from the ESA Land Cover Product User Guide (ESA, 2017). This table originated from Poulter et al. (2015) and was further adjusted for some classes due to improved understanding of how the LC class descriptions can be interpreted to estimate fractional cover of PFTs from each LC class, in particular for mosaic classes and sparsely vegetated regions. PFTs were grouped into major vegetation types: forest, shrub, grassland and cropland. The tree PFTs and shrub PFTs (Table S1) were summed to obtain the forest and shrub area respectively; thus, the shrub PFTs are excluded from tree PFTs in our analyses. The net area change was calculated by comparing two annual PFT maps at $0.5^\circ \times 0.5^\circ$ resolution.

2.3 Gross PFT changes and transitions

Gross changes need to be considered differently because it is only possible to derive the net change by comparing the annual 300 m maps sequentially. Gross changes may be far larger than the net changes, and thus may show different magnitudes or even directions of LULCC carbon fluxes when simulated in LSMs. To document all the bidirectional LC transitions at $0.5^\circ \times 0.5^\circ$ resolution, high-resolution LC transition data are needed. Therefore, the annual ESA CCI LC maps are compared year by year at 300 m resolution to record the gross loss and gain of each original LC class over the whole period from 1992 to 2015. There are 23 original LC classes that experienced gross changes (classes with stars in Table S1).

In order to derive the gross transitions, all possible transitions (506 in total) between the 23 original LC classes with gross changes were calculated at 300 m resolution. There are a total of 422 gross transitions between these 23 original LC classes. These gross changes in the original classes were then translated into gross changes in PFTs using the LC-to-PFT cross-walking table (Table S1) and grouped into the major vegetation types (forest, shrub, grassland, cropland). For example, a LC transition from class "50", corresponding to 90 % tree PFT in Table S1, to class "30" (10 % tree PFT) is taken as a forest loss of 80 % in that 300 m grid cell. Finally, the converted transitions were aggregated into fractions in each $0.5^\circ \times 0.5^\circ$ grid cell.

Table 1. Description and comparison of different land-use/land-cover datasets used in this study. The term of "dataset" in this study can also involve some model output (e.g. forest area from Land Use Harmonization dataset, LUH2v2h).

	PFTs from annual ESA CCI maps	LUH2v2h (Hurtt et al., 2011)	HYDE 3.2 (Klein Goldewijk et al., 2016)	Hansen et al. (2013)	Houghton and Nassikas (2017)
Time span	1992–2015	850–2100	10 000 BC-2015	2000–2014	1850–2015
Time step	annual	annual	1000 year for the BCE period, then 100 year till 1700, 10 year till 2000, and from 2000 to 2015 annual	gross loss, annual; gross gain for one period (2000–2012)	annual
Spatial resolution	300 m	0.25°	5 arcmin	30 m	country
Land-use/ land-cover type	forest, shrub, grassland, cropland, bare soil, water and urban	forest, cropland, pasture, rangeland, urban and non-forested	cropland, grazing lands and urban	forest	forest
Gross or net	gross and net	gross and net	net	gross and net	net at country level
Data source	satellite (MERIS, SPOT-VGT, AVHRR, and PROBA-V)	urban, cropland, pasture and rangeland from HYDE 3.2 (Klein Goldewijk et al., 2016); forest and transitions based on model	cropland and grazing land are based on the FAO categories for "Arable land and permanent crops" and "Permanent meadows and pastures" (FAOSTAT, 2015); Spatial distribution based on ESA CCI epoch LC map 2010	satellite (Landsat)	FAO FRA (FAO, 2015), based on country reports
Advantage	full land cover types; relatively long time series; relatively high resolution; full gross transitions	full gross transitions; long time series	long time series; inventory-based	high resolution	inventory-based
Disadvantage	no specific pasture; uncertainty in cross-walking table	no separation of deciduous and evergreen forest; model-based forest areas; model-based temporal changes in historical cropland and grazing land (HYDE 3.2)	no forest; coarse time steps	short time period; no annual forest gain, but only for the whole period of 2000–2012; no other LC types	not grid-cell explicit; no other LC types; inconsistency of data sources and forest definitions between different countries

2.4 Comparison with other datasets

Three land-use and land-cover datasets (Table 1) were used for comparison: (i) forest, grassland and cropland area from the Land Use Harmonization (LUH2v2h) dataset (Hurtt et al., 2011); (ii) forest cover data from Hansen et al. (2013) and (iii) national forest area data from Houghton and Nassikas (2017). The cropland and pasture areas in the LUH2v2h dataset are from HYDE 3.2 (Klein Goldewijk et al., 2016), in which the ESA CCI epoch LC map in 2010 (representing 2008–2012) was used as a spatial reference map for the area allocation and the national cropland and grazing land were adjusted to match the FAO STAT data (FAOSTAT, 2015) as close as possible. The national forest areas from Houghton and Nassikas (2017) are based on FAO Forest Resources Assessment (FRA) data (FAO, 2015; also see Keenan et al. ,2015, for the main findings of FAO FRA 2015). Thus, these two additional sources of data, HYDE 3.2 (Klein Goldewijk et al., 2016) and FAO FRA (FAO, 2015) were not shown in the figures.

It should be noted that land use data are not necessarily the same as land cover data, and the exact definitions and categorization of forest (cropland and grassland) are different for each dataset (see details in Discussion). Nevertheless, these represent the best datasets available for the use in LSMs for comparison, and we have tried to harmonize the definitions where possible (see below), but to some degree this is an ongoing discussion between the modeling and data communities. Furthermore, all the LSMs have to use these datasets for deriving PFT changes back through time, so it is a very worthwhile exercise to determine if the broad groupings differ, and to what extent.

Absolute areas, net changes and gross transitions from 1992 to 2015 in the LUH2v2h dataset (Hurtt et al., 2011) were used for comparison. Forest used in this study from LUH2v2h (Hurtt et al., 2011) refers to the total of primary and secondary forest, cropland refers to all crop types and grassland refers to the total of pasture and rangeland. Because LUH2v2h data use cropland and grazing land areas from HYDE 3.2 as an input (Hurtt et al., 2011), the spatial distributions are mainly determined by HYDE 3.2. The gross transitions in LUH2v2h data are calculated from the Global Land Use Model (Hurtt et al., 2006) that tracks sub-grid cell loss and gain in land use categories. They first determined the urban area in each grid cell proportionally from cropland, pasture and secondary lands, and if these areas cannot fulfill the urban increase, primary lands were cleared. The minimum transition rates between cropland, pasture and other (sum of primary and secondary) lands were then calculated to identify the gross transitions between these land use categories (Hurtt et al., 2011). Transitions related to shifting cultivation and wood harvest were determined last (Hurtt et al., 2011).

Only annual gross forest loss each year during 2000–2014 and total gross forest gain during 2000–2012 are available in the dataset of Hansen et al. (2013). Thus, the net forest area change from this dataset only refers to the period of 2000–2012. The national forest area data from 1992 to 2015 in the dataset of Houghton and Nassikas (2017) were used to calculate the forest area changes.

A land mask with nine regions (Fig. 1) defined by Houghton (1999) was used to derive the regional values.

3 Results

3.1 PFT areas in the year 2000

After translating the original ESA CCI LC classes into PFTs using the cross-walking table (Table S1), the global and regional areas of forest, cropland and grassland PFTs in the year 2000 are shown in Fig. 1. Global areas of forest (excluding shrub), cropland and grassland PFTs are 30.4, 19.2 and 35.7 million km^2, respectively Global forest area is 6.7, 11.2 or 10.1 million km^2 lower than that from LUH2v2h (Hurtt et al., 2011), Hansen et al. (2013) or Houghton and Nassikas (2017), respectively. It is also much lower than the recently reported global forest area of 43.3 million km^2 with increased forest area estimate in dryland biomes using Google Earth images (Bastin et al., 2017). Global cropland area from ESA CCI is 4.2 million km^2 larger than that from LUH2v2h, while the difference in global grassland area is relatively small.

Forest area from ESA CCI is lower than that from Hansen et al. (2013) in all regions except North Africa and Middle East and the Pacific developed region. Here, the regions refer to the defined regions in Fig. 1. Forest area from LUH2v2h (Hurtt et al., 2011) is larger than that from ESA CCI in most regions except in South and Central America, tropical African and the Pacific developed region. Forest area from Houghton and Nassikas (2017), however, is systematically higher than that from ESA CCI in all regions. Cropland area from ESA CCI matches that from LUH2v2h (Hurtt et al., 2011) in North America but is higher in all the other regions. Although the global grassland area is similar between ESA CCI and LUH2v2h (Hurtt et al., 2011), larger differences are seen on a regional scale. Grassland area from ESA CCI was found to be much higher than that from LUH2v2h (Hurtt et al., 2011) in North America and the former Soviet Union (4.0 and 3.5 million km^2 higher, respectively) but much lower (2.4 million km^2) in North Africa and Middle East.

3.2 Gross area change

3.2.1 Time series of gross PFT change

After translating all of the 422 gross transitions detected between the original ESA LC classes into PFTs, the time series of gross changes in PFTs are shown in Fig. 2, and the mean annual change rates are shown in Table S2. Generally, the gross changes are related to the net, i.e. where there are more gross changes, more net changes can be found. Major gross changes occur in forest, cropland and grassland PFTs, with a global gross gain of 0.91, 1.2 and 1.1 million km^2 and a global gross loss of 1.5, 0.56 and 0.98 million km^2 respectively, from 1992 to 2015. The magnitudes of gross changes in these three PFTs are larger before 2005 than after 2005. Especially during the late 1990s, both intensive gross forest loss and gain occurred but overall resulted in net forest loss. Accordingly, both gross and net cropland area expands during this period. Two other peaks of net forest loss were found in 1995 and 2004, during which net cropland area increased. Although grassland experienced large gross loss and gross gain, the net area remains stable, except in 2004 where a net increase was found.

The temporal correlations of gross and net changes between ESA CCI PFTs, Hansen et al. (2013) and LUH2v2h (Hurtt et al., 2011) are not significant ($p > 0.05$, Table S3). The magnitudes of gross changes in forest from LUH2v2h (Hurtt et al., 2011) and Hansen et al. (2013) and cropland from LUH2v2h (Hurtt et al., 2011) are much larger than those detected from ESA CCI PFT maps (Fig. 2). In contrast to gross changes of forest and cropland from ESA CCI maps, annual gross changes from LUH2v2h (Hurtt et al., 2011) show larger variations after 2005 than before 2005. Especially before 2000, the annual gross changes in forest and cropland from LUH2v2h (Hurtt et al., 2011) are constant because HYDE 3.2 provides cropland and pasture

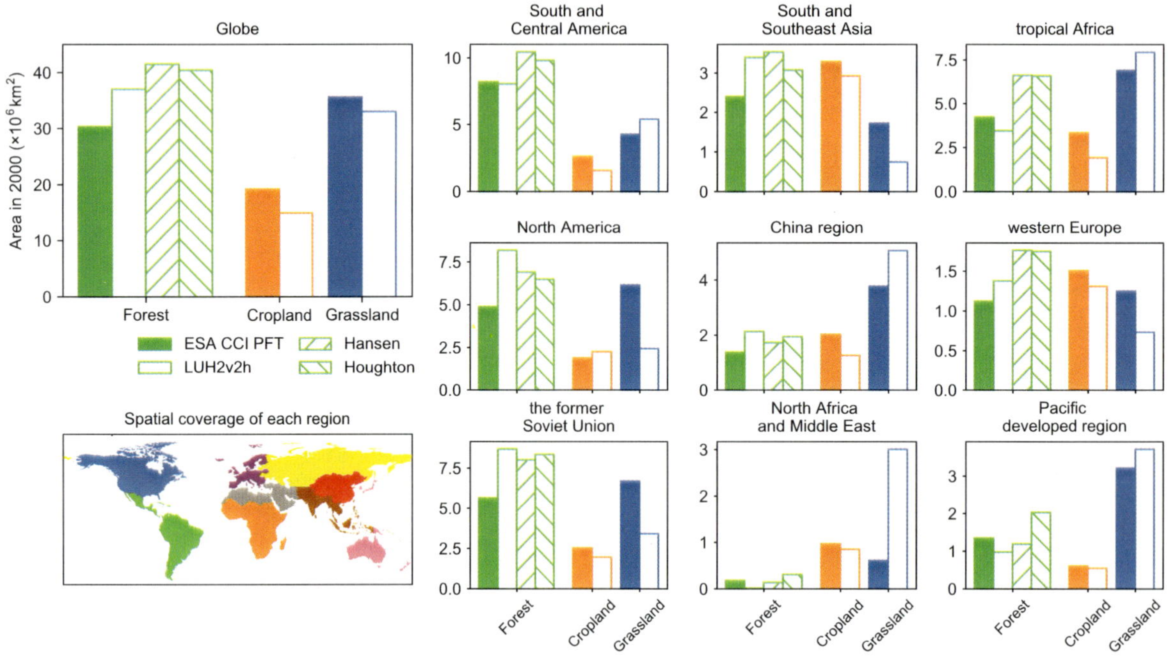

Figure 1. Global and regional areas of forest, cropland and grassland PFTs in the year 2000 in comparison with data from LUH2v2h (Hurtt et al., 2011), Hansen et al. (2013) and Houghton and Nassikas (2017). Different colours indicate different PFTs.

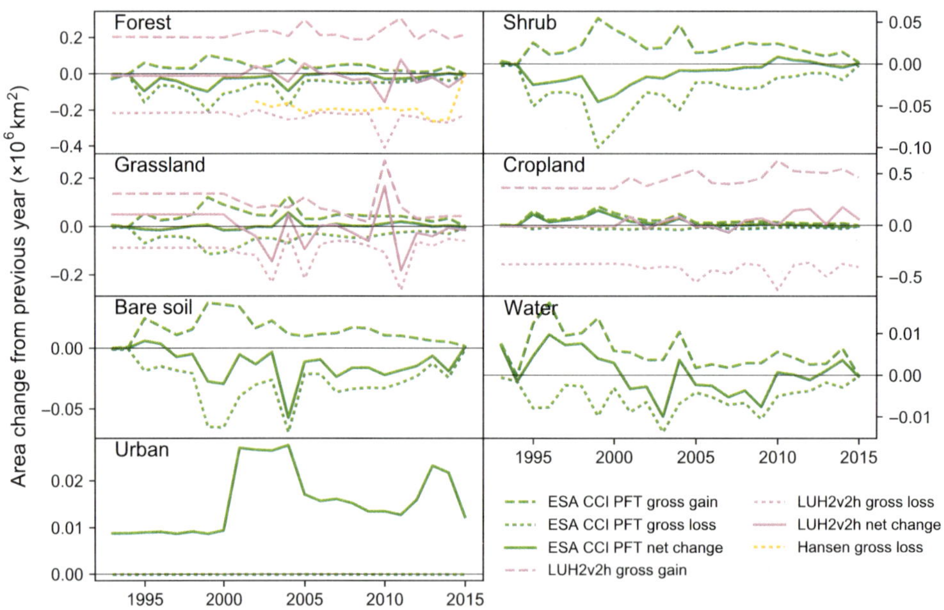

Figure 2. Gross changes in PFTs from 1992 to 2015 after translating gross transitions between original ESA land cover classes. Gross changes from LUH2v2h (Hurtt et al., 2011) and Hansen et al. (2013) are also shown for comparison. The red line indicates the zero line.

area only at a 10-year time step before 2000 (at an annual time step after 2000), and a linear interpolation was used in LUH2v2h (Hurtt et al., 2011) to produce the annual maps from HYDE 3.2 before 2000. The net forest loss and corresponding net cropland gain in 2004 coincides with the ESA CCI PFT maps and LUH2v2h (Hurtt et al., 2011) but the years of cropland gain in HYDE 3.2 are rather different from in ESA CCI PFTs during the other periods. Although the difference in the magnitude of gross grassland changes between ESA CCI PFTs and LUH2v2h (Hurtt et al., 2011) is relatively smaller than that of forest and cropland, the net grassland changes are not consistent over time.

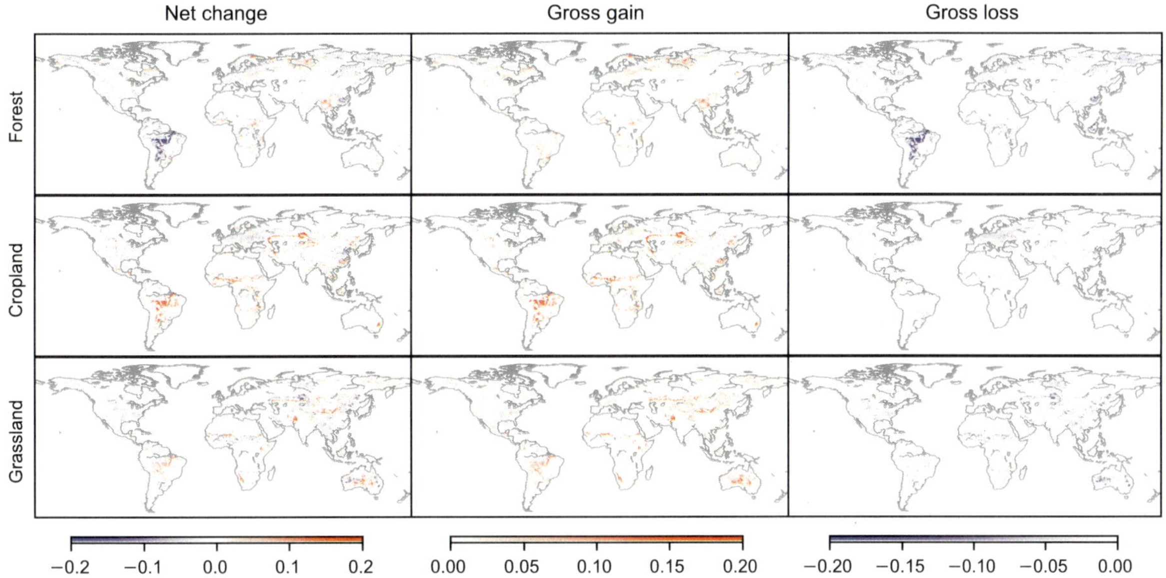

Figure 3. Spatial distributions of net and cumulative gross changes in forest, cropland and grassland PFTs between 1992 and 2015 derived from the ESA CCI data. The color scales indicate the changed fraction in each half-degree grid cell.

Gross changes in shrub and bare soil are also detected over the whole period, and the net change in these PFTs is generally a loss in area. The magnitudes of gross water body area changes are small compared to other PFTs. There is a relatively large net increase during 1995–2000 and a moderate net decrease during 2000–2010. Urban areas keep expanding over the whole period, and the increasing rates are high during 2001–2004 and 2012–2014.

3.2.2 Spatial distributions of gross PFT changes

The spatial distributions of net and cumulative gross changes in forest, cropland and grassland PFTs between 1992 and 2015 are shown in Fig. 3, and the distributions of the other PFTs are shown in Fig. S1. Intensive gross forest loss and sparse gross forest gain in South America result in a strong net decrease in forest area (Fig. 3). There are also considerable gross and net forest losses in South and Southeast Asia and in some regions of tropical Africa. Gross forest gain occurs pervasively in boreal regions. Some regions of intensive gross forest gain were found in south Asia, tropical Africa and South America, but with a small extent. Gross cropland gain occurs all over the world, and especially in South America, tropical Africa (particularly in the Sahel), South and Southeast Asia, and central Asia. By contrast, gross cropland loss is only observed in Europe and across the North China Plain. The cropland loss in these two regions is mainly caused by urbanization and thus an increase in urban area was found (Fig. S1). Overall, the net cropland change is an increase in most regions except Europe and the North China Plain. Grassland in temperate and tropical regions experienced extensive gross gain and gross loss, but the gross gain and loss are not fully coincident, leading to a pattern of coexisting net gain and loss everywhere (Fig. 3). The changes in grassland are relatively small in boreal regions.

The changes to shrubs are largely distributed in tropical regions, with a net gain in South America and net loss in tropical Africa and south Asia (Fig. S1). Intensive gross changes in bare soil were found in north China, central Asia, Australia and the south edge of the Sahara, mainly caused by the gross transitions between original ESA LC classes “200” (bare areas) and “150” (sparse vegetation; tree, shrub, herbaceous cover $< 15\,\%$). Water body changes are relatively small compared to other PFTs. In addition to the urban area increase over cropland in Europe and the North China Plain, there is also urban expansion to cropland in the United States (Fig. S1).

3.3 Net area change of PFTs

3.3.1 Global change

The global and regional net area changes in forest, cropland and grassland PFTs from ESA CCI LC maps since 1992 are shown in Fig. 4 (solid lines). Global net forest loss and net cropland gain between 1992 and 2015 are 0.60 and 0.67 million km^2, respectively. Global forest area decreased fast from 1992 to 2004 accompanied by fast increases in cropland. Forest area stayed stable between 2004 and 2009 and then decreased again, although by a smaller magnitude than in 1992–2004, during the recent period from 2009 to 2015. Meanwhile, cropland area remains relatively stable since 2004. Net grassland changes are small compared to forest and cropland changes.

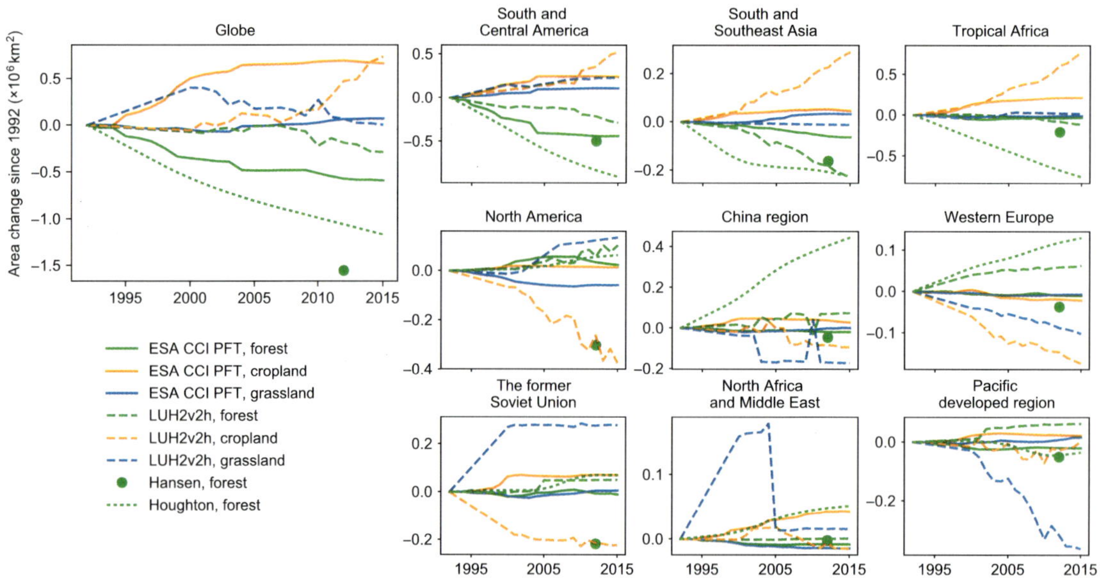

Figure 4. Global and regional net area changes in forest, cropland and grassland PFTs derived from ESA CCI land cover maps since 1992. Data from LUH2v2h (Hurtt et al., 2011), Hansen et al. (2013) and Houghton and Nassikas (2017) are also shown for comparison. Note that net forest area change from Hansen et al. (2013) is corresponding to the period of 2000–2012, and thus the forest area change between 1992 and 2000 from ESA CCI was added as Hansen et al. (2013) data in the plot.

The magnitudes of net forest area change from LUH2v2h (Hurtt et al., 2011) are much smaller than those from ESA CCI, mainly because the forest area decrease between 1992 and 2009 (Fig. 4) is not reflected in the LUH2v2h dataset (Hurtt et al., 2011). Although the net increased cropland areas from 1992 to 2015 are similar between ESA CCI and LUH2v2h (Hurtt et al., 2011), the temporal trajectories are rather different. The increase in cropland in ESA CCI data happened between 1992 and 2004, while cropland area in LUH2v2h (Hurtt et al., 2011) mainly increased since 2007 (Fig. 4). Grassland area changes in LUH2v2h (Hurtt et al., 2011) display more variations than those from ESA CCI. There was an increase in grassland in LUH2v2h (Hurtt et al., 2011) in the earlier period (1992–2004) where ESA CCI had the increase in cropland. Globally, net forest area loss between 1992 and 2015 from both Hansen et al. (2013) and Houghton and Nassikas (2017) is much larger than that from ESA CCI and LUH2v2h data (Hurtt et al., 2011).

3.3.2 Regional change

Consistent with the spatial distributions of net forest change in Fig. 3, net forest loss in South and Central America dominates the global net forest loss (Fig. 4), accounting for 75 % of the global total. The magnitude of net forest loss is close to that observed by Hansen et al. (2013) in this region. However, the magnitudes of net forest loss from ESA CCI PFTs in other regions are generally smaller than those from Hansen et al. (2013). Net forest area change from Houghton and Nassikas (2017) also shows a stronger loss in all three tropical regions than that in other datasets, especially in South and Central America and tropical Africa. It should be noted that the net forest loss in South and Southeast Asia is consistent between LUH2v2h (Hurtt et al., 2011), Hansen et al. (2013) and Houghton and Nassikas (2017), and all these datasets have much larger net forest area loss than ESA CCI data. All datasets demonstrate net forest gain in North America, except Hansen et al. (2013) which has a strong forest loss. The forest area in LUH2v2h data (Hurtt et al., 2011) and inventory-based data from Houghton and Nassikas (2017) shows a net increase in the China region and western Europe. In contrast, forest area in the satellite-based datasets of ESA CCI PFTs and Hansen et al. (2013) is stable or slightly decreasing.

South and Central America, tropical Africa and the former Soviet Union are the regions with the largest contributions to the global total net cropland increase, representing 37, 33 and 11 % of the global total. The regional patterns of temporal net cropland area change are rather different between ESA CCI PFTs and LUH2v2h (Hurtt et al., 2011) although the global net changes from 1992 to 2015 are similar. Cropland from LUH2v2h (Hurtt et al., 2011) expands more in tropical regions but decreases more in other regions than in ESA CCI PFTs (Fig. 4).

Grassland area from ESA CCI PFTs slightly increases in South and Central America and South and Southeast Asia, and slightly decreases in North America, the former Soviet Union, North Africa and Middle East. Differences in grassland change are large between ESA CCI PFTs and LUH2v2h (Hurtt et al., 2011) in all regions other than tropical regions.

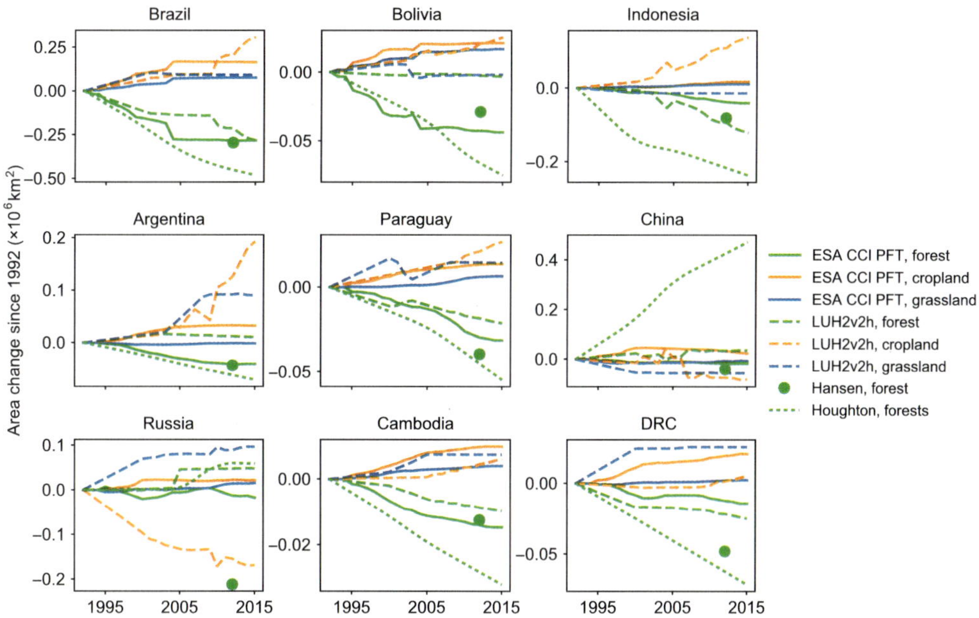

Figure 5. Net area changes in forest, cropland and grassland PFTs derived from ESA CCI land cover maps since 1992 in countries with largest net forest area loss between 1992 and 2015. Data from LUH2v2h (Hurtt et al., 2011), Hansen et al. (2013) and Houghton and Nassikas (2017) are also shown for comparison.

3.3.3 Countries with largest net forest area loss and gain

The countries with the largest net forest PFT area loss between 1992 and 2015 from ESA CCI maps are shown in Fig. 5, and countries with the largest net forest PFT gain in Fig. 6. Brazil, Bolivia and Indonesia are the three countries with largest net forest losses during 1992–2015 with a net loss of 0.28, 0.044 and 0.042 million km^2, respectively. The net forest loss in Brazil during the whole period is consistent between ESA CCI PFTs, LUH2v2h (Hurtt et al., 2011) and Hansen et al. (2013), despite the fact that temporal patterns are different between ESA CCI and LUH2v2h (Hurtt et al., 2011). Net forest changes between ESA CCI PFTs and Hansen et al. (2013) are also similar in Indonesia, Argentina and Cambodia, while net forest loss in Russia and the Democratic Republic of the Congo (DRC) from Hansen et al. (2013) is much larger than that from ESA CCI. Net forest loss from Houghton and Nassikas (2017) is always higher than the loss from other datasets in all these countries except in China and Russia, where a net forest gain was found in Houghton and Nassikas (2017).

The overall net cropland gain from 1992 to 2015 between ESA CCI and LUH2v2h (Hurtt et al., 2011) is similar in Bolivia but is rather different in all other countries in Fig. 5. Larger cropland gain from LUH2v2h (Hurtt et al., 2011) compared to ESA CCI was found in Brazil, Indonesia, Argentina and Paraguay, while lower cropland gain was found in Cambodia and the DRC. The cropland area change in China and Russia from LUH2v2h (Hurtt et al., 2011) even shows a net loss rather than gain. Grassland area increased in Argentina, Paraguay, Russia, Cambodia and the DRC in LUH2v2h (Hurtt et al., 2011), which was not captured by ESA CCI maps.

The magnitudes of forest change in the countries with the largest forest gain in Fig. 6 are much smaller than those with largest forest loss (Fig. 5). For example, the net forest gain from 1992 to 2015 is 0.019 million km^2 in Canada, compared with a forest loss of 0.28 million km^2 in Brazil. In these largest forest gain countries, forest area change from Hansen et al. (2013) indicates a net forest gain only in Uruguay, and a net loss or stable in other countries. Again, contrary to ESA CCI PFTs, Houghton and Nassikas (2017) forest area data show large magnitudes of net forest loss in Myanmar, North Sudan and Nigeria, and greater magnitudes of net forest gain than other datasets in Uruguay. Cropland changes from LUH2v2h (Hurtt et al., 2011) display larger magnitudes, more variations and even different directions than those from ESA CCI in these nine countries in Fig. 6. Grassland area changes from ESA CCI are rather flat, which is different from those in LUH2v2h (Hurtt et al., 2011).

4 Discussion

4.1 Differences in total area of forest, cropland and grassland

The forest, cropland and grassland areas from different datasets do not match on global or regional scales (Fig. 1),

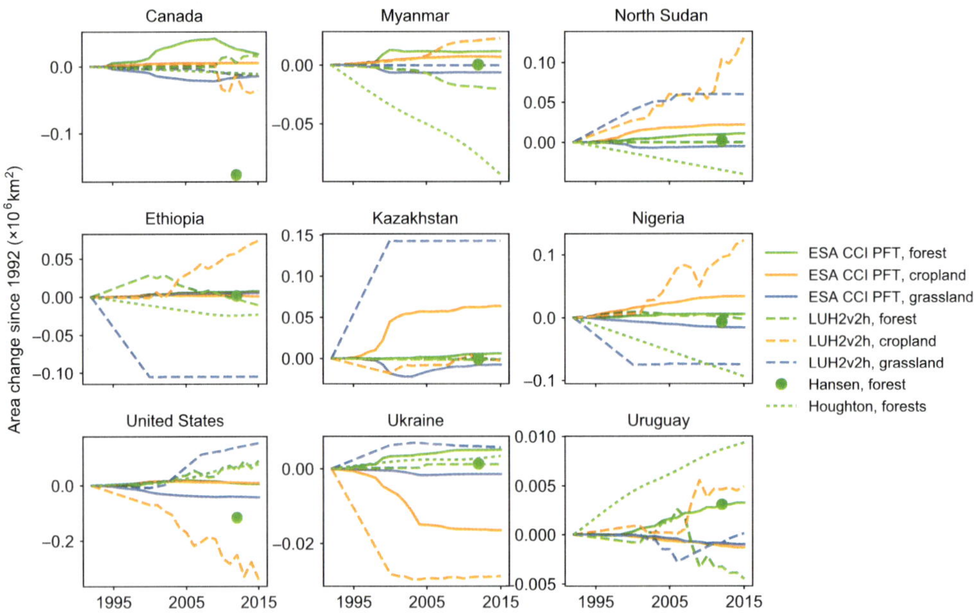

Figure 6. Net area changes in forest, cropland and grassland PFTs derived from ESA CCI land cover maps since 1992 in countries with largest net forest area gain between 1992 and 2015. Data from LUH2v2h (Hurtt et al., 2011), Hansen et al. (2013) and Houghton and Nassikas (2017) are also shown for comparison.

mainly caused by the differences in land cover definitions and data sources (Table 1), as well as the uncertainties in the cross-walking table used for translating original ESA CCI LC classes into PFTs. The canopy cover of forest varies in different ESA CCI LC classes with defined ranges such as > 15 %, 15–40 % and > 40 % depending on the "openness" of the canopy and according to the UN-LCCS framework provided by the FAO (Di Gregorio, 2005). Although continuous tree cover fractions are provided in data from Hansen et al. (2013), the forest cover is defined as > 25 % canopy closure for trees higher than 5 m (Hansen et al., 2010). Forest areas in Hansen et al. (2013) are obtained from NASA's Landsat instruments with a spatial resolution of 30 m that can capture the small-scale forest areas. This partly explains the larger forest extent in Hansen et al. (2013) than ESA CCI PFT maps. The definition of forest by FAO, which is the data source of Houghton and Nassikas (2017), is a canopy cover > 10 %. FAO's forest areas are based on reports from the member countries (FAO, 2015) and the methods of compiling data in each country may vary largely, e.g. from field survey or from satellite-imagery-based estimation (Grainger, 2008; Harris et al., 2012). Furthermore, in the definition of forest by FAO, natural disturbance suppressing forests do not change the land remaining a forest, but from satellite, they are not detected as forest cover. Keenan et al. (2015) also compared the forest area from FAO FRA 2015 with remote sensing data and attributed their differences to five factors, the major one of which is the different definitions of "forest". It seems that forest area from ESA CCI PFTs is lower than that from Hansen et al. (2013) in humid regions (Fig. S2). Additionally, in the drylands like tropical Africa, it is difficult to map and estimate forest area using medium- (e.g. Landsat; Hansen et al., 2013) or coarse-resolution satellite data (e.g. ESA CCI LC; Achard et al., 2014; Bastin et al., 2017; Gross et al., 2017), in particular when tree cover is below 30 % (Achard et al., 2014). Bastin et al. (2017) recently reported a forest estimate in drylands using very high spatial resolution satellite imagery, which is 40–47 % more than previous forest assessments. The difficulty of detecting forest in these sparse tree cover regions could partly be responsible for the lower forest area from ESA CCI PFT maps than those from Hansen et al. (2013) and Houghton and Nassikas (2017) in tropical Africa (Fig. 1).

Forest area estimates in LUH2v2h (Hurtt et al., 2011) are based on aboveground biomass density from the Miami-LU ecosystem model (Hurtt et al., 2006), and cropland and pasture areas are based on HYDE 3.2 (Klein Goldewijk et al., 2016). HYDE 3.2 uses the cropland and pasture areas from FAO STAT (FAOSTAT, 2015) as the main land-use input data and the ESA CCI epoch LC map of 2010 as a spatial reference map (Klein Goldewijk et al., 2016). Thus, the grasslands in LUH2v2h refer to the sum of intensively managed pastures and less intensively used rangelands (Klein Goldewijk et al., 2016), while the grassland PFT from ESA CCI maps also includes natural grassland, which may be the reasons for less grassland in LUH2v2h (Hurtt et al., 2011) than ESA CCI, especially in the former Soviet Union, western Europe and North America (Fig. 1).

Similarly, the underestimate in cropland area in ESA CCI maps (Fig. 1) is likely due to differences in definitions of what constitutes a cropland based on remote sensing datasets used to derive the ESA maps vs. land use statistics and country-dependent reporting used to derive FAO statistics that are used to define croplands in HYDE 3.2 (Klein Goldewijk et al., 2016), in addition to differences in spatial resolution. For example, the attribution of oil palm plantations is an important factor for the differences in area changes between different datasets, especially in Indonesia. Oil palm is taken as cropland rather than forest in the FAO definitions (FAOSTAT, 2015) but detected as tree covers from the remote sensing (Carlson et al., 2012, 2013; Hansen et al., 2013; Koh et al., 2011; Tropek et al., 2014), including in the CCI LC products. This partly explains that the larger cropland increase in LUH2v2h (Hurtt et al., 2011) and larger forest decrease in Houghton and Nassikas (2017) than those in ESA CCI PFTs and Hansen et al. (2013) in Indonesia (Fig. 4). Furthermore, the classification of cropland in ESA CCI is also based on remote sensing temporal analysis. In the ESA CCI algorithm, for example, spectral features at key moments during the year were used to optimize the discriminations between all major crop classes: differentiating between cropland and natural vegetation (typically harvesting dates). Cropland in LUH2v2h that is essentially from FAO statistics (Klein Goldewijk et al., 2016), on the other hand, depends on country reporting and therefore comprises different definitions and data sources.

Pérez-Hoyos et al. (2017) provide an extensive comparison of multiple cropland datasets, including ESA CCI epoch and annual maps, for the purposes of cropland monitoring, and they found that the ESA CCI 2015 annual map is more suitable for cropland monitoring than the epoch map because of the reduction in cropland area over the Congo basin. They also showed that spatial resolution is a key driver of product suitability for agriculture monitoring (Pérez-Hoyos et al., 2017). However, the specific framework of their study is the suitability for agriculture monitoring for early warning and they focus on a limited number of countries selected to be "with high risk of food insecurity" (Pérez-Hoyos et al., 2017). The issues of cropland area from the ESA CCI LC maps discussed in their study is fully justified for a study addressing the challenge of cropland monitoring, but it does not allow generalizing the conclusion to all domains (e.g. to derive PFT maps for use in global land surface modeling in this study). As Pérez-Hoyos et al. (2017) and our study show, the agreement (or lack thereof) is country-dependent, further implying that more consistent definitions of LC classes are required and/or regional LC satellite mapping algorithms (or cross-walking table, see below) are needed. Cropland mapping issues, including those discussed in Pérez-Hoyos et al. (2017) are being addressed in upcoming versions of the ESA CCI maps. Additionally, Waldner et al. (2016) have produced a product that aims to combine the "fittest" LC maps on the country level into a unified 250 m resolution cropland product, but again this is dependent upon a specific definition (the JECAM, Joint Experiment of Crop Assessment and Monitoring, cropland definition for the purposes of cropland monitoring).

The final spatial area of each PFT in this study is derived from a combination of the ESA LC map and the cross-walking table (Table S1) used for translating original ESA LC classes into PFTs. The range in tree cover canopy openness (as discussed above) and percent of each type of vegetation for mosaic LC classes in the LC description contributes to uncertainty in the conversion fractions used to translate the LC classes into PFTs in the cross-walking table. Thus, uncertainties in the cross-walking table contribute to the differences in forest, cropland and grassland PFT areas when comparing with other datasets. Only one value is used to prescribe the fraction of each PFT for a given class, e.g. class "50" corresponds to 90 % of broadleaf evergreen trees in Table S1. This hinders an explicit representation of spatially heterogeneous tree cover fractions. In the absence of other information, the approximate mid-point of the range in the LC class description is used when calculating the fraction of forest PFT from a given LC class. For example, class "61" represents a closed canopy (> 40 %) and therefore we use a LC-to-tree PFT conversion fraction of 70 % (0.7) as the mid-point between 40 and 100 % (Table S1). Class "62", on the other hand, is an open canopy (15–40 % cover) and therefore we use a LC-to-tree PFT conversion factor of 30 %. Some exceptions to this general rule are made when we have a better understanding of the species or biomes included in a given LC class. For example, class "50" (broadleaved evergreen trees) encompasses tropical rainforests. Although the class description states that the canopies in this class can be closed to open (> 15 %), we know that the tree cover fraction is much higher than a mid-point of ∼ 60 %, therefore we use a conversion factor of 90 %. However, this level of knowledge is not available for all LC classes. This is particularly true for mosaic and sparse vegetation classes (e.g. classes "100", "110" and "150") that span different regions/biomes that may contain different fractional coverages of vegetation.

Likewise, an explicit regional classification is required for cropland. For example, class "10" (cropland, rainfed) is separated well in North America, i.e. mainly partitioning into class "11" (herbaceous cover), and thus the cropland area in this region is highly consistent with LUH2v2h data (Hurtt et al., 2011; Fig. 1). In tropical Africa where class "10" is not separated into a more detailed classification, the difference in cropland areas between these two datasets are large (Fig. 1). This is because if most of the cropland in this region belongs to class "12", using the corresponding value for class "10" in the cross-walking table (90 % for class "10" vs. 30 % for class "12", Table S1) overestimates cropland areas.

Hartley et al. (2017) also investigated the uncertainty in simulations of carbon, water and energy fluxes from three LSMs as a result of cross-walking table uncertainty. This study found that the spread in model outputs due to cross-walking uncertainty was higher than uncertainty due to the underlying LC maps (mapping algorithm; Hartley et al., 2017). Despite these uncertainties, satellites provide the only plausible way to derive the global maps of vegetation distribution needed to drive LSMs and validate dynamic global vegetation models. Future efforts by the ESA CCI LC project and collaborators will focus on reducing the uncertainty introduced when translating from LC to PFT, including using optimized and regionally based cross-walking tables.

4.2 Differences in area changes

The ESA CCI LC magnitudes of gross changes for all PFTs are lower than those of all three products considered. This is explained by the effect of spatial resolution combined with a change consolidation approach. Using Earth observation time series of 1 km spatial resolution to annually detect the land cover change for the ESA CCI maps does not allow capturing small-scale LC changes, which is part of the reason for smaller gross and net forest changes than those in Hansen et al. (2013). On the other hand, this is the only way to have a consistent method of LC change detection over the whole period. In spite of the consolidation strategy, confirming the change over several years, ESA CCI LC trends of area change mitigate only partly the impact of the heterogeneous quality of the data acquired by the various sensors. For instance, larger change variations for forest and cropland in the 1990s result from poorer radiometric and spectral quality of the AVHRR input data. This instrument, first designed for meteorological observation, is, however, the only one recording the land surface systematically before 1999.

The large magnitude of gross changes in forest and cropland in LUH2v2h (Hurtt et al., 2011; Fig. 2) mostly distributes in the tropical regions (Figs. S3 and S4) where gross changes reflect shifting agriculture (Heinimann et al., 2017). The gross gain and loss of forest (or cropland) in the tropics from LUH2v2h maintains a similar constant rate with other small variations (Figs. S3 and S4). This is because the gross changes in LUH2v2h are mainly generated from the shifting cultivation in the tropics by assuming a turnover rate of 6.7 % $year^{-1}$ (i.e. a residence time of 15 year) of all agricultural lands (Hurtt et al., 2011) and based on a spatial distribution map from Butler (1980). The Butler (1980) map is a hand-drawn map indicating the presence or absence (no precise area or fraction) of both shifting cultivation and other non-shifting farming systems based on some regional studies and "general knowledge" (Heinimann et al., 2017). The estimate of shifting cultivation extent from LUHv2h (Hurtt et al., 2011) is thus highly uncertain because of the simple assumptions and the old reference map (representing 1960s–1970s) but strongly affects the gross land use change areas. Heinimann et al. (2017) recently estimated the global extent of shifting cultivation visually using Landsat 30 m forest data and very-high-resolution satellite imagery from Bing and Google. They found that shifting cultivation area decreases over the last 40 to 50 years, in particular in Southeast Asia (Heinimann et al., 2017). This, however, is not reflected in the LUH2v2h dataset (Figs. S3 and S4). From LUH1 to LUH2v2h, The area of shifting agriculture is reduced (see an example in tropical Africa; Fig. S5) because of the separation of forest from natural vegetation in LUH2 (Hurtt et al., 2011). However, the gross forest changes in LUH2v2h (Hurtt et al., 2011) are still much higher than those in ESA CCI PFTs and Hansen et al. (2013). Especially in the ESA 300 m resolution data, the gross change area seems very small (Figs. S3 to S5). Therefore, the shifting cultivation area in LUH2v2h may be overestimated due to (1) the binary (presence/absence) indication rather than a precise extent of shifting cultivation in the Butler (1980) map and (2) no temporal change (missing the decreasing trend) of the reference map. Still, it should be noted that the coarse spatial resolution of ESA CCI products cannot detect small-scale LC changes, resulting in an underestimation of gross changes. The shifting cultivation today remains extensive and is very important for land carbon modeling, but there are only very limited studies on the regional or national extent estimates (Heinimann et al., 2017). More research and developments in the mapping and change detection of shifting cultivation are urgently desired.

The discrepancies in temporal PFT net area changes between ESA CCI maps and FAO data (cropland and pasture area changes in LUH2v2h, Hurtt et al., 2011, and forest area changes in Houghton and Nassikas, 2017; Figs. 4–6) are mainly caused by the different approaches for estimating LC change used by different countries in FAO reports (FAO, 2015; FAOSTAT, 2015). Some countries like Canada distinguish land use and land cover when compiling forest statistics. For example, a forest cleared for wood harvest is not taken as a forest loss because new secondary forest will be planted on this land, thus no change in land use (Keenan et al., 2015). However, remote sensing can easily detect such land cover change and treat it as forest loss. Cropland and pasture in HYDE 3.2 (Klein Goldewijk et al., 2016) adopted the FAO categories for "Arable land and permanent crops" and "Permanent meadows and pastures", respectively, as the main data source. In the ESA CCI LC maps, pastures are mapped as grassland and translated into 100 % "Natural Grass" PFT (Table S1). The different trajectories of temporal cropland changes between ESA CCI and LUH2v2h (the former shows increasing from 1992 to 2004 while the latter increases after 2007; Fig. 4) are probably caused by the time lag between the real changes and country reporting to FAO. Finally, the trends of cropland area change from FAO STAT data may contradict those from national statistics (Li et al., 2016), e.g. comparing FAO STAT data (FAOSTAT, 2015) with USDA estimates (Nickerson et al., 2011) for the United States or with NBSC estimates (NBSC, 2015) for China.

There are also many other land cover and land use datasets that can be used for comparisons to assess the accuracy of land cover or land cover change in ESA CCI LC products. However, they are either regional maps (e.g. the maps for Europe from Fuchs et al., 2015) or global epoch maps (e.g. the GlobeLand30 maps for 2000 and 2010; Chen et al., 2014) and not suitable for the application in LSMs. Thus, we did not include them in this study. In fact, there have already been studies on the detailed comparisons of different datasets in a region (e.g. Fuchs et al., 2015, for Europe; Yang et al., 2017, for China; and Achard et al., 2014, for the tropics). In addition to the accuracy assessments conducted in the ESA CCI project (ESA, 2017), a systematic comparison with all other land cover datasets in future will help to validate the land cover classification and land cover change detection in the ESA CCI LC products. Instead of using a single dataset, combining a sample of several datasets is reported to be considerably more efficient and accurate to estimate land cover area and change (Olofsson et al., 2014; Sannier et al., 2016) and has been adopted as technical guidelines (GOFC-GOLD, 2016; GFOI, 2016) in the remote sensing community, especially for forest monitoring to reduce emissions from deforestation and forest degradation in developing countries (the REDD+ programme).

5 Data availability

The ESA CCI LC maps can be viewed online using http://maps.elie.ucl.ac.be/CCI/viewer/index.ph, and the data products can be download from http://maps.elie.ucl.ac.be/CCI/viewer/download.php. After entering some basic information, the land cover maps with a specific version number are available for download in the Climate Research Data Package (CRDP) section. In this study, we used the version "Land Cover Maps – v2.0.7". A protocol for translating the original ESA CCI LC maps into PFT maps and an example of a LC map and PFT map in 2000 used in this study can be downloaded from https://doi.org/10.5281/zenodo.834229 (Li et al., 2017a). The annual ESA CCI PFT maps from 1992 to 2015 at $0.5^\circ \times 0.5^\circ$ resolution can also be downloaded from https://doi.org/10.5281/zenodo.1048163 (Li et al., 2017b).

6 Conclusions

In this study, we compare the absolute areas and areal changes between PFTs from annual ESA CCI LC products and other datasets. In the intensive LULCC regions like South and Central America, both forest area and net forest change are consistent with those from other datasets. The detection of LC changes has significantly improved from the last version of 5-year epoch ESA CCI maps (Li et al., 2016). The detailed annual cropland changes from 1992 to 2000 fill the gaps of HYDE 3.2 data for this period, in which only decadal changes are available (Klein Goldewijk et al., 2016).

Considering the discrepancies, advantages and disadvantages among different datasets (Table 1), we propose different choices of these datasets for application in LSMs depending on research purposes. For example, if we would like all LSMs to share the same historical and future maps in a model intercomparison project (e.g. using LUH2v2h data in CMIP6), annual ESA CCI data products should be cautiously harmonized considering the large differences between ESA CCI and LUH2v2h (Hurtt et al., 2011). On the other hand, if we want to analyse recent carbon and water budgets with LSMs, ESA CCI maps are definitely an appropriate choice. The detailed LC classes in ESA CCI products provide a valuable reference map for modellers to partition land covers into PFTs, e.g. separating the generic forest in LUH2v2h dataset (Hurtt et al., 2011) into different forest PFTs (Table S1). LSMs can also benefit from the 300 m spatial resolution changes in ESA CCI products when accounting for gross land use changes to simulate the LULCC carbon fluxes. Therefore, the current annual ESA CCI land cover maps with full land cover classes, 300 m spatial resolution and relatively long-term time series are sufficient to be implemented in LSMs and help better characterize the recent global and regional carbon cycles.

Competing interests. The authors declare that they have no conflict of interest.

Acknowledgements. Wei Li was supported by the European-Commission-funded project LUC4C (grant no. 603542). Wei Li and Philippe Ciais were supported by the European Research Council through Synergy grant ERC-2013-SyG-610028 "IMBALANCE-P". Natasha MacBean, Sophie Bontemps, Céline Lamarche and Pierre Defourny were supported by the Climate Change Initiative program supported by the European Space Agency.

Edited by: David Carlson
Reviewed by: two anonymous referees

References

Achard, F., Beuchle, R., Mayaux, P., Stibig, H.-J., Bodart, C., Brink, A., Carboni, S., Desclée, B., Donnay, F., Eva, H. D., Lupi, A., Raši, R., Seliger, R., and Simonetti, D.: Determination of tropical deforestation rates and related carbon losses from 1990 to 2010, Glob. Change Biol., 20, 2540–2554, https://doi.org/10.1111/gcb.12605, 2014.

Alkama, R. and Cescatti, A.: Biophysical climate impacts of recent changes in global forest cover, Science, 351, 600–604, https://doi.org/10.1126/science.aac8083, 2016.

Bartholomé, E. and Belward, A. S.: GLC2000: a new approach to global land cover mapping from Earth observation data, Int. J. Remote Sens., 26, 1959–1977, https://doi.org/10.1080/01431160412331291297, 2005.

Bastin, J.-F., Berrahmouni, N., Grainger, A., Maniatis, D., Mollicone, D., Moore, R., Patriarca, C., Picard, N., Sparrow, B., Abraham, E. M., Aloui, K., Atesoglu, A., Attore, F., Bassüllü, Ç., Bey, A., Garzuglia, M., García-Montero, L. G., Groot, N., Guerin, G., Laestadius, L., Lowe, A. J., Mamane, B., Marchi, G., Patterson, P., Rezende, M., Ricci, S., Salcedo, I., Diaz, A. S.-P., Stolle, F., Surappaeva, V., and Castro, R.: The extent of forest in dryland biomes, Science, 356, 635–638, https://doi.org/10.1126/science.aam6527, 2017.

Bayer, A. D., Lindeskog, M., Pugh, T. A. M., Anthoni, P. M., Fuchs, R., and Arneth, A.: Uncertainties in the land-use flux resulting from land-use change reconstructions and gross land transitions, Earth Syst. Dynam., 8, 91–111, https://doi.org/10.5194/esd-8-91-2017, 2017.

Bonan, G. B.: Forests and climate change: forcings, feedbacks, and the climate benefits of forests., Science, 320, 1444–1449, https://doi.org/10.1126/science.1155121, 2008.

Bontemps, S., Defourny, P., van Bogaert, E., Kalogirou, V., and Arino, O.: GlobCover 2009: Products description and validation report, available at: http://due.esrin.esa.int/files/GLOBCOVER2009_Validation_Report_2.2.pdf (last access: 10 November 2017), 2011.

Bontemps, S., Herold, M., Kooistra, L., van Groenestijn, A., Hartley, A., Arino, O., Moreau, I., and Defourny, P.: Revisiting land cover observation to address the needs of the climate modeling community, Biogeosciences, 9, 2145–2157, https://doi.org/10.5194/bg-9-2145-2012, 2012.

Bontemps, S., Defourny, P., Radoux, J., Van Bogaert, E., Lamarche, C., Achard, F., Mayaux, P., Boettcher, M., Brockmann, C., and Kirches, G.: Consistent global land cover maps for climate modelling communities: current achievements of the ESA's land cover CCI, in: Proceedings of the ESA Living Planet Symposium, 9–13 September 2013, Edimburgh, UK, 9–13, 2013.

Butler, J. H.: Economic Geography: Spatial and Environmental Aspects of Economic Activity, John Wiley & Sons., New York, USA, 1980.

Carlson, K. M., Curran, L. M., Asner, G. P., Pittman, A. M., Trigg, S. N., and Adeney, J. M.: Carbon emissions from forest conversion by Kalimantan oil palm plantations, Nat. Clim. Change, 3, 283–287, https://doi.org/10.1038/Nclimate1702, 2013.

Carlson, K. M., Curran, L. M., Ratnasari, D., Pittman, A. M., Soares-Filho, B. S., Asner, G. P., Trigg, S. N., Gaveau, D. A., Lawrence, D., and Rodrigues, H. O.: Committed carbon emissions, deforestation, and community land conversion from oil palm plantation expansion in West Kalimantan, Indonesia, P. Natl. Acad. Sci. USA, 109, 7559–7564, https://doi.org/10.1073/pnas.1200452109, 2012.

Chen, J., Ban, Y., and Li, S.: China: open access to Earth land-cover map, Nature, 514, 434–434, https://doi.org/10.1038/514434c, 2014.

Defourny, P., Mayaux, P., Herold, M., and Bontemps, S.: Global land-cover map validation experiences: toward the characterization of quantitative uncertainty, in: Remote Sensing of Land Use and Land Cover: Principles and Applications, 207–223, CRC Press, Boca Raton, FL, USA, 2012.

Di Gregorio, A.: Land Cover Classification System: Classification Concepts And User Manual: LCCS, Food & Agriculture Org., Rome, Italy, 2005.

ESA: Land Cover CCI Product User Guide Version 2.0, available at: http://maps.elie.ucl.ac.be/CCI/viewer/download/ESACCI-LC-Ph2-PUGv2_2.0.pdf, last access: 10 November 2017.

FAO: Global Forest Resources Assessment 2015, Desk Reference, Rome, available at: http://www.fao.org/3/a-i4808e.pdf (last access: 10 November 2017), 2015.

FAOSTAT: Food and agriculture data, Food Agric. Organ. United Nations, available at: http://www.fao.org/faostat/en/\T1\textbackslash#home (last access: 19 May 2017), 2015.

Friedl, M. A., Sulla-Menashe, D., Tan, B., Schneider, A., Ramankutty, N., Sibley, A., and Huang, X.: MODIS Collection 5 global land cover: algorithm refinements and characterization of new datasets, Remote Sens. Environ., 114, 168–182, https://doi.org/10.1016/j.rse.2009.08.016, 2010.

Fuchs, R., Herold, M., Verburg, P. H., Clevers, J. G. P. W., and Eberle, J.: Gross changes in reconstructions of historic land cover/use for Europe between 1900 and 2010, Glob. Change Biol., 21, 299–313, https://doi.org/10.1111/gcb.12714, 2015.

GFOI: Integration of remote-sensing and ground-based observations for estimation of emissions and removals of greenhouse gases in forests: Methods and Guidance from the Global Forest Observations Initiative, Edition 2.0, Food and Agriculture Organization, Rome, Italy, 2016.

GOFC-GOLD: A sourcebook of methods and procedures for monitoring and reporting anthropogenic greenhouse gas emissions and removals associated with deforestation, gains and losses of carbon stocks in forests remaining forests, and forestation. GOFC-GOLD Report version COP22-1, GOFC-GOLD Land Cover Project Office, Wageningen University, the Netherlands, 2016.

Grainger, A.: Difficulties in tracking the long-term global trend in tropical forest area, P. Natl. Acad. Sci. USA, 105, 818–823, https://doi.org/10.1073/pnas.0703015105, 2008.

Gross, D., Achard, F., Dubois, G., Brink, A., and Prins, H. H. T.: Uncertainties in tree cover maps of Sub-Saharan Africa and their implications for measuring progress towards CBD Aichi Targets, Remote Sens. Ecol. Conserv., online first, https://doi.org/10.1002/rse2.52, 2017.

Hansen, M. C., Stehman, S. V. and Potapov, P. V.: Quantification of global gross forest cover loss, P. Natl. Acad. Sci. USA, 107, 8650–8655, https://doi.org/10.1073/pnas.0912668107, 2010.

Hansen, M. C., Potapov, P. V., Moore, R., Hancher, M., Turubanova, S. A., Tyukavina, A., Thau, D., Stehman, S. V, Goetz, S. J., Loveland, T. R., Kommareddy, A., Egorov, A., Chini, L., Justice, C. O., and Townshend, J. R. G.: High-resolution global maps of 21st-century forest cover change., Science, 342, 850–853, https://doi.org/10.1126/science.1244693, 2013.

Harris, N. L., Brown, S., Hagen, S. C., Saatchi, S. S., Petrova, S., Salas, W., Hansen, M. C., Potapov, P. V and Lotsch, A.: Baseline map of carbon emissions from deforestation in tropical regions., Science, 336, 1573–1576, https://doi.org/10.1126/science.1217962, 2012.

Hartley, A. J., MacBean, N., Georgievski, G., and Bontemps, S.: Uncertainty in plant functional type distributions and its impact on land surface models, Remote Sens. Environ., 203, 71–89, https://doi.org/10.1016/j.rse.2017.07.037, 2017.

Heinimann, A., Mertz, O., Frolking, S., Egelund Christensen, A., Hurni, K., Sedano, F., Parsons Chini, L., Sahajpal, R., Hansen, M., and Hurtt, G.: A global view of shifting cultivation: Recent, current, and future extent, edited by B. Poulter, Plos One, 12, e0184479, https://doi.org/10.1371/journal.pone.0184479, 2017.

Houghton, R. A.: The annual net flux of carbon to the atmosphere from changes in land use 1850–1990, Tellus B, 51, 298–313, https://doi.org/10.1034/j.1600-0889.1999.00013.x, 1999.

Houghton, R. A. and Nassikas, A. A.: Global and regional fluxes of carbon from land use and land cover change 1850–2015, Global Biogeochem. Cy., 31, 456–472, https://doi.org/10.1002/2016GB005546, 2017.

Hurtt, G. C., Chini, L. P., Frolking, S., Betts, R. A., Feddema, J., Fischer, G., Fisk, J. P., Hibbard, K., Houghton, R. A., Janetos, A., Jones, C. D., Kindermann, G., Kinoshita, T., Klein Goldewijk, K., Riahi, K., Shevliakova, E., Smith, S., Stehfest, E., Thomson, A., Thornton, P., van Vuuren, D. P., and Wang, Y. P.: Harmonization of land-use scenarios for the period 1500–2100: 600 years of global gridded annual land-use transitions, wood harvest, and resulting secondary lands, Climatic Change, 109, 117–161, https://doi.org/10.1007/s10584-011-0153-2, 2011.

Hurtt, G. C., Frolking, S., Fearon, M. G., Moore, B., Shevliakova, E., Malyshev, S., Pacala, S. W., and Houghton, R. A.: The underpinnings of land-use history: three centuries of global gridded land-use transitions, wood-harvest activity, and resulting secondary lands, Glob. Change Biol., 12, 1208–1229, https://doi.org/10.1111/j.1365-2486.2006.01150.x, 2006.

Keenan, R. J., Reams, G. A., Achard, F., de Freitas, J. V., Grainger, A., and Lindquist, E.: Dynamics of global forest area: results from the FAO Global Forest Resources Assessment 2015, Forest Ecol. Manag., 352, 9–20, https://doi.org/10.1016/j.foreco.2015.06.014, 2015.

Klein Goldewijk, K., Beusen, A., Doelman, J., and Stehfest, E.: Anthropogenic land use estimates for the Holocene – HYDE 3.2, Earth Syst. Sci. Data, 9, 927–953, https://doi.org/10.5194/essd-9-927-2017, 2017.

Koh, L. P., Miettinen, J., Liew, S. C., and Ghazoul, J.: Remotely sensed evidence of tropical peatland conversion to oil palm, P. Natl. Acad. Sci. USA, 108, 5127–5132, https://doi.org/10.1073/pnas.1018776108, 2011.

Le Quéré, C., Andrew, R. M., Canadell, J. G., Sitch, S., Korsbakken, J. I., Peters, G. P., Manning, A. C., Boden, T. A., Tans, P. P., Houghton, R. A., Keeling, R. F., Alin, S., Andrews, O. D., Anthoni, P., Barbero, L., Bopp, L., Chevallier, F., Chini, L. P., Ciais, P., Currie, K., Delire, C., Doney, S. C., Friedlingstein, P., Gkritzalis, T., Harris, I., Hauck, J., Haverd, V., Hoppema, M., Klein Goldewijk, K., Jain, A. K., Kato, E., Körtzinger, A., Landschützer, P., Lefèvre, N., Lenton, A., Lienert, S., Lombardozzi, D., Melton, J. R., Metzl, N., Millero, F., Monteiro, P. M. S., Munro, D. R., Nabel, J. E. M. S., Nakaoka, S.-I., O'Brien, K., Olsen, A., Omar, A. M., Ono, T., Pierrot, D., Poulter, B., Rödenbeck, C., Salisbury, J., Schuster, U., Schwinger, J., Séférian, R., Skjelvan, I., Stocker, B. D., Sutton, A. J., Takahashi, T., Tian, H., Tilbrook, B., van der Laan-Luijkx, I. T., van der Werf, G. R., Viovy, N., Walker, A. P., Wiltshire, A. J., and Zaehle, S.: Global Carbon Budget 2016, Earth Syst. Sci. Data, 8, 605–649, https://doi.org/10.5194/essd-8-605-2016, 2016.

Li, W., Ciais, P., MacBean, N., Peng, S., Defourny, P., and Bontemps, S.: Major forest changes and land cover transitions based on plant functional types derived from the ESA CCI Land Cover product, Int. J. Appl. Earth Obs., 47, 30–39, https://doi.org/10.1016/j.jag.2015.12.006, 2016.

Li, W., MacBean, N., Ciais, P., Defourny, P., Lamarche, C., Bontemps, S., Houghton, R., and Peng, S.: Derivation of plant functional type (PFT) maps from the ESA CCI Land Cover product, version 1, Zenodo, https://doi.org/10.5281/zenodo.834229, 2017a.

Li, W., MacBean, N., Ciais, P., Defourny, P., Lamarche, C., Bontemps, S., Houghton, R., and Peng, S.: Derivation of plant functional type (PFT) maps from the ESA CCI Land Cover product, version 2, Zenodo, https://doi.org/10.5281/zenodo.1048163, 2017b.

NBSC: National data from National Bureau of Statistics of China, available at: http://data.stats.gov.cn/easyquery.htm?cn=C01, last access: 16 September 2015, 2015.

Nickerson, C., Ebel, R., Borchers, A., and Carriazo, F.: Major uses of land in the United States, 2007, US Department of Agriculture, Economic Research Service, USA, 2011.

Olofsson, P., Foody, G. M., Herold, M., Stehman, S. V., Woodcock, C. E., and Wulder, M. A.: Good practices for estimating area and assessing accuracy of land change, Remote Sens. Environ., 148, 42–57, https://doi.org/10.1016/j.rse.2014.02.015, 2014.

Peng, S.-S., Piao, S., Zeng, Z., Ciais, P., Zhou, L., Li, L. Z. X., Myneni, R. B., Yin, Y., and Zeng, H.: Afforestation in China cools local land surface temperature., P. Natl. Acad. Sci. USA, 111, 2915–2919, https://doi.org/10.1073/pnas.1315126111, 2014.

Pérez-Hoyos, A., Rembold, F., Kerdiles, H., and Gallego, J.: Comparison of global land cover datasets for cropland monitoring, Remote Sens., 9, 1118, https://doi.org/10.3390/rs9111118, 2017.

Poulter, B., MacBean, N., Hartley, A., Khlystova, I., Arino, O., Betts, R., Bontemps, S., Boettcher, M., Brockmann, C., Defourny, P., Hagemann, S., Herold, M., Kirches, G., Lamarche, C., Lederer, D., Ottlé, C., Peters, M., and Peylin, P.: Plant functional type classification for earth system models: results from the European Space Agency's Land Cover Climate Change Initiative, Geosci. Model Dev., 8, 2315–2328, https://doi.org/10.5194/gmd-8-2315-2015, 2015.

Sannier, C., McRoberts, R. E., and Fichet, L.-V.: Suitability of Global Forest Change data to report forest cover estimates at national level in Gabon, Remote Sens. Environ., 173, 326–338, https://doi.org/10.1016/j.rse.2015.10.032, 2016.

Shevliakova, E., Pacala, S. W., Malyshev, S., Hurtt, G. C., Milly, P. C. D., Caspersen, J. P., Sentman, L. T., Fisk, J. P., Wirth, C., and Crevoisier, C.: Carbon cycling under 300 years of land use change: importance of the secondary vegetation sink, Global Biogeochem. Cy., 23, GB2022, https://doi.org/10.1029/2007GB003176, 2009.

Stocker, B. D., Feissli, F., Strassmann, K. M., Spahni, R., and Joos, F.: Past and future carbon fluxes from land use change, shifting cultivation and wood harvest, Tellus B, 66, 23188, https://doi.org/10.3402/tellusb.v66.23188, 2014.

Tropek, R., Sedláček, O., Beck, J., Keil, P., Musilová, Z., Símová, I., and Storch, D.: Comment on "High-resolution global maps of 21st-century forest cover change", Science, 344, 981, https://doi.org/10.1126/science.1248753, 2014.

Tsendbazar, N.-E., de Bruin, S., Fritz, S., and Herold, M.: Spatial accuracy assessment and integration of global land cover datasets, Remote Sens., 7, 15804–15821, https://doi.org/10.3390/rs71215804, 2015.

Waldner, F., Fritz, S., Di Gregorio, A., Plotnikov, D., Bartalev, S., Kussul, N., Gong, P., Thenkabail, P., Hazeu, G., Klein, I., Löw, F., Miettinen, J., Dadhwal, V., Lamarche, C., Bontemps, S., and Defourny, P.: A unified cropland layer at 250 m for global agriculture monitoring, Data, 1, 3, https://doi.org/10.3390/data1010003, 2016.

Wilkenskjeld, S., Kloster, S., Pongratz, J., Raddatz, T., and Reick, C. H.: Comparing the influence of net and gross anthropogenic land-use and land-cover changes on the carbon cycle in the MPI-ESM, Biogeosciences, 11, 4817–4828, https://doi.org/10.5194/bg-11-4817-2014, 2014.

Yang, Y., Xiao, P., Feng, X., and Li, H.: Accuracy assessment of seven global land cover datasets over China, ISPRS J. Photogramm. Remote Sens., 125, 156–173, https://doi.org/10.1016/j.isprsjprs.2017.01.016, 2017.

Yue, C., Ciais, P., Luyssaert, S., Li, W., McGrath, M. J., Chang, J., and Peng, S.: Representing anthropogenic gross land use change, wood harvest and forest age dynamics in a global vegetation model ORCHIDEE-MICT (r4259), Geosci. Model Dev. Discuss., https://doi.org/10.5194/gmd-2017-118, in review, 2017.

Zhao, K. and Jackson, R. B.: Biophysical forcings of land-use changes from potential forestry activities in North America, Ecol. Monogr., 84, 329–353, https://doi.org/10.1890/12-1705.1, 2014.

11

Global manure nitrogen production and application in cropland during 1860–2014: a 5 arcmin gridded global dataset for Earth system modeling

Bowen Zhang[1], **Hanqin Tian**[1,2], **Chaoqun Lu**[3,1], **Shree R. S. Dangal**[1], **Jia Yang**[1], **and Shufen Pan**[1,2]

[1]International Center for Climate and Global Change Research, Auburn University, Auburn, AL 36849, USA
[2]State Key Laboratory of Urban and Regional Ecology, Research Center for Eco-Environmental Sciences, Chinese Academy of Sciences, Beijing 100085, China
[3]Department of Ecology, Evolution, and Organismal Biology, Iowa State University, Ames, IA 50011, USA

Correspondence to: Hanqin Tian (tianhan@auburn.edu)

Abstract. Given the important role of nitrogen input from livestock systems in terrestrial nutrient cycles and the atmospheric chemical composition, it is vital to have a robust estimation of the magnitude and spatiotemporal variation in manure nitrogen production and its application to cropland across the globe. In this study, we used the dataset from the Global Livestock Impact Mapping System (GLIMS) in conjunction with country-specific annual livestock populations to reconstruct the manure nitrogen production during 1860–2014. The estimated manure nitrogen production increased from 21.4 Tg N yr^{-1} in 1860 to 131.0 Tg N yr^{-1} in 2014 with a significant annual increasing trend (0.7 Tg N yr^{-1}, $p < 0.01$). Changes in manure nitrogen production exhibited high spatial variability and concentrated in several hotspots (e.g., Western Europe, India, northeastern China, and southeastern Australia) across the globe over the study period. In the 1860s, the northern midlatitude region was the largest manure producer, accounting for ~ 52 % of the global total, while low-latitude regions became the largest share (~ 48 %) in the most recent 5 years (2010–2014). Among all the continents, Asia accounted for over one-fourth of the global manure production during 1860–2014. Cattle dominated the manure nitrogen production and contributed ~ 44 % of the total manure nitrogen production in 2014, followed by goats, sheep, swine, and chickens. The manure nitrogen application to cropland accounts for less than one-fifth of the total manure nitrogen production over the study period. The 5 arcmin gridded global dataset of manure nitrogen production generated from this study could be used as an input for global or regional land surface and ecosystem models to evaluate the impacts of manure nitrogen on key biogeochemical processes and water quality. To ensure food security and environmental sustainability, it is necessary to implement proper manure management practices on cropland across the globe.

1 Introduction

Human-induced nitrogen flow, mainly driven by the increasing needs for food production, has a tremendous impact on the Earth's biogeochemical cycles (Bouwman et al., 2013; Galloway et al., 2008; Liu et al., 2010). Chemical fertilizer use began to play an important role in enhancing crop yield in the 1960s (Lu and Tian, 2017; Potter et al., 2010), and manure has long been recognized as a traditional source of soil nutrients for centuries, contributing up to ~ 37–61 % of the total nitrogen input to the land surface (Bouwman et al., 2013). Manure nitrogen production is expected to increase in the coming decades due to the growing demand for livestock populations as a result of the ever-increasing human population and shifts in diet structure with more meat consumption (Herrero and Thornton, 2013). The resultant changes have been suggested to surpass the sustainability threshold (Pelletier and Tyedmers, 2010) with a substantial impact on biogeochemical processes and greenhouse gas balance in terrestrial ecosystems (Tian et al., 2016).

The increasing application of manure nutrients has contributed to an increase in crop production and, at the same time, has been identified as one of the major causes for a litany of environmental problems that impinge on the land, the aquatic ecosystem, and even the atmospheric composition (Bouwman et al., 2013; Burkart and James, 1999; Davidson and Kanter, 2014; Potter et al., 2010). To maintain high yield, farmers tend to apply large amounts of nitrogen fertilizer and organic manure, especially in intensive crop-producing systems. A recent study revealed that only 38 % of total reactive nitrogen input was finally transferred into harvested crop yield (Liu et al., 2016). Part of the surplus nitrogen can be accumulated in soil nitrogen pools. Manure-derived nitrous oxide (N_2O) accounts for 44 % of total anthropogenic N_2O emissions, which is the largest anthropogenic stratospheric ozone-depleting substance and the third most important anthropogenic greenhouse gas (Davidson, 2009; Davidson and Kanter, 2014; Tian et al., 2016). It has been suggested that manure was the single largest source of the anthropogenic emission of N_2O in the 2000s (Davidson, 2009; Davidson and Kanter, 2014; Syakila and Kroeze, 2011). At the same time, manure also acted as the dominant source of ammonia (NH_3), which played a vital role in the formation of atmospheric particulate matter (PM), such as $PM_{2.5}$, and atmospheric nitrogen deposition (Behera et al., 2013; Sutton et al., 2013). Manure production contributed over 66 % of NH_3 emissions from the agricultural system (Beusen et al., 2008). Thus, increasing manure production could lead to an increase in NH_3 emissions, which impairs public and environmental health (Sutton et al., 2013). The rest of the surplus nitrogen can leach through the soil profile and contaminate groundwater in the form of nitrate (Ju et al., 2006). Excess nitrogen together with phosphorous can stimulate the eutrophication of inland water (Conley et al., 2009), be transported far away from original sources, exacerbate degrading coastal water quality, and even lead to hypoxia (Burkart and James, 1999; Yang et al., 2015).

To determine the status of unevenly distributed nitrogen at large scales, it is critical to have a good understanding of the geographic distribution of nitrogen inputs from different sectors. In spite of extensive studies on the development of nitrogen fertilizer data at both regional and global scales (FAOSTAT, 2014; Lu and Tian, 2017; Matthews, 1994; Nishina et al., 2017; Potter et al., 2010), most previous datasets for manure nitrogen production at the global scale either relied on a livestock population dataset with coarse resolution or were only available for limited time periods without consecutive inter-annual variation; e.g., Herrero and Thornton (2013), Holland et al. (2005), Liu et al. (2010), and Potter et al. (2010). Recent research has expanded the estimation of manure nutrient production in the conterminous United States during 1930–2012 and in China during 2002–2008 (Ouyang et al., 2013; Yang et al., 2016). In the conterminous United States, manure nitrogen has increased by 46 % from 1930 to 2012 with substantial spatial heterogeneity (Yang et al., 2016). In China, manure nutrients are unevenly distributed with seven provinces contributing over half of the total manure nitrogen (Ouyang et al., 2013).

Although these datasets have expanded our recognition of manure nitrogen estimates, spatially explicit estimates of manure nitrogen production on a global scale are still lacking. To reduce the uncertainty in estimating several key biogeochemical processes at the global scale, such as the continuously increased emission of nitrous oxide and the occurrences of inland and coastal hypoxia due to nutrient enrichment at large scales, it is necessary to quantify the spatial and temporal variations in manure nitrogen production over a long period. Together with other data, quantification of manure nitrogen production could also be used to generate a comprehensive assessment for livestock sectors and design sustainable options for the sector's development (Herrero and Thornton, 2013). At the same time, it could quantify the uncertainties in analyzing the key nutrient cycles in terrestrial ecosystems and their feedback to the climate over a century-long period.

The original Gridded Livestock of the World (GLW) database (Wint and Robinson, 2007) was further revised and improved through the collection of more up-to-date livestock statistics and the application of finer-resolution predictor variables and more reasonable analytical procedures to develop the Global Livestock Impact Mapping System (GLIMS, also called GLW2; Robinson et al., 2014). GLIMS offers an exceptional opportunity to improve manure data from earlier studies and extend our knowledge of manure production over a century-long period (Robinson et al., 2014). Thus, the major objective of this study is to produce global gridded maps of manure nitrogen production at a 5 arcmin resolution in latitude by longitude during 1860–2014. More specifically, we (1) estimate the magnitude and spatial and temporal variation in manure nitrogen production, (2) quantify the relative contribution of major livestock groups on the manure nitrogen production, (3) investigate the spatial and temporal variation in manure nitrogen applied to cropland, and (4) discuss the impacts of manure nitrogen production on terrestrial biogeochemical cycles.

2 Methods

2.1 Manure nitrogen production

To develop the gridded annual nitrogen production rate from manure during 1860–2014, we used the dataset from GLIMS (GLW2), which provided the spatial distribution of different livestock at a spatial resolution of 0.00833° (a nominal pixel resolution of approximately 1 × 1 km at the Equator) for cattle (dairy and other cattle), swine,

Table 1. Summary of data sources used in this study.

Data source	Dataset	Units	Reference
Global Livestock Impact Mapping System (GLIMS)	Spatial distribution of different livestock	Head	Robinson et al. (2014)
FAOSTAT 2014	Annual stock of country-specific livestock	Head	FAOSTAT (2014)
History Database of the Global Environment (HYDE)	Fills the gaps for years without livestock populations from FAOSTAT	n/a	Mitchell (1998a, b, 1993)
IPCC 2006 guidelines	Regional excretion rate from livestock	$kg\,N\,animal^{-1}\,yr^{-1}$	IPCC (2006)
IPCC 2006 guidelines	Typical animal mass from livestock	$kg\,animal^{-1}$	IPCC (2006)
Holland et al., 2005	Manure nitrogen production from 1860 to 1960	$Tg\,N\,yr^{-1}$	Holland et al. (2005)
Livestock production systems	Manure management for different livestock production systems	n/a	Herrero et al. (2013)
History Database of the Global Environment (HYDE 3.2)	Global cropland distributions from 1860 to 2014	n/a	Klein Goldewijk et al. (2016)
Siebert et al., 2013	Spatial distribution of global irrigated area (expressed as the percentage of area equipped for irrigation)	n/a	Siebert et al. (2013)
FAOSTAT 2014	Country-level area equipped for irrigation	km^2	FAOSTAT (2014)

chickens, goats, sheep, and ducks with a partial distribution (https://livestock.geo-wiki.org/home-2/; Robinson et al., 2014). The annual variation in national livestock stock from 1961 to 2014 was obtained from FAOSTAT (FAOSTAT, 2014; http://faostat.fao.org/site/291/default.aspx). For the countries (including the United States, Australia, Brazil, Canada, China, and Mongolia) with subregional (province- or state-level) cattle populations, we disaggregated FAO country-level populations into subregions (see detailed description in Dangal et al., 2017). For the missing data in FAOSTAT, the annual trend was generated by linear interpolation of the five time periods (1960, 1970, 1980, 1990, and 1998) of livestock populations from the History Database of the Global Environment (HYDE; Table S1 in the Supplement) to fill the gaps (Mitchell, 1998a, b, 1993). Default values for the regional nitrogen excretion rates of different livestock were obtained from the Intergovernmental Panel on Climate Change (IPCC) 2006 guidelines (Tier 1; IPCC, 2006; Table S2). By using the livestock population (including cattle, swine, sheep, and poultry) from FAO Production Yearbooks, Holland et al. (2005) applied the default excretion rate suggested by Souchu and Etchanchu (1989) to generate the annual manure nitrogen production from 1860 to 1960. We obtained the change in manure nitrogen production before 1960 from Holland et al. (2005) and applied it to each grid cell to estimate the amount of manure nitrogen production from 1860 to 1960.

The development of the time series on the nitrogen excretion rate from livestock is provided below in more detail (Table 1). To distribute the yearly country-level livestock population from FAOSTAT (after 1960) or Holland et al. (before 1960), we standardized the livestock distribution with spatially explicit gridded information from GLIMS to match the annual country-level livestock records from FAOSTAT:

$$D(\mathrm{FAO})_{i,j,k} = D(\mathrm{GLIMS})_{i,j} \times \frac{\mathrm{NTH(FAO)}_{i,j,k}}{\mathrm{NTH(GLIMS)}_{i,j}}, \quad (1)$$

where NTH indicates the national total head of animal j from a specific country i (unit: head) in year k (k indicates 1961–2014). D indicates the density of animal j from a specific country i (unit: head km^{-2} land in each grid) in year k.

Then we calculated the average nitrogen excretion rate by applying the IPCC 2006 guidelines (Tier 1; IPCC, 2006):

$$N_{\mathrm{ex}(i,j)} = N_{\mathrm{rate}(i,j)} \times \frac{\mathrm{TAM}_{(i,j)}}{1000} \times 365, \quad (2)$$

where $N_{\mathrm{ex}(i,j)}$ indicates annual N excretion for livestock category j from a specific country i (unit: $kg\,N\,animal^{-1}\,yr^{-1}$), $N_{\mathrm{rate}(i,j)}$ indicates the default N excretion rate for livestock category j from a specific country i (unit: kg N (1000 kg animal mass)$^{-1}$ day^{-1}), and $\mathrm{TAM}_{(i,j)}$ indicates typical animal mass for livestock category j from a specific region i (unit: $kg\,animal^{-1}$). For cattle, we collected information for $N_{\mathrm{rate}(i,j)}$ and $\mathrm{TAM}_{(i,j)}$ (Table S2.1) at the continent, country, and subregional level. For other livestock, we use region-specific values from IPCC (2006; Tables S2.2–2.3).

We calculated the gridded average nitrogen excretion rate with

$$N_{\mathrm{man}(i,j,k)} = N_{\mathrm{ex}(i,j)} \times D(\mathrm{FAO})_{(i,j,k)}, \quad (3)$$

where $N_{\mathrm{man}(i,j,k)}$ indicates gridded average nitrogen excretion rates for livestock category j from a specific country i in year k (unit: $kg\,N\,km^{-2}\,yr^{-1}$).

2.2 Manure nitrogen applied to cropland

We further developed the gridded map of the manure nitrogen applied to cropland at 5 arcmin of resolution based on

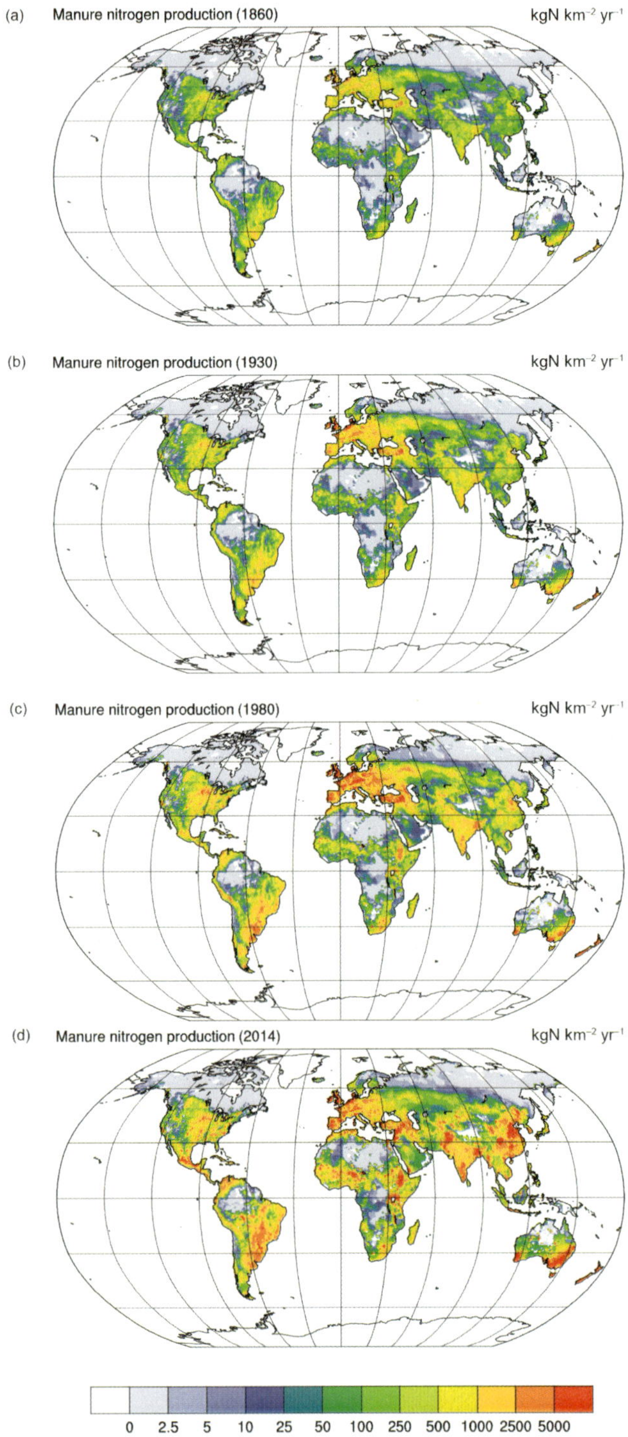

Figure 1. Spatial distribution of manure nitrogen production across the global land surface in the four years 1860, 1930, 1980, and 2014.

manure management in three livestock production systems, including rangeland-based systems, mixed rainfed farming systems, and mixed irrigated farming systems for cattle (dairy and other cattle), goats and sheep, and smallholder and industrial systems for poultry and swine (Herrero et al., 2013). Herrero et al. (2013) further classified the livestock systems into different agroecological zones (arid and semi-arid, humid and subhumid, and temperate and tropical highland areas) based on temperature and the length of the growing period (LGP). The data on the spatial distribution for livestock production systems for ruminants, swine, and chickens were obtained from GLIMS (https://livestock.geo-wiki.org/download/), which represents the status around 2006. Thus,

$$F_{\mathrm{M}(j,\,\mathrm{ProSys})} = F_{\mathrm{MT}(j,\mathrm{ProSys})} \cdot \left(1 - F_{\mathrm{MO}(j,\mathrm{ProSys})}\right) \cdot \left(1 - F_{\mathrm{Loss}(j,\mathrm{ProSys})}\right), \tag{4}$$

where $F_{\mathrm{M}(j,\mathrm{ProSys})}$ indicates the fraction of manure from livestock category j applied to cropland, and $F_{\mathrm{MT}(j,\mathrm{ProSys})}$ indicates the fraction of total manure managed for different livestock production systems. $F_{\mathrm{MO}(j,\mathrm{ProSys})}$ indicates the fraction of managed manure to other use, e.g., the production of biogas. $F_{\mathrm{Loss}(j,\mathrm{ProSys})}$ indicates the fraction of managed manure lost through volatilization as NH_3 and NO_x. All the parameters used in Eq. (4) can be found in Table S3. ProSys indicated livestock production systems for cattle (dairy and other cattle) and small ruminants, including rangeland-based systems (LGY: livestock-only systems in hyperarid areas; LGA: livestock-only systems in arid areas; LGH: livestock-only systems in humid areas; and LGT: livestock-only systems in temperate areas or tropical highlands), mixed rainfed farming systems (MRY: mixed rainfed systems in hyperarid areas; MRA: mixed rainfed systems in arid areas; MRH: mixed rainfed systems in humid areas; and MRT: mixed rainfed systems in temperate areas or tropical highlands) and mixed irrigated farming systems (MIY: mixed irrigated systems in hyperarid areas; MIA: mixed irrigated systems in arid areas; MIH: mixed irrigated systems in humid areas; and MIT: mixed irrigated systems in temperate areas or tropical highlands), and smallholder (POsm) and industrial (POin) for poultry and swine.

To develop the spatial maps for manure nitrogen applied to soils on cropland during 1860–2014, we made several assumptions due to absence of the appropriate data and calculated as

$$N_{\mathrm{man}_{\mathrm{CR}(j,k)}} = N_{\mathrm{man}_{(i,j,k)}} \times \begin{cases} F_{\mathrm{M}(j,\mathrm{ProSys_{rl}})} & \text{(5a)} \\ F_{\mathrm{M}(j,\mathrm{ProSys_{rd}})} \times \dfrac{f_{\mathrm{crp}(k)}}{f_{\mathrm{crp}(2006)}} + F_{\mathrm{M}(j,\mathrm{ProSys_{rl}})} \times \left(1 - \dfrac{f_{\mathrm{crp}(k)}}{f_{\mathrm{crp}(2006)}}\right) & \text{(5b)} \\ F_{\mathrm{M}(j,\mathrm{ProSys_{irri}})} \times \dfrac{f_{\mathrm{irri}(k)}}{f_{\mathrm{irri}(2006)}} + \left\{F_{\mathrm{M}(j,\mathrm{ProSys_{rd}})} \times \dfrac{f_{\mathrm{crp}(k)}}{f_{\mathrm{crp}(2006)}} + F_{\mathrm{M}(j,\mathrm{ProSys_{rl}})} \times \left(1 - \dfrac{f_{\mathrm{crp}(k)}}{f_{\mathrm{crp}(2006)}}\right)\right\} \times \left(1 - \dfrac{f_{\mathrm{irri}(k)}}{f_{\mathrm{irri}(2006)}}\right), & \text{(5c)} \end{cases}$$

where $N_{\mathrm{man}_{\mathrm{CR}(j,k)}}$ indicates the manure nitrogen applied to soils on cropland, and $F_{\mathrm{M}(j,\mathrm{ProSys_{rd}})}$ indicates the fraction of

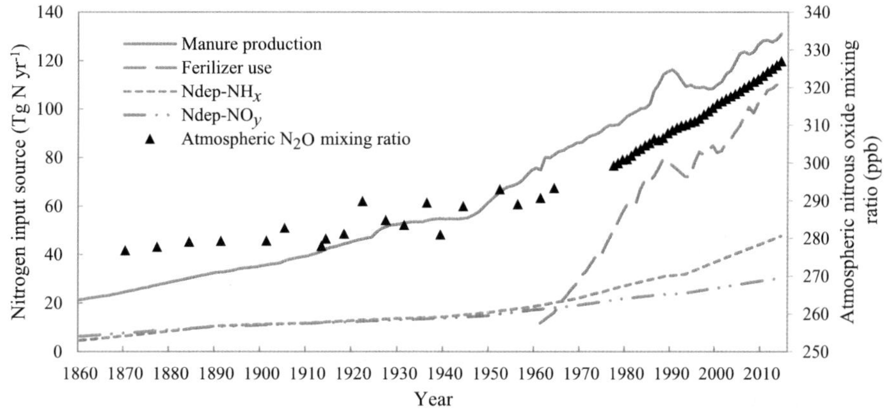

Figure 2. Comparison of nitrogen input from global manure production, fertilizer use, and atmospheric nitrogen deposition with atmospheric nitrous oxide mixing ratio during 1860–2014.

manure applied to mixed rainfed farming systems, including MRY, MRA, MRH, and MRT. $F_{\mathrm{M}(j,\mathrm{ProSys_{irri}})}$ indicates the fraction of manure applied to mixed irrigated farming systems, including MIY, MIA, MIH, and MIT. $F_{\mathrm{M}(j,\mathrm{ProSys_{rl}})}$ indicates the fraction of manure applied to rangeland-based systems, $f_{\mathrm{irri}(k)}$ indicates the fraction of irrigated area to the total area in year k in each grid cell, and $f_{\mathrm{crp}(k)}$ indicates the fraction of cropland area to the total area in year k in each grid cell.

The spatial distribution of livestock production systems in 2006 serves as a baseline map to characterize the change in livestock production system during 1860–2005. We also assumed that the spatial distribution of livestock production systems remained the same during 2006–2014. We assumed the following: if the grid cell was identified as a rangeland-based system, the livestock production system remained the same during the study period (See Eq. 5a); if the grid cell was identified as a mixed rainfed farming system, the percent change in the livestock production system would be proportional to the changes in the cropland area in that grid cell before 2006, and the mixed rainfed farming system was converted from a rangeland-based system (See Eq. 5b); if the grid cell was identified as a mixed irrigated farming system, the percent change in the livestock production system would be proportional to the changes in the irrigated area in that grid cell before 2006, and the mixed irrigated farming systems were converted from mixed rainfed farming systems (See Eq. 5c).

The gridded cropland distribution map during 1860–2014 was obtained from HYDE 3.2 (Klein Goldewijk et al., 2016). We spatialized the country-level area equipped for irrigation from FAOSTAT during 1961–2014 by adopting the gridded irrigated area (expressed as the percentage of area equipped for irrigation; Siebert et al., 2013) to create the gridded irrigation map during 1961–2014. We assumed the irrigated area did not change before 1961.

We assumed that if the grid cell was identified as smallholder for poultry and swine, the livestock production system remained the same during the study period; if the grid cell was identified as industrial, the fraction of the industrial livestock production system was assumed to be 0 in 1860 and 1 in 2006, and it linearly increased from 1860 to 2006 for swine and chickens.

Previous studies suggested that the intensive duck production system first came out in the early 1950s (Ahuja, 2013; Raud and Faure, 1994). Thus, the intensive duck production system was assumed to be 0 in 1950 and 81.6 % in 2008, and it linearly increased from 1950 to 2008. The rest was occupied by extensive duck production systems (Ahuja, 2013; Duc and Long, 2008; MOA, 2013; Raud and Faure, 1994).

3 Results

3.1 Temporal changes in manure nitrogen production

In this study, we quantified the total manure nitrogen production from six livestock categories, including cattle (dairy and other cattle), chickens, ducks, goats, swine, and sheep at a global scale during 1860–2014 (Fig. 1). We referred to the total mass of nitrogen excreted by livestock for the manure nitrogen production. The estimated global manure nitrogen production increased about 5 times from 21.4 Tg N $\mathrm{yr^{-1}}$ in 1860 to 131.0 Tg N $\mathrm{yr^{-1}}$ in 2014 with an overall significant increasing trend during 1860–2014 (0.7 Tg N $\mathrm{yr^{-2}}$, $p < 0.01$; Fig. 2). In 1990, there was near peak manure production ($\sim$ 116.3 Tg N $\mathrm{yr^{-1}}$) followed by a decrease until 1998 (108.4 Tg N $\mathrm{yr^{-1}}$) and then an increase afterward.

3.2 Spatial patterns of manure nitrogen production

Manure nitrogen production exhibited large spatial variation over the study period. In the 1860s, the northern mid-latitudes (NM, 30–60° N) accounted for over half of the

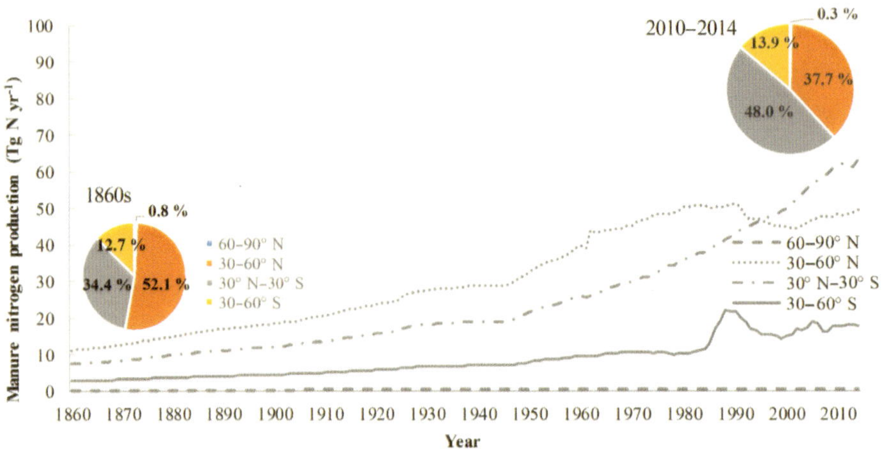

Figure 3. Estimation of global manure nitrogen production in northern high latitudes (60–90° N), northern midlatitudes (30–60° N), tropical regions (30° N–30° S), and southern midlatitudes (30–60° S).

global total manure production ($\sim 12.0 \pm 0.5\,\mathrm{Tg\,N\,yr^{-1}}$, average ±1 SD, same hereafter). Low-latitude regions (30° N–30° S) contributed one-third of total manure nitrogen production, followed by the southern midlatitudes (SM, 30–60° S; ~12.7 %) and the northern high latitudes (NH, 60–90° N; ~0.8 %). However, the dominant regions of the total manure nitrogen production have changed in recent years. During the most recent 5 years (2010–2014), low-latitude regions took the largest share, which was around 48.0 % of the estimated global manure production ($\sim 61.9 \pm 0.9\,\mathrm{Tg\,N\,yr^{-1}}$), followed by NM (~37.7 %), SM (~13.9 %), and NH, which contributed the least to the global manure nitrogen production (Fig. 3).

From a continental perspective, manure nitrogen production in Europe ($\sim 6.2 \pm 0.3\,\mathrm{Tg\,N\,yr^{-1}}$) appeared to be similar to that in Asia ($\sim 6.0 \pm 0.2\,\mathrm{Tg\,N\,yr^{-1}}$) in the 1860s, which was much higher than in any other continent, including South America ($\sim 3.6 \pm 0.1\,\mathrm{Tg\,N\,yr^{-1}}$), Africa ($\sim 2.8 \pm 0.1\,\mathrm{Tg\,N\,yr^{-1}}$), North America ($\sim 2.6 \pm 0.1\,\mathrm{Tg\,N\,yr^{-1}}$), and Oceania ($\sim 1.9 \pm 0.1\,\mathrm{Tg\,N\,yr^{-1}}$). During 2010–2014, however, Asia accounted for the largest single share (~34.2 %), followed by Africa (~17.6 %), South America (~14.2 %), Oceania (~13.3 %), Europe (~11.6 %), and North America (~9.2 %; Table 2).

Changes in manure nitrogen production showed high spatial variability and revealed several hotspots over the globe due to imbalances in global economic development and population growth (Fig. 4). Western Europe experienced an increase in the annual changing trend of manure nitrogen production from 1860 to the late 1980s and a decline thereafter. Southern Mexico, Central America, Columbia, southern Brazil, southeastern Australia, and India showed a continuing increasing trend for manure nitrogen production during 1860–2014. Western and eastern Africa and northeastern China experienced an increase in manure nitrogen production during recent decades.

3.3 Relative contribution of different livestock categories

At the global level, cattle dominated the manure nitrogen production among different livestock categories and contributed around 55.5 and 43.7 % of the total manure nitrogen production in 1860 and 2014, respectively (Figs. 5 and 6). Goats and sheep together contributed another one-third of the total manure nitrogen production during the study period, followed by swine and chickens. Ducks contributed the least to manure nitrogen production. However, at the regional level in terms of the dominant livestock species to the total manure nitrogen production, ducks were the dominant contributor in Alaska and Canada, while cattle played a dominant role in the conterminous United States, Mexico, India, and most areas in South America and Europe (Figs. 5 and 6). Goats contributed the most in North Africa, Australia, and central and northeastern Asia, while chickens and swine dominated in Russia. Over the study period, the relative contribution of cattle (dairy and other cattle) and sheep showed a significant decreasing trend to the total manure production, while for goats and chickens a significant increasing trend was found ($p < 0.001$; Fig. 6).

3.4 Spatial and temporal variation in manure nitrogen applied to cropland

At the global scale, the manure nitrogen applied to cropland increased from 3.6 Tg N in 1860 to 24.5 Tg N in 2014 with a significant increasing trend ($0.14\,\mathrm{Tg\,N\,yr^{-2}}$, $p < 0.01$) during 1860–2014. The application to cropland only accounted for 16.9–19.1 % of the total manure nitrogen production over the study period. Among different livestock categories, cattle (dairy and other cattle) contributed around half (42.4–

Table 2. Estimates of manure nitrogen production at the continental scale.

Manure nitrogen production ($Tg\,N\,yr^{-1}$)	1860s	1900s	1940s	1980s	2010s
Asia	6.0 ± 0.2	9.8 ± 0.4	14.8 ± 0.7	29.3 ± 1.2	44.3 ± 0.7
North America	2.6 ± 0.1	4.2 ± 0.2	6.3 ± 0.3	10.7 ± 0.2	11.8 ± 0.04
Europe	6.2 ± 0.3	10.1 ± 0.4	15.3 ± 0.7	25.7 ± 0.1	14.9 ± 0.1
Africa	2.8 ± 0.1	4.6 ± 0.2	6 ± 0.4	13.3 ± 0.6	22.6 ± 0.8
South America	3.6 ± 0.1	5.9 ± 0.2	8.9 ± 0.4	14.4 ± 0.4	18.3 ± 0.1
Oceania	1.9 ± 0.1	3.1 ± 0.1	4.7 ± 0.2	13.0 ± 5.4	17.2 ± 0.1
Global	23.1 ± 1.0	37.5 ± 1.4	57.0 ± 2.8	106.4 ± 7.3	129.0 ± 1.5

Note: values indicate mean $\pm$ standard deviation of 10-year estimates.

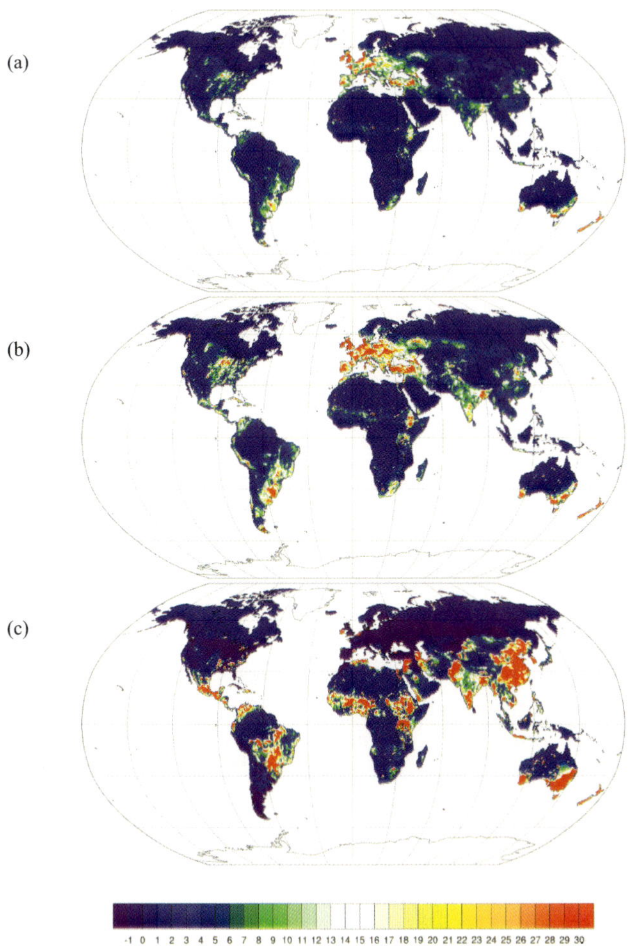

Figure 4. Spatial variation in the annual changing trend of manure nitrogen production ($kg\,N\,km^{-2}\,yr^{-1}$) during **(a)** 1860–1910, **(b)** 1911–1960, and **(c)** 1961–2014.

58.7 %) of the total manure nitrogen applied to cropland. Other ruminants (goats and sheep) only accounted for 14.5–22.1 % over the study period, which was similar to the contribution from swine (16.9–23.3 %). At the continental scale,

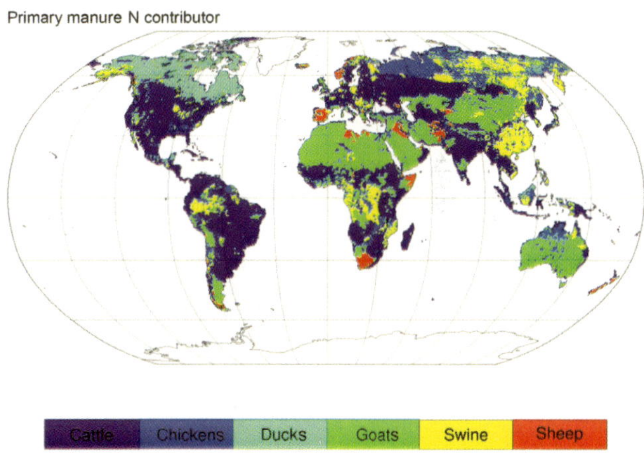

Figure 5. Spatial distribution of the primary contributors to manure nitrogen production in the year 2014.

Europe was the dominant contributor (27.8–37.3 % of the global total) before the 1990s; however, its manure production has been reduced dramatically since the early 1990s (Fig. 7). Asia accounted for 24.4–37.7 % of the global manure nitrogen applied to cropland over the study period with the fastest growing rate of $0.47\,Tg\,N\,decade^{-1}$ compared to other continents.

4 Discussion

4.1 Comparison with previous studies

Over the last 2 decades due to the recognition of the importance of manure nitrogen production in nitrogen cycles, various previous studies have estimated the manure nitrogen production at both regional and global levels. At the global scale, it has been suggested that manure nitrogen production increased from $26.3\,Tg\,N\,yr^{-1}$ in 1860 to $142.5\,Tg\,N\,yr^{-1}$ in

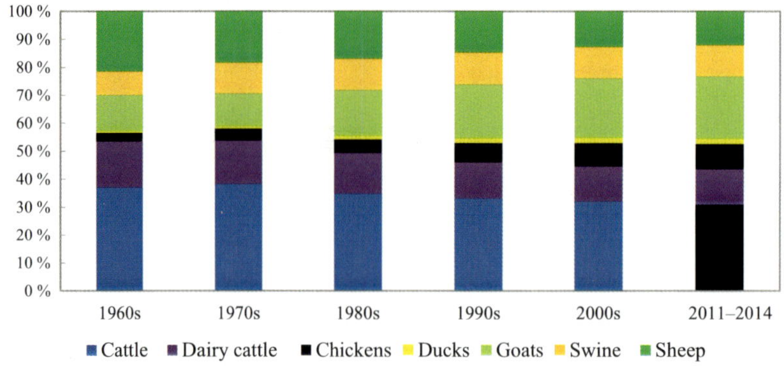

Figure 6. Relative contributions of different livestock animals to the total manure production.

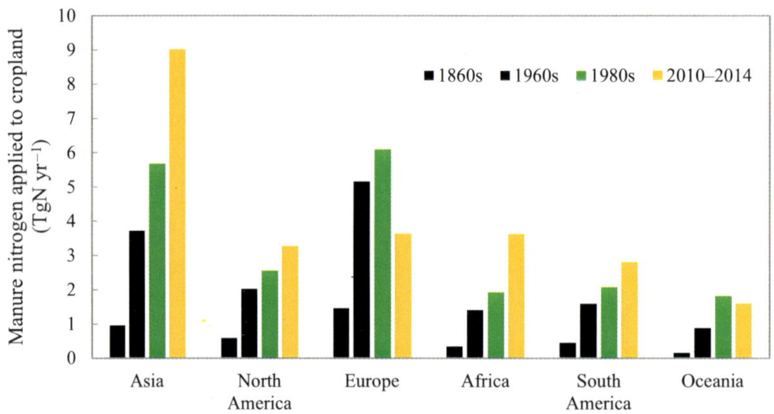

Figure 7. Changes in manure nitrogen amount applied to cropland at the continental level.

2004 with an increasing trend of $0.84\,\text{Tg N yr}^{-1}$ (Holland et al., 2005), which was 18.5 % higher than our estimate from 1860 ($\sim 21.4\,\text{Tg N yr}^{-1}$) to 2004 ($\sim 119.1\,\text{Tg N yr}^{-1}$). However, our result during the 1990s ($\sim 110.0 \pm 1.9\,\text{Tg N yr}^{-1}$) was more consistent with estimates from other studies, ranging from 101.4 to $128.3\,\text{Tg N yr}^{-1}$ (Bouwman et al., 2009; Potter et al., 2010; Van der Hoek et al., 1999). There were some spatial differences between the estimated manure nitrogen application in this study and Bouwman et al. (2013) and manure nitrogen production in this study and Potter et al. (2010; Fig. 8), partly due to the difference in calculation processes. Bouwman's estimate for manure nitrogen applied to cropland is higher than our estimate, mainly due to the consideration of more refined manure management in different livestock production systems from our study. Gerber et al. (2016) and Carlson et al. (2017) suggested that only 7.4–$7.8\,\text{Tg N yr}^{-1}$ was applied to cropland, which is lower than our estimate. One big difference between our study and Gerber et al. (2016) and Carlson et al. (2017) is that we include the managed manure lost through leaching. Since their studies tried to estimate greenhouse gas emissions, it might be appropriate to remove all the lost N through different pathways, including leaching. However, here we try to estimate the total manure applied to cropland, so it may be more reasonable to account for this portion since the leaching processes occurred after manure was applied to soils.

Our analyses indicated that the total amount of manure production in different continents was close to other estimates with a difference of around ±4 % (difference = [estimate from this study − estimate from Potter et al. (2010)]/estimate from this study). Our results showed that manure nitrogen production in Europe started to decline in the early 1990s, which was mainly due to the reduction of livestock populations in Europe (FAOSTAT, 2014). At the country scale, our estimation of manure nitrogen production ($\sim 5.3 \pm 0.8\,\text{Tg N yr}^{-1}$) was close to the previous estimation for the conterminous United States ($\sim 5.9 \pm 0.7\,\text{Tg N yr}^{-1}$) during 1930–2012 (Yang et al., 2016). Meanwhile, both studies identified cattle as the dominant contributor to the manure nitrogen production in the conterminous United States. For the manure nitrogen applied to cropland and grassland in China, our estimation (3.0–$3.6\,\text{Tg N yr}^{-1}$) was lower than previous studies (5.1–$6.2\,\text{Tg N yr}^{-1}$) from 2002 to 2008 (Ouyang et al., 2013), which might be due to our consideration of livestock-specific and region-specific manure management factors to calculate the amount applied to cropland.

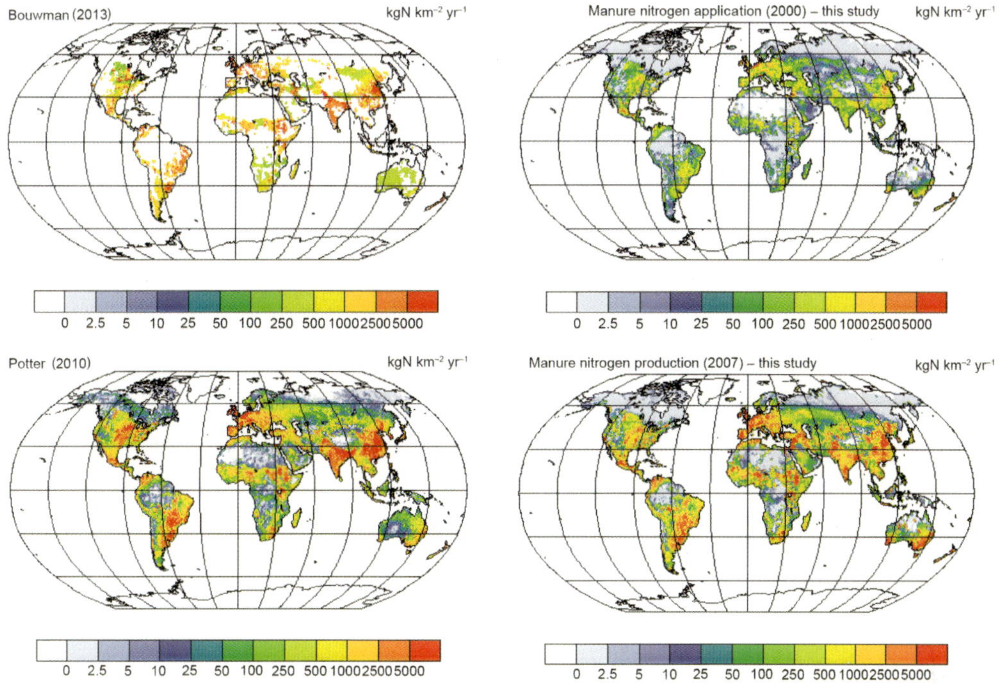

Figure 8. Comparison of manure nitrogen production estimated by Bouwman et al. (2013), Potter et al. (2010), and this study.

4.2 Manure production in the context of global environmental changes

During the past 155 years, the nitrogen input from atmospheric deposition has increased constantly with a significantly increasing rate of $0.36\,Tg\,N\,yr^{-1}$ (Dentener, 2006; Wei et al., 2014). Nitrogen fertilizer use began and has altered the global nitrogen cycle since the early 1960s. Fertilizer use increased by 835 % during 1961–2013 with a significant increasing trend of $1.8\,Tg\,N\,yr^{-1}$ (EPI, 2016). The magnitude of nitrogen production from manure was always higher than fertilizer consumption (Fig. 2) despite the fact that only 16.9–19.1 % of the total produced manure nitrogen could be applied to cropland. Previous studies suggested that manure nitrogen production is the single largest source of nitrous oxide emissions (Davidson, 2009; Davidson and Kanter, 2014). By using the regression equation derived by Davidson (2009), we could roughly estimate manure-induced N_2O emissions at around $2.7\,Tg\,N_2O\text{-}N\,yr^{-1}$ in 2014, which accounted for 21.1 and 17.5 % of the total biogenic N_2O emissions estimated by a top-down approach and bottom-up approach, respectively (Tian et al., 2016).

4.3 Uncertainties

Our study estimates the magnitude and spatiotemporal distribution of manure nitrogen production over the globe during 1860–2014. There are several uncertainties that need to be considered while interpreting the results of this study. First, the livestock distribution was generated by using one-phase static GLIMS. Thus, the changes in the spatial distribution of livestock at the subnational level over time might not be accurate. For instance, free-grazing livestock may migrate due to the availability of food, especially in the early stage of the study period. Therefore, the spatial distribution of different livestock at the subnational scale, such as cattle, sheep, and goats, might be different considering livestock migration. Second, we assumed that the typical mass for different livestock from specific regions did not change over time. However, other studies have suggested that the carcass weights of chickens and beef cattle increased by about 30 % and the carcass weights of swine increased by about 20 % from the early 1960s to the mid-2000s (Thornton, 2010). Thus, manure nitrogen production may be overestimated in the past, and the relative contribution of cattle and chickens to the total manure production may be overestimated since their carcass weight grows much faster than other livestock. Third, the uniform excretion rate for specific livestock types at the regional scale could bring some uncertainties without considering the feed availability and quality across different seasons and various regions (Ouyang et al., 2013; Rufino et al., 2014). For example, Ouyang et al. (2013) provided the provincial N excretion rate in China, ranging from 53 to $94\,kg\,N\,animal^{-1}\,yr^{-1}$ for dairy cattle and 17 to $36\,kg\,N\,animal^{-1}\,yr^{-1}$ for other cattle. Velthof et al. (2015) suggested that the N excretion factors for EU countries using the gross N excretions in the Nitrates Directive reports was 75–184 and $20\text{–}90\,kg\,N\,animal^{-1}\,yr^{-1}$ for dairy and other cattle, respectively. However, for most countries in the world, the N excretion rate at the regional scale is not available. In addition, we made several other assumptions to develop global datasets for manure nitrogen production and manure nitrogen applied to cropland due to the absence of appropriate datasets, which could introduce some uncertainties. For instance, we assumed that the spatial distribution of livestock production systems remained the same during 2006–2014. If more rangeland has been converted to cropland, the total amount of manure applied to cropland might be underestimated globally, except in North America. Additionally, we assumed that the irrigation area did not change before 1961. If less cropland was irrigated before, the manure application to cropland might be underestimated globally, except in Asia. When using this dataset for a specific purpose, further analyses or assumptions need to be made to fulfill the objectives of different studies (Yang et al., 2016). Reducing the associated uncertainty seems straightforward but is hard to accomplish at the current stage; it requires more available data, e.g., the detailed excretion rates for different livestock groups in a specific region over time or the spatial distribution of livestock and livestock systems at a finer scale with temporal variation. In addition, system thinking is another way to unravel complexity and explore options for sustainable development.

5 Data availability

The 5 arcmin gridded global dataset of manure nitrogen production and application in cropland is available at https://doi.org/10.1594/PANGAEA.871980 (Zhang et al., 2017). Data are in text/ASCII format. A supplemental file is added to the list of all other parameters used in this study to calculate the manure nitrogen production and the application in cropland.

6 Conclusion

In this study, we quantified the spatially explicit global manure nitrogen production across the globe during 1860–2014. The estimated total manure nitrogen production increased from 21.4 Tg N yr^{-1} in 1860 to 131.0 Tg N yr^{-1} in 2014 with an overall significant increasing trend during 1860–2014 (0.7 Tg N yr^{-1}, $p < 0.01$). Along the latitudinal gradient, the low latitudes and northern middle latitudes dominated the estimated global manure nitrogen production. From a continental perspective, Asia contributed the largest portion of global manure nitrogen production during recent decades. Southern Mexico, Central America, Columbia, southern Brazil, Uruguay, Western Europe, India, northeastern China, and southeastern Australia increased most rapidly in manure nitrogen production during 1860–2014. We estimated that the manure nitrogen applied to cropland only accounted for 16.9–19.1 % of the total manure nitrogen production over the study period. Further studies are expected to comprehensively evaluate the tradeoff between food production, climate mitigation, and environmental pollution caused by the application of manure to further improve manure management. Together with other data, this 5 arcmin gridded dataset could be used as an input for ecosystem and Earth system models to assess the impact of manure production on global biogeochemical processes, water resources, and climate change.

Competing interests. The authors declare that they have no conflict of interest.

Acknowledgements. This work has been supported by the National Key Research and Development Program of China (no. 2017YFA0604700), the US National Science Foundation (NSF; grants 1243232, 121036), the US National Oceanic and Atmospheric Administration (NOAA grants: NA16NOS4780207, NA16NOS4780204), the Chinese Academy of Sciences STS Program (KFJ-STS-ZDTP-0), a SKLURE grant (SKLURE2017-1-6), and the Auburn University IGP Program.

Edited by: David Carlson
Reviewed by: three anonymous referees

References

Ahuja, V.: Asian livestock: Challenges, opportunities and the response. Food and Agriculture Organization of the United Nations, Regional Office for Asia and the Pacific, Bangkok, 2013.

Behera, S. N., Sharma, M., Aneja, V. P., and Balasubramanian, R.: Ammonia in the atmosphere: a review on emission sources, atmospheric chemistry and deposition on terrestrial bodies, Environ. Sci. Pollut. R., 20, 8092–8131, 2013.

Beusen, A. H. W., Bouwman, A. F., Heuberger, P. S. C., Van Drecht, G., and Van Der Hoek, K. W.: Bottom-up uncertainty estimates of global ammonia emissions from global agricultural production systems, Atmos. Environ., 42, 6067–6077, 2008.

Bouwman, A., Beusen, A., and Billen, G.: Human alteration of the global nitrogen and phosphorus soil balances for the period 1970–2050, Global Biogeochem. Cy., 23, GB0A04, https://doi.org/10.1029/2009GB003576, 2009.

Bouwman, L., Goldewijk, K. K., Van Der Hoek, K. W., Beusen, A. H. W., Van Vuuren, D. P., Willems, J., Rufino, M. C., and Stehfest, E.: Exploring global changes in nitrogen and phosphorus cycles in agriculture induced by livestock production over the 1900–2050 period, P. Natl. Acad. Sci. USA, 110, 20882–21195, 2013.

Burkart, M. R. and James, D. E.: Agricultural-nitrogen contributions to hypoxia in the Gulf of Mexico, J. Environ. Qual., 28, 850–859, 1999.

Carlson, K. M., Gerber, J. S., Mueller, N. D., Herrero, M., MacDonald, G. K., Brauman, K. A., Havlik, P., Oconnell, C. S., Johnson, J. A., Saatchi, S., and West, P. C.: Greenhouse gas emissions intensity of global croplands, Nature Climatc Change, 7, 63–68, 2017.

Conley, D. J., Paerl, H. W., Howarth, R. W., Boesch, D. F., Seitzinger, S. P., Havens, K. E., Lancelot, C., and Likens, G. E.: Controlling eutrophication: nitrogen and phosphorus, Science, 323, 1014–1015, 2009.

Dangal, S. R. S., Tian, H., Zhang, B., Pan, S., Lu, C., and Yang, J.: Methane emission from global livestock sector during 1890–2014: Magnitude, trends and spatiotemporal patterns, Glob. Change Biol., https://doi.org/10.1111/gcb.13709, online first, 2017.

Davidson, E. A.: The contribution of manure and fertilizer nitrogen to atmospheric nitrous oxide since 1860, Nat. Geosci., 2, 659–662, 2009.

Davidson, E. A. and Kanter, D.: Inventories and scenarios of nitrous oxide emissions, Environ. Res. Lett., 9, 105012, https://doi.org/10.1088/1748-9326/9/10/105012, 2014.

Dentener, F. J.: Global Maps of Atmospheric Nitrogen Deposition, 1860, 1993, and 2050, Data set, Oak Ridge National Laboratory Distributed Active Archive Center, Oak Ridge, Tennessee, USA, https://doi.org/10.3334/ORNLDAAC/830, 2006.

Duc, N. V. and Long, T.: Poultry production systems in Vietnam, Animal genetics and breeding department, National Institute of Animal Husbandry, Vietnam, 2008.

EPI: World Fertilizer Consumption. Earth Policy Institude (EPI) from Worldwatch, IFA, USDA, 2016.

FAOSTAT: Food and Agriculture Organization of the United Nations Statistics Division, available at: http://www.fao.org/faostat/en/#data (last access: 1 September 2017), 2014.

Galloway, J. N., Townsend, A. R., Erisman, J. W., Bekunda, M., Cai, Z., Freney, J. R., Martinelli, L. A., Seitzinger, S. P., and Sutton, M. A.: Transformation of the nitrogen cycle: recent trends, questions, and potential solutions, Science, 320, 889–892, 2008.

Gerber, J. S., Carlson, K. M., Makowski, D., Mueller, N. D., Garcia de Cortazar-Atauri, I., Havlík, P., Herrero, M., Launay, M., O'Connell, C. S., and Smith, P.: Spatially explicit estimates of N_2O emissions from croplands suggest climate mitigation opportunities from improved fertilizer management, Glob. Change Biol., 22, 3383–3394, https://doi.org/10.1111/gcb.13341, 2016.

Herrero, M. and Thornton, P. K.: Livestock and global change: Emerging issues for sustainable food systems, P. Natl. Acad. Sci. USA, 110, 20878–20881, 2013.

Herrero, M., Havlík, P., Valin, H., Notenbaert, A., Rufino, M. C., Thornton, P. K., Blümmel, M., Weiss, F., Grace, D., and Obersteiner, M.: Biomass use, production, feed efficiencies, and greenhouse gas emissions from global livestock systems, P. Natl. Acad. Sci. USA, 110, 20888–20893, 2013.

Holland, E. A., Lee-Taylor, J., Nevison, C., and Sulzman, J.: Global N Cycle: Fluxes and N_2O Mixing Ratios Originating from Human Activity, Data set, Oak Ridge National Laboratory Distributed Active Archive Center, Oak Ridge, Tennessee, USA, https://doi.org/10.3334/ORNLDAAC/797, 2005.

IPCC: 2006 IPCC Guidelines for National Greenhouse Gas Inventories, prepared by: National Greenhouse Gas Inventories Programme, edited by: Eggleston, H. S., Buendia, L., Miwa, K., Ngara, T., and Tanabe, K., Published:IGES, Japan, 2006.

Ju, X., Kou, C., Zhang, F., and Christie, P.: Nitrogen balance and groundwater nitrate contamination: Comparison among three intensive cropping systems on the North China Plain, Environ. Pollut., 143, 117–125, 2006.

Klein Goldewijk, K., Beusen, A., Doelman, J., and Stehfest, E.: New anthropogenic land use estimates for the Holocene; HYDE 3.2, Earth Syst. Sci. Data Discuss., https://doi.org/10.5194/essd-2016-58, in review, 2016.

Liu, J., Ma, K., Ciais, P., and Polasky, S.: Reducing human nitrogen use for food production, Scientific Reports, 6, 30104, https://doi.org/10.1038/srep30104, 2016.

Liu, J., You, L., Amini, M., Obersteiner, M., Herrero, M., Zehnder, A. J., and Yang, H.: A high-resolution assessment on global nitrogen flows in cropland, P. Natl. Acad. Sci. USA, 107, 8035–8040, 2010.

Lu, C. and Tian, H.: Global nitrogen and phosphorus fertilizer use for agriculture production in the past half century: shifted hot spots and nutrient imbalance, Earth Syst. Sci. Data, 9, 181–192, https://doi.org/10.5194/essd-9-181-2017, 2017.

Matthews, E.: Nitrogenous fertilizers: global distribution of consumption and associated emissions of nitrous oxide and ammonia, Global Biogeochem. Cy., 8, 411–439, 1994.

Mitchell, B. R.: International historical statistics Europe: 1750–1993, 4th Edn., New York, Stockton Press, London, MacMillan Reference Ltd., 959 pp., 1998a.

Mitchell, B. R.: International Historical Statistics, Africa, Asia & Oceania: 1750–1993, 3rd Edn., New York, Stockton Press, London, MacMillan Reference Ltd., 1113 pp., 1998b.

Mitchell, B. R.: International Historical Statistics, The Americas: 1750–1988, New York, Stockton Press, London, MacMillan Publishers Ltd., 817 pp., 1993.

MOA: Ministry of Agriculture of the People's Republic of China, Oview on China's Animal Husbandry, available at: http://english.agri.gov.cn/hottopics/fa/ahic/201307/t20130729_20030.htm (last access: 1 September 2017), 2013.

Nishina, K., Ito, A., Hanasaki, N., and Hayashi, S.: Reconstruction of spatially detailed global map of NH_4^+ and NO_3^- application in synthetic nitrogen fertilizer, Earth Syst. Sci. Data, 9, 149–162, https://doi.org/10.5194/essd-9-149-2017, 2017.

Ouyang, W., Hao, F., Wei, X., and Huang, H.: Spatial and temporal trend of Chinese manure nutrient pollution and assimilation capacity of cropland and grassland, Environ. Sci. Pollut. Res., 20, 5036–5046, 2013.

Pelletier, N. and Tyedmers, P.: Forecasting potential global environmental costs of livestock production 2000–2050, P. Natl. Acad. Sci. USA, 107, 18371–18374, 2010.

Potter, P., Ramankutty, N., Bennett, E. M., and Donner, S. D.: Characterizing the spatial patterns of global fertilizer application and manure production, Earth Interact., 14, 1–22, 2010.

Raud, H. and Faure, J.: Welfare of ducks in intensive units, Revue scientifique et technique, International Office of Epizootics, 13, 119–129, 1994.

Robinson, T. P., Wint, G. R. W., Conchedda, G., Van Boeckel, T. P., Ercoli, V., Palamara, E., Cinardi, G., D'Aietti, L., Hay, S. I., and Gilbert, M.: Mapping the Global Distribution of Livestock, Plos One, 9, https://doi.org/10.1371/journal.pone.0096084, 2014.

Rufino, M., Brandt, P., Herrero, M., and Butterbach-Bahl, K.: Reducing uncertainty in nitrogen budgets for African livestock systems, Environ. Res. Lett., 9, 105008, https://doi.org/10.1088/1748-9326/9/10/105008, 2014.

Siebert, S., Henrich, V., Frenken, K., and Burke, J.: Global Map of Irrigation Areas version 5. Rheinische Friedrich-Wilhelms-University, Bonn, Germany/Food and Agriculture Organization of the United Nations, Rome, Italy, 2013.

Souchu, P. and Etchanchu, D.: The Environmental Effects of the Intensive Application of Nitrogen Fertilizers in Western Europe: Past Problems and Future Prospects, 1989.

Sutton, M. A., Reis, S., Riddick, S. N., Dragosits, U., Nemitz, E., Theobald, M. R., Tang, Y. S., Braban, C. F., Vieno, M., and Dore, A. J.: Towards a climate-dependent paradigm of ammonia emission and deposition, Philos. T. R. Soc. Lond. B, 368, 20130166, https://doi.org/10.1098/rstb.2013.0166, 2013.

Syakila, A. and Kroeze, C.: The global nitrous oxide budget revisited, Greenhouse Gas Measurement and Management, 1, 17–26, 2011.

Thornton, P. K.: Livestock production: recent trends, future prospects, Philos. T. R Soc. B, 365, 2853–2867, 2010.

Tian, H., Lu, C., Ciais, P., Michalak, A. M., Canadell, J. G., Saikawa, E., Huntzinger, D. N., Gurney, K. R., Sitch, S., Zhang, B., Yang, J., Bousquet, P., Bruhwiler, L., Chen, G., Dlugokencky, E., Friedlingstein, P., Melillo, J., Pan, S., Poulter, B., Prinn, R., Saunois, M., Schwalm, C. R., and Wofsy, S. C.: The terrestrial biosphere as a net source of greenhouse gases to the atmosphere, Nature, 531, 225–228, 2016.

Van der Hoek, K., Bouwman, A., Smaling, E., Oenema, O., and Fresco, L.: Upscaling of nutrient budgets from agroecological niche to global scale, Nutrient disequilibria in agroecosystems: concepts and case studies, 57–73, Wallingford, Oxon, UK, New York, NY, CABI Pub., 1999.

Velthof, G. L., Hou, Y., and Oenema, O.: Nitrogen excretion factors of livestock in the European Union: a review, J. Sci. Food Agr., 95, 3004–3014, 2015.

Wei, Y., Liu, S., Huntzinger, D. N., Michalak, A. M., Viovy, N., Post, W. M., Schwalm, C. R., Schaefer, K., Jacobson, A. R., Lu, C., Tian, H., Ricciuto, D. M., Cook, R. B., Mao, J., and Shi, X.: The North American Carbon Program Multi-scale Synthesis and Terrestrial Model Intercomparison Project – Part 2: Environmental driver data, Geosci. Model Dev., 7, 2875–2893, https://doi.org/10.5194/gmd-7-2875-2014, 2014.

Wint, W. and Robinson, T.: Gridded livestock of the world, Food and Agriculture Organization of the United Nations, Rome, 131 pp., 2007.

Yang, Q., Tian, H., Friedrichs, M. A., Hopkinson, C. S., Lu, C., and Najjar, R. G.: Increased nitrogen export from eastern North America to the Atlantic Ocean due to climatic and anthropogenic changes during 1901–2008, J. Geophys. Res.-Biogeo., 120, 1046–1068, 2015.

Yang, Q., Tian, H., Li, X., Ren, W., Zhang, B., Zhang, X., and Wolf, J.: Spatiotemporal patterns of livestock manure nutrient production in the conterminous United States from 1930 to 2012, Sci. Total Environ., 541, 1592–1602, 2016.

Zhang, B., Tian, H., Lu, C., Dangal, R. S. S., Yang, J., and Pan, S.: Manure nitrogen production and application in cropland during 1860–2014: A 5-minute gridded global data set for Earth system modeling, PANGAEA, https://doi.org/10.1594/PANGAEA.871980, 2017.

12

Radiocarbon measurements of ecosystem respiration and soil pore-space CO_2

Lydia J. S. Vaughn[1,2] and Margaret S. Torn[2,3]

[1]Integrative Biology, University of California, Berkeley, Berkeley, CA 94720, USA
[2]Lawrence Berkeley National Laboratory, Berkeley, CA 94720, USA
[3]Energy and Resources Group, University of California, Berkeley, Berkeley, CA 94720, USA

Correspondence: Lydia J. S. Vaughn (lydiajsvaughn@gmail.com)

Abstract. Radiocarbon measurements of ecosystem respiration and soil pore space CO_2 are useful for determining the sources of ecosystem respiration, identifying environmental controls on soil carbon cycling rates, and parameterizing and evaluating models of the carbon cycle. We measured flux rates and radiocarbon content of ecosystem respiration, as well as radiocarbon in soil profile CO_2 in Utqiaġvik (Barrow), Alaska, during the summers of 2012, 2013, and 2014. We found that radiocarbon in ecosystem respiration ($\Delta^{14}C_{Reco}$) ranged from $+60.5$ to -160 ‰ with a median value of $+23.3$ ‰. Ecosystem respiration became more depleted in radiocarbon from summer to autumn, indicating increased decomposition of old soil organic carbon and/or decreased CO_2 production from fast-cycling carbon pools. Across permafrost features, ecosystem respiration from high-centered polygons was depleted in radiocarbon relative to other polygon types. Radiocarbon content in soil pore-space CO_2 varied between -7.1 and -280 ‰, becoming more negative with depth in individual soil profiles. These pore-space radiocarbon values correspond to CO_2 mean ages of 410 to 3350 years, based on a steady-state, one-pool model. Together, these data indicate that deep soil respiration was derived primarily from old, slow-cycling carbon, but that total CO_2 fluxes depended largely on autotrophic respiration and heterotrophic decomposition of fast-cycling carbon within the shallowest soil layers. The relative contributions of these different CO_2 sources were highly variable across microtopographic features and time in the sampling season. The highly negative $\Delta^{14}C$ values in soil pore-space CO_2 and autumn ecosystem respiration indicate that when it is not frozen, very old soil carbon is vulnerable to decomposition. Radiocarbon data and associated CO_2 flux and temperature data are stored in the data repository of the Next Generation Ecosystem Experiments (NGEE-Arctic) at http://dx.doi.org/10.5440/1364062 and https://doi.org/10.5440/1418853.

1 Introduction

The flux of CO_2 from ecosystems to the atmosphere is a critical component of the global carbon budget. This flux is highly heterogeneous in space and time, so extensive datasets are needed to evaluate the carbon balance within ecosystems. Many measurements have been made of soil surface CO_2 emissions using soil chambers and eddy covariance towers (e.g., Baldocchi, 2008; Bond-Lamberty and Thomson, 2010; Davidson et al., 2002; Norman et al., 1997; Xu and Baldocchi, 2004). Such measurements reveal spatial and temporal patterns in soil and ecosystem respiration that are important for scaling soil carbon emissions across the landscape or identifying drivers of respiration rates (Wainwright et al., 2015). Bulk fluxes of soil or ecosystem respiration, however, do not provide information on the cycling rates or relative contributions of different soil carbon pools to total CO_2 emissions. CO_2 emitted from the soil surface includes autotrophic respiration of rapidly cycling carbon as well as heterotrophic decomposition of carbon that cycles on a broad range of timescales. Shifts in these carbon pool distributions can have large long-term consequences for soil carbon stocks, but may be impossible to detect in bulk CO_2

flux rates (Hopkins et al., 2012; Schuur et al., 2009; Torn et al., 2009; Trumbore, 2000, 2009).

Linking CO_2 emissions with carbon cycling rates requires a tracer of carbon dynamics that can differentiate between source carbon pools. Natural abundance radiocarbon provides such a tracer, as the radiocarbon content of CO_2 reflects the age (since photosynthesis) and decomposition rates of its component sources (Trumbore, 2000). Accordingly, radiocarbon in surface CO_2 emissions and soil pore gas can be used to quantify carbon cycling rates and assess their variability across space, time, carbon pools, and environmental factors such as soil moisture, temperature, and vegetation (Gaudinski et al., 2000; Trumbore, 2000). Within and across sites, variations in the radiocarbon abundance of respired CO_2 can indicate differences in substrate utilization by microbial decomposers (Borken et al., 2006; Chasar et al., 2000) and shifts between more fast-cycling and slow-cycling substrates (Hicks Pries et al., 2013; Hopkins et al., 2012; Schuur et al., 2009). Radiocarbon measurements of soil respiration may be particularly useful for models of the carbon cycle. Carbon pool structures and decomposition dynamics inferred from radiocarbon data may be used to parameterize pool-specific respiration rates; radiocarbon measurements in surface CO_2 emissions and soil pore gas can be used to benchmark model performance; and the radiocarbon signature of respiration can be used to constrain the terrestrial signal in top-down carbon cycle analyses (He et al., 2016; Randerson et al., 2002).

In the high latitudes in particular, increased decomposition rates from large, slow-cycling soil carbon stores have the potential to generate an important long-term climate change feedback (Schuur et al., 2015). There, an estimated 1300 Pg of soil carbon has been protected from decomposition by cold temperatures and often frozen or anoxic conditions (Hugelius et al., 2014). With climate change, these controls on decomposition rates are expected to change as soils warm, hydrological changes occur, and permafrost degradation intensifies (ACIA, 2004). In the lower depths of the seasonally thawed active layer, carbon that has historically cycled on millennial timescales may be highly decomposable under aerobic and thawed conditions (Elberling et al., 2013; Mueller et al., 2015; Strauss et al., 2015; Waldrop et al., 2010). Our current understanding of this old carbon's decomposability, however, is based primarily on CO_2 production rates from laboratory incubations, which may not accurately reflect in situ decomposition dynamics. To quantify in situ decomposition rates, field radiocarbon measurements can be used to differentiate between slow-cycling and fast-cycling carbon and link decomposition dynamics to their environmental controls.

In spite of their use in tracing carbon cycle dynamics, only a limited set of radiocarbon data have been published for Arctic ecosystem respiration (e.g., Hardie et al., 2009; Hicks Pries et al., 2013; Lupascu et al., 2014a; Nowinski et al., 2010; Phillips et al., 2015; Schuur et al., 2009) or

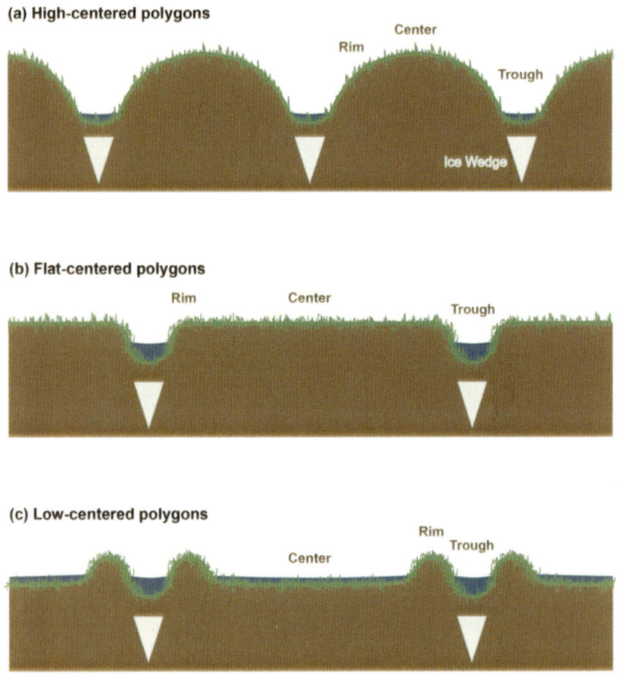

Figure 1. Schematic diagram of polygon morphology. Ice wedge locations and the relative scales of features are approximate.

soil pore-space CO_2 (Czimczik and Welker, 2010; Lupascu et al., 2014a, b), from only a few sites and tundra types. To our knowledge, no data are currently available on radiocarbon in CO_2 from Arctic polygon tundra. Polygon tundra, which covers an estimated 250 000 km^2 throughout the Arctic (Donner et al., 2007), is characterized by microtopographic features $\sim$ 10–32 m in diameter known as ice wedge polygons, which form by the growth and subsequent degradation of subsurface ice formations (ice wedges) (Billings and Peterson, 1980; Brown et al., 1980). Polygons are separated by low-lying drainage channels called troughs, and can be classified according to the elevation of their centers into low-centered, flat-centered, and high-centered categories. Low-centered polygons (LC) have raised, dry rims at their perimeter surrounding low, saturated or inundated centers; flat-centered polygons (FC) have flat, even surfaces with minimal rim delineation; and high-centered polygons (HC) have raised centers with dry surface soils that slope steeply into troughs (Fig. 1). Among LC, FC, and HC polygons, differences in active layer depth, oxygen availability, soil carbon distribution, and thermal conductivity (Liljedahl et al., 2016; Lipson et al., 2012; Ping et al., 1998) create strong differences in surface carbon fluxes (Vaughn et al., 2016; Wainwright et al., 2015). Given these steep spatial gradients in carbon cycling rates and their environmental controls, radiocarbon in respiration from polygon tundra ecosystems offers unique information on subsurface carbon dynamics across space, environmental factors, and the progression of seasonal thaw.

Here, we present measurements of radiocarbon in ecosystem respiration, soil pore-space CO_2, and soil organic carbon from a polygon tundra ecosystem in Utqiaġvik, Alaska. We measured rates and radiocarbon contents of soil surface CO_2 emissions across three summer seasons, from a range of ice wedge polygon features. Using soil pore-space $^{14}CO_2$ profiles, we quantified vertical distributions of carbon cycling rates within the soil profile. Together, these measurements reveal temporal patterns and spatial relationships in ecosystem respiration and its source carbon pools.

2 Study site

Field sampling was conducted at the Barrow Environmental Observatory (BEO), ~ 6 km east of Utqiaġvik (formerly Barrow), Alaska (71.3° N, 156.5° W), at the northern end of the Alaskan Arctic coastal plain. Utqiaġvik has a mean annual temperature of −9.8 °C and mean annual precipitation of 146 mm, with long, dry winters and short, moist, cool summers (NOAA, 2018). The land surface has low topographic relief reaching a maximum elevation of 5 m (Brown et al., 1980; Hubbard et al., 2013). The seasonally thawed active layer ranges in depth from 20 to 60 cm, underlain by continuous ice-rich permafrost to depths greater than 400 m (Hinkel and Nelson, 2003). Formed from the late Pleistocene Gubic formation (Black, 1964), soils in the region are dominated by Typic Aquiturbels (53 %), Typic Histoturbels (22 %), and Typic Aquorthels (8.6 %) (Bockheim et al., 1999). Across the Utqiaġvik peninsula, 24 % of the ground surface is covered by LC polygons, 17 % by FC polygons, 11 % by HC polygons, and 5 % by their surrounding troughs (Lara et al., 2014). Within the specific study region, polygons cover the majority of the land surface, with the remainder comprised of a polygon-free drainage region. In the study area, FC polygons occupy the greatest percentage of the total ground surface, HC polygons occupy the smallest percentage, and LC polygons have intermediate percentage cover (Wainwright et al., 2015). Sample collection was distributed across centers, rims, and troughs of the three polygon types to capture variations in polygon microtopography and associated carbon cycle controls.

3 Field measurements and sample collection

3.1 Surface CO_2 emissions

In August and October 2012, July and September 2013, and September 2014, we collected soil surface CO_2 emissions for radiocarbon analysis using a static chamber method modified from Hahn et al. (2006). Samples were collected from a total of 19 locations within 11 polygons (4 LC, 4 FC, 3 HC). Opaque chambers (25 cm diameter) were seated on circular PVC chamber bases extending to depths of ~ 10 cm below the soil surface, with aboveground chamber height varying between ~ 15 and 20 cm. We installed all bases at least 2 days prior to sampling to limit the influence of disturbance on gas flux rates and radiocarbon values. Chambers blocked light transmission and were tall enough to enclose surface vegetation, so CO_2 emissions were equivalent to ecosystem respiration. For sample collection, the chamber was placed in a 3 cm deep channel on the top rim of each base, which was filled with water to create an airtight seal. We then circulated chamber gas through soda lime for 20 min at a flow rate of 1 L min^{-1} to remove ambient CO_2. CO_2 was allowed to accumulate in the chamber over 2 to 48 h, depending on the rate of CO_2 accumulation, which we monitored periodically by injecting a 30 mL sample of chamber gas into a LI-820 CO_2 gas analyzer (LI-COR) at a flow rate of ~ 1 L min^{-1}. For all samples, the final chamber CO_2 concentration was more than twice its initial concentration, important for accuracy in chamber radiocarbon measurements (Egan et al., 2014). Based on this concentration measurement, a volume of chamber gas sufficient for radiocarbon analysis was collected in one or more 500–1000 mL evacuated stainless steel canisters connected by capillary tubing to a chamber sampling port. High-concentration samples were collected with a syringe and needle through a septum in the sampling port and immediately injected into evacuated glass vials sealed with 14 mm thick chlorobutyl septa (Bellco Glass, Inc.). To correct chamber gas samples for atmospheric contamination, we collected local air samples in 3000 mL stainless steel canisters on 12 August 2012, 13 July 2013, and 2 and 7 September 2014.

In July and September 2013 and September 2014, we measured rates of ecosystem respiration from radiocarbon sampling locations. CO_2 fluxes were measured within 2 days of radiocarbon sample collection, using opaque static chambers seated on bases described above and vented according to Xu et al. (2006) to minimize pressure changes due to the Venturi effect. For each measurement, we measured CO_2 concentrations within the chamber over a period of 4–8 min using a Los Gatos Research, Inc. Portable Greenhouse Gas Analyzer. We calculated the CO_2 flux rate (equivalent to ecosystem respiration) as the slope of the linear portion of its concentration vs. time curve, converted to units of $\mu mol\ m^{-2}\ s^{-2}$ according to chamber volume and temperature. Endpoints of this linear region were determined manually for each curve, and curves lacking a clear linear range were not included in the dataset. Corresponding to each CO_2 flux measurement, we measured thaw depth with a tile probe.

3.2 Soil pore-space CO_2

In August 2012 and July 2013, we collected soil pore gas from a total of six soil profiles within five polygons (one LC, two FC, two HC). Samples were collected with a method similar to Czimczik and Welker (2010). Briefly, 1 / 4″ diameter stainless steel probes were inserted into the soil at 45° angles to vertical depths of 10, 20, and 30 cm, or to 2 cm above the frost table if thaw depth was less than 30 cm. Probes were capped with gastight septa and allowed to remain in place throughout the sampling season. Before collecting each sample, we purged 10 mL of gas from the probe and measured the CO_2 concentration with a LI-820 CO_2 gas analyzer as described above. Radiocarbon samples were collected by connecting evacuated 500–1000 mL stainless steel canisters to probes via flow-restricting tubing (Upchurch scientific, 0.01″ ID ×10 cm length), which allowed canisters to fill slowly over 4 h with minimal disturbance to the soil CO_2 concentration gradient (Gaudinski et al., 2000). In-line Drierite water traps were used during sample collection to prevent moisture accumulation in canisters. As with surface respiration samples, high-concentration samples were collected from probes with a syringe and needle and immediately injected into evacuated glass vials. Due to water-saturated soils or clogged soil probes, we were unable to obtain samples from all profiles and depths. For this reason, the final sample set represents only a subset of depths and sampling locations and is not a random sample of the entire landscape.

3.3 Soil organic matter

On 8 September 2014, we collected 1″ diameter soil cores from the centers of the nine polygons used for September 2014 soil surface CO_2 sampling (3 LC, 3 FC, 3 HC). Soil cores were collected from within 0.5 m of each static chamber. Cores were collected using a manual soil recovery probe (AMS Samplers Inc.) to the base of the thawed soil layer (16–44 cm depth) and stored intact for 8 days at 5 °C before further processing. Following this 8-day period, cores from FC and HC polygons were divided into 2 equal depth increments, which ranged from 8 to 15.5 cm in length. In the case of LC polygons, where standing water was present at the time of sampling, the shallow and deep increments were defined as surface water and sediment, respectively. For a separate experiment, core increments were incubated at 3 °C for 387 days, during which CO_2 production rates were monitored with a LI-820 CO_2 analyzer (LI-COR). Following this 387-day period, we collected evolved CO_2 for radiocarbon analysis using stainless steel air sampling canisters or glass serum vials. Incubated soil, water, and sediment samples were freeze-dried and ground for carbon content and radiocarbon analyses.

3.4 Sample purification and radiocarbon analysis

CO_2 from gas samples was cryogenically purified under vacuum, divided for ^{14}C and ^{13}C analysis, and sealed in 9 mm quartz tubes. For radiocarbon analysis, we sent samples to Lawrence Livermore National Laboratory's Center for Accelerator Mass Spectrometry (CAMS) or the Carbon, Water, and Soils Research Lab at the USDA-FS Northern Research Station, where CO_2 was reduced to graphite on iron powder under H_2. ^{14}C abundance was then measured at CAMS using an HVEC FN Tandem Van de Graaff accelerator mass spectrometer or at UC Irvine's Keck Carbon Cycle AMS facility. $^{13}C/^{12}C$ in CO_2 splits was analyzed on the UC Davis Stable Isotope Laboratory GVI Optima Stable Isotope Ratio Mass Spectrometer.

Two quality control measures were employed to test for contamination by non-sample carbon during gas sample storage and preparation. First, we evaluated the air sampling canisters with repeated leak testing and an ethanol-derived certified CO_2 reference gas standard with empirically determined ^{14}C (Airgas, USA) to confirm that no contamination was introduced during sample storage. Second, the vacuum line used for CO_2 purification was tested with USGS coal and oxalic acid II (NIST SRM 4990C) standards for contamination or leakage. Standards were well characterized with respect to ^{14}C.

Subsamples of bulk soil from the 2014 incubation were sent to the Carbon, Water, and Soils Research Lab, where they were combusted to CO_2 and analyzed for radiocarbon as above. Carbon concentrations and ^{13}C abundances were measured on a Costech analytical Technologies ECS 4010 elemental analyzer coupled to a Thermo Fischer Scientific Delta V^{plus} isotope ratio mass spectrometer. Given the acidic pH range of soils at this site (Zona et al., 2011), we consider bulk soil carbon to be equivalent to soil organic carbon.

Following the conventions of Stuiver and Polach (1977), radiocarbon results are presented as fraction modern relative to the NBS Oxalic Acid I (OX1) standard ($F^{14}C$), and deviations in parts per thousand (‰) from the absolute (decay-corrected) OX1 standard ($\Delta^{14}C$). All results have been corrected for mass-dependent isotopic fractionation using ^{13}C measurements.

4 Calculations and data quality control

4.1 Correction of surface CO_2 emissions for atmospheric contamination

We corrected surface-chamber radiocarbon measurements for atmospheric contamination using the method described in Schuur and Trumbore (2006). Briefly, we determined fractional contributions of background atmosphere and ecosystem respiration to total chamber gas using ^{13}C values in a two-pool mixing model:

$$^{13}C_S = f_{Reco} \times {}^{13}C_{Reco} + f_{atm} \times {}^{13}C_{atm}, \tag{1}$$

$$f_{Reco} + f_{atm} = 1, \tag{2}$$

where f_{Reco} and f_{atm} are the fractional contributions of ecosystem respiration and background atmosphere, $^{13}C_S$ and $^{13}C_{atm}$ are the measured ^{13}C abundances in the sample and background atmosphere in units of at. %, and $^{13}C_{Reco}$ is the ^{13}C abundance in ecosystem respiration, approximated separately for each polygon type as the mean ^{13}C of chamber

CO_2 samples with $[CO_2] > 4000$ ppm. Mean $\delta^{13}C_{Reco}$ values calculated this way were −24.6, −26.5, and −26.2‰ for LC, FC, and HC polygons. To minimize error due to large proportions of atmospheric CO_2, we omitted samples with $f_{Reco} < 0.5$. For each sample, we calculated $\Delta^{14}C$ of ecosystem respiration ($\Delta^{14}C_{Reco}$) according to Eq. (3):

$$\Delta^{14}C_S = f_{Reco} \times \Delta^{14}C_{Reco} + f_{atm} \times \Delta^{14}C_{atm}, \quad (3)$$

where $\Delta^{14}C_S$ and $\Delta^{14}C_{atm}$ are the measured $\Delta^{14}C$ values of the sample and background atmosphere, and f_{Reco} and f_{atm} were calculated from Eqs. (1) and (2). Errors due to analytical precision and variations in source isotopic signatures were propagated through this correction according to the formulation in Phillips and Gregg (2001).

Soil surface CO_2 flux measurements were assessed for quality using two criteria, the standard error of the slope (SE) and the per cent relative standard error (PRSE), defined as $100 \times SE_{slope}/Estimate_{slope}$. Flux measurements with $SE > 0.05$ and $PRSE > 5$ were omitted from the dataset. This set of dual criteria avoided biasing the dataset toward low fluxes (if SE alone were used) or high fluxes (if PRSE alone or R^2 were used).

4.2 Modeled mean age of soil pore-space CO_2

With soil pore-space radiocarbon data, we omitted measurements with CO_2 yields below the expected yield for atmospheric measurements due to possible leakage and atmospheric contamination during sampling, with the exception of one sample from 31 cm depth with highly negative $\Delta^{14}C$, indicating a low proportion of atmospheric CO_2. At this field site, ^{13}C abundances in pore-space CO_2 vary greatly due to isotopic fractionation from methane production and consumption (Vaughn et al., 2016). For this reason, we could not correct subsurface samples for atmospheric CO_2. Reported radiocarbon values thus represent the total CO_2 present in the soil pore space, sourced from heterotrophic respiration, root respiration, and downward atmospheric diffusion.

With soil-profile CO_2 samples, we used radiocarbon measurements to model the mean age of carbon in respired CO_2 using the time-dependent steady-state turnover time model described in Torn et al. (2009), modified to account for carbon residence time in plant tissues:

$$F'_{C,t} C_t = I F'_{atm,\, t-T_R} + C_{t-1} F'_{C,\, t-1} \left(1 - \frac{1}{\tau} - \lambda\right), \quad (4)$$

where $F' = \frac{\Delta^{14}C}{1000} - 1$, $F'_C = F'$ of the given carbon pool, equal to F' of the CO_2 sample, $F'_{atm} = F'$ of CO_2 in the local atmosphere, I = input rate of carbon from the atmosphere to the given carbon pool (g C y^{-1}), C = stock of carbon in the given carbon pool (g), τ = turnover time of the given carbon pool, equivalent to the mean age of carbon in its decomposition flux (y) at steady state, λ = radioactive decay rate of ^{14}C (1/8267 y), and T_R = mean residence time of carbon in plants before entering soil organic matter (y).

At steady state, $C_t = C_{t-1} = I \times \tau$, so Eq. (4) reduces to

$$F'_{C,t} = \frac{1}{\tau} F'_{atm,\, t-T_R} + F'_{C,\, t-1} \left(1 - \frac{1}{\tau} - \lambda\right). \quad (5)$$

Following Eq. (5), the $\Delta^{14}C$ value of CO_2 at time t thus depends on the turnover time of carbon in the decomposing carbon pool, the mean residence time of carbon in plant material, and the $\Delta^{14}C$ of atmospheric CO_2 in the current and previous years, which has changed continuously since the release of radiocarbon into the atmosphere from nuclear weapons testing between 1950 and the mid 1960s (Trumbore, 2000). This model assumes that CO_2 is derived from a homogeneous pool of decomposing carbon, such that the turnover time of this pool is equal to the mean age of its decomposition flux at steady state (Sierra et al., 2017).

The mean residence time of carbon in plants (T_R) reflects a mixture of materials with varying transfer rates. In the Arctic vegetation dominant at our site, some photosynthates enter the soil within 1 day of fixation (Loya et al., 2002); leaf tissues with transit times of < 1 year represent the majority of annual production (Shaver and Kummerow, 1991); roots and shoots are estimated to live for ranges of 3–7 and 1–8 years, respectively (Chapin III et al., 1980; Shaver and Billings, 1975), and rhizomes and stem bases, the longest-lived of belowground tissues, have estimated lifetimes of 2.7–15.6 years (Dennis, 1977). Overall, T_R represents the flux-weighted average of these various pools. Given the large percentage of carbon flux dedicated to annual leaf production and rapid transport to soils, we estimate that across plant organs, plant species, and seasons, the mean value of T_R lies between 0 and 5 years.

Annual atmospheric $\Delta^{14}C$ values were compiled from our data and three other sources: the IntCal13 dataset (Reimer et al., 2013), measurements from Fruholmen, Norway, between 1962 and 1991 (Nydal and Lövseth, 1996), measurements from Utqiaġvik between 1999 and 2007 (Graven et al., 2012), and our measurements from the BEO in 2012, 2013, and 2014. From roughly the same latitude as Utqiaġvik, Fruholmen $\Delta^{14}C$ measurements provide a close approximation for missing Utqiaġvik data (Meijer et al., 2008). Because ecosystem CO_2 uptake occurs primarily during the growing season, we averaged June–August $\Delta^{14}C$ values to produce an annualized summer dataset. Data gaps from 1992–1998 and 2008–2011 were filled using an exponential interpolation constrained by the available data from the 10 years surrounding each gap.

Using our measured $\Delta^{14}C$ values and annually resolved atmospheric $\Delta^{14}C$ data, we iteratively solved for the mean age of each CO_2 sample. We performed this calculation twice, using T_R values of 0 and 5 to bracket the likely T_R range. Samples containing a large percentage of recently fixed carbon yielded two possible solutions for each T_R value

(Trumbore, 2000). In such cases, we chose the appropriate solution either by comparing the two values to other (unique) mean age values within the same profile, or by comparing the measured carbon stock with the carbon stocks calculated from the CO_2 production rate and the candidate solutions (Torn et al., 2009).

4.3 Correction of soil organic carbon for respiration losses

To calculate $\Delta^{14}C$ values of soil organic carbon, we corrected bulk soil $\Delta^{14}C$ measurements for loss of CO_2 respired during the 387-day incubation. We used the post-incubation masses of non-respired soil carbon (C_{soil}, calculated from measured soil carbon contents and sample masses) and carbon respired as CO_2 (C_{CO_2}, calculated from CO_2 production rates) to calculate pre-incubation $\Delta^{14}C$ values of bulk soil organic carbon ($\Delta^{14}C_{SOC}$) according to

$$\Delta^{14}C_{SOC} = \frac{C_{soil}}{C_{soil} + C_{CO_2}} \times \Delta^{14}C_{soil} + \frac{C_{CO_2}}{C_{soil} + C_{CO_2}} \times \Delta^{14}C_{CO_2}, \quad (6)$$

where $\Delta^{14}C_{soil}$ and $\Delta^{14}C_{CO_2}$ are the radiocarbon abundances measured from bulk soil and respired CO_2 following the incubation.

5 Results and discussion

5.1 Radiocarbon in ecosystem respiration

Radiocarbon contents of soil surface CO_2 emissions, equivalent to ecosystem respiration ($\Delta^{14}C_{Reco}$), ranged from +60.5 to −160 ‰, with a median value of +23.3 ‰ (Table S1 in the Supplement). The positive $\Delta^{14}C_{Reco}$ values measured in 28 of the 37 samples indicate high proportions of carbon fixed since 1950, likely sourced from both autotrophic respiration and decomposition of rapidly cycling soil carbon in shallow soil layers. In contrast, the negative values measured in the other nine samples show that the sources of CO_2 at those times and places were dominated by carbon that cycles on centennial to millennial timescales. $\Delta^{14}C_{Reco}$ followed a left-skewed distribution, with notably negative $\Delta^{14}C_{Reco}$ values measured from LC1-center in October 2012 (−159.5 ‰) and HC1-center in September 2013 (−115.5 ‰). The variations in these $\Delta^{14}C_{Reco}$ values reflect differences in the relative CO_2 production and transport rates from both autotrophic and heterotrophic source pools, whose decomposition rates and radiocarbon contents vary spatially and temporally (Nowinski et al., 2010). Together, seasonal variations in soil thaw depths and spatial variations in soil organic carbon ^{14}C depth profiles are known to affect the radiocarbon signature of surface CO_2 emissions.

From data compiled over the 3 measurement years, the lowest $\Delta^{14}C_{Reco}$ values were measured in September, when

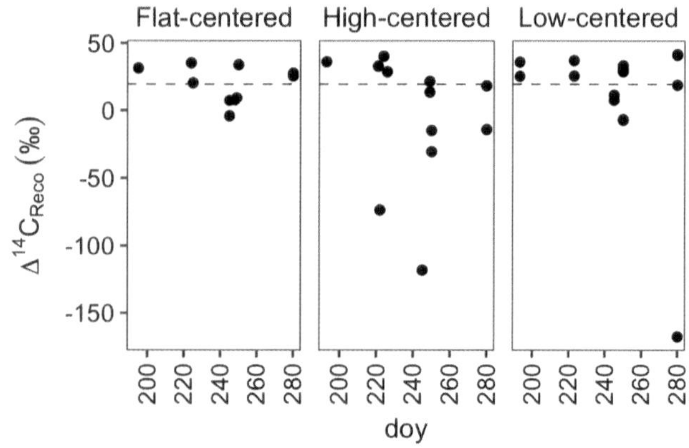

Figure 2. Radiocarbon content of ecosystem respiration, sampled from flat-centered, high-centered, and low-centered polygons in 2012, 2013, and 2014. All samples have been corrected for atmospheric contamination using ^{13}C and ^{14}C abundances of the local atmosphere at the time of sample collection. Dashed horizontal line indicates the mean $\Delta^{14}C$ of local atmospheric CO_2 during the 2012–2014 summer seasons.

the depth of soil thaw was near its annual maximum (Fig. 2). This observation is consistent with other data from high-latitude sites (Hicks Pries et al., 2013; Trumbore, 2000) and reflects seasonal changes in thaw depth, soil temperature, and vegetation activity. Surface soils at this site begin to thaw in June, typically reaching their maximum temperatures in July (Hinkel et al., 2001; Torn, 2015). As shallow soils warm and plant activity increases in this early summer period, ecosystem respiration includes high proportions of ^{14}C-enriched CO_2 from autotrophic respiration and heterotrophic decomposition of shallow, rapid-cycling soil carbon (Fig. 2). Autotrophic respiration peaks in July or August, and decreases substantially into the fall after plants senesce (Hicks Pries et al., 2013), shifting the balance of respiration toward soil carbon decomposition. As thaw depth increases into October (Table S1) (Zona et al., 2016), decomposition of deep, ^{14}C-depleted soil carbon can become an increasing fraction of total soil respiration. The effect of these changes is a seasonal shift in respiration from primarily shallow, fast-cycling source carbon pools to deeper, more ^{14}C-depleted soil organic matter.

Whereas September $\Delta^{14}C_{Reco}$ values were notably low, this was not the case with October measurements (Fig. 2). Although it is possible that this observation reflects the relative carbon mineralization rates of the different soil depths, the relatively enriched $\Delta^{14}C_{Reco}$ values measured in October may instead be due to physical limitations to gas transport within the soil profile or bias in the distribution of samples across polygon types and features. Surface soils were frozen at the time of October sample collection, possibly trapping deeper, ^{14}C-depleted CO_2 below the surface soil layers. This effect has been proposed as a mechanism underlying pulses

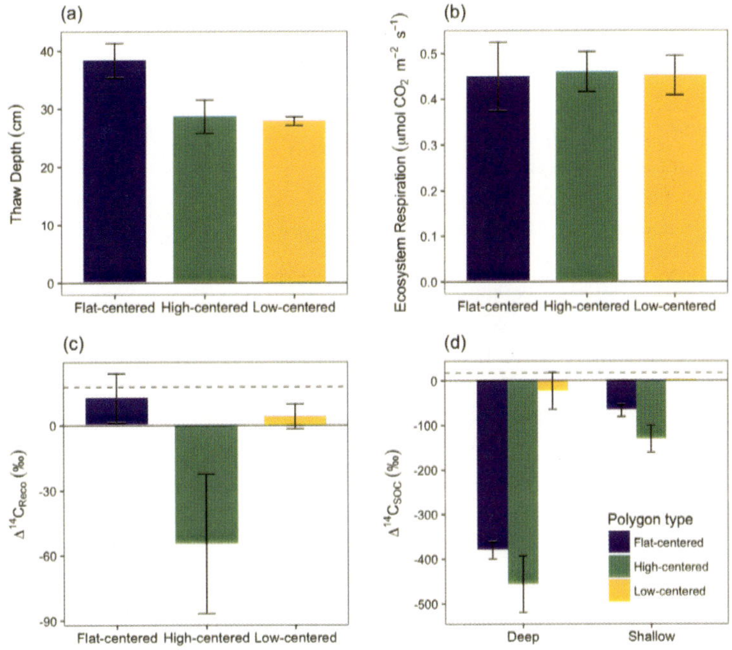

Figure 3. Depth of soil thaw **(a)**, ecosystem respiration rate **(b)**, radiocarbon content of ecosystem respiration **(c)**, and radiocarbon content of bulk soil organic carbon **(d)**, measured from polygon centers in September 2014. Error bars represent standard error with $n = 3$ with the exception of $\Delta^{14}C_{SOC}$ from shallow low-centered polygons ($n = 2$; no standard error calculated). Dashed horizontal line indicates $\Delta^{14}C$ of local atmospheric CO_2 at the time of sampling. In panel **(d)**, deep and shallow samples represent the lower and upper halves of the thawed soil layer for flat-centered and high-centered polygons. For low-centered polygons, shallow samples were collected as standing water and deep samples span the full depth of the thawed sediment layer.

of CO_2 release during the autumn freeze-up and spring thaw periods (Bubier et al., 2002; Oechel et al., 1997; Raz-Yaseef et al., 2017). In the absence of such a pulse, surface CO_2 fluxes may be derived largely from shallow soil decomposition, yielding relatively enriched $\Delta^{14}C_{Reco}$ values late in the autumn season. Additionally, because of snow and frozen surface soils, we were unable to obtain CO_2 samples from a number of locations. Notably, measurements from the centers of HC and FC polygons are absent from the October dataset. If the seasonal decrease in $\Delta^{14}C_{Reco}$ were greatest in polygon centers, the lack of ^{14}C-depleted October measurements may thus be due to this sampling bias.

Time-series measurements from individual profiles or polygon types indicate that $\Delta^{14}C_{Reco}$ varied not only with time in the sampling season, but also with polygon type (Fig. 2, Table S1). HC polygons displayed particularly strong seasonal trends in $\Delta^{14}C_{Reco}$; in each HC polygon from which multiple measurements were made, $\Delta^{14}C_{Reco}$ decreased throughout the summer months to a minimum value in September or October. The magnitude of this seasonal decrease varied greatly among individual HC polygons and microtopographic positions within the polygons, reaching a minimum $\Delta^{14}C_{Reco}$ value of -115 ‰ from the center of polygon HC1, but only $+13$ ‰ from the trough of HC3 (Table S1). In contrast with HC polygons, $\Delta^{14}C_{Reco}$ from FC and LC polygons remained closer to atmospheric values throughout the sampling season (Fig. 2). One exception was one highly negative $\Delta^{14}C_{Reco}$ value measured in October from the center of polygon LC1. In this instance, with October thaw depths close to their annual maximum (Zona et al., 2016), thaw may have penetrated into the transition layer at the top of the permafrost, exposing old, previously frozen carbon to decomposition. Alternatively, this isolated measurement of ^{14}C-depleted CO_2 may reflect heterogeneity within the active layer due to cryoturbation, the vertical transport of soil material due to freeze–thaw processes (Kaiser et al., 2007; Ping et al., 1998).

In September 2014, we measured $\Delta^{14}C_{Reco}$, $\Delta^{14}C_{SOC}$, and ecosystem respiration rates from the centers of three polygons of each type (Fig. 3, Table S2). The influence of microtopography on old carbon emissions was particularly apparent in this measurement set, which offers an even distribution of measurements across polygon types. At this time, decomposition of old carbon consistently dominated the respiration flux from HC polygon centers ($\Delta^{14}C_{Reco} = -51.6 \pm 32.4$ ‰), whereas fast-cycling pools dominated respiration from FC and LC polygon centers (13.0 ± 10.9 and 5.9 ± 5.6 ‰, respectively; Fig. 3c). Ecosystem respiration rates were comparable across the three polygon types (Fig. 3b), which indicates that absolute rates of old carbon decomposition were greater from HC polygons than from LC or FC polygons. The spatial pattern in old carbon decomposition was not tightly linked to $\Delta^{14}C_{SOC}$, which was ^{14}C-depleted relative to the respiration flux in both shallow and deep samples (Fig. 3c, d, Tables S1, S2). Differences in ^{14}C between bulk soil carbon and the respired fraction are commonly observed across ecosystems and soils, as soil carbon is comprised of components with a range of cycling rates (Dioumaeva et al., 2002; Hicks Pries et al., 2013; Schuur and Trumbore, 2006; Trumbore, 2000). Additionally, spatial and temporal variations in autotrophic respiration obscure relationships between soil-respired CO_2 and ecosystem respiration.

We saw no correlation in September 2014 between $\Delta^{14}C_{Reco}$ and thaw depth, which was greatest at this time in FC polygons (Fig. 3a). In contrast, repeated measurements from individual soil locations showed a general decrease in $\Delta^{14}C_{Reco}$ as thaw depth increased and exposed deeper soil layers to unfrozen conditions (Table S1). These findings suggest that the relationship between the depth of thaw and old carbon mineralization depends on the spatial and temporal scales of observation. At the scale of an individual soil profile, seasonal variations in $\Delta^{14}C_{Reco}$ correspond with changes in thaw depth. $\Delta^{14}C$ of soil organic matter tends to increase with depth in the soil, particularly in permafrost environments where the brevity of summer thaw strongly limits carbon turnover at depth (Hicks Pries et al., 2013; McFarlane et al., 2013; Torn et al., 2002; Trumbore, 2000; Fig. 3d). Accordingly, increasingly ^{14}C-depleted soil organic matter becomes available to decomposers as soil thaw deepens over the summer season. At the site scale, however, thaw depth may not be a useful predictor of spatial variations in $\Delta^{14}C_{Reco}$. Other differences – including variations in soil temperature profiles, organic layer thickness, vegetation productivity and rooting depth, the presence of cryoturbation, and oxygen availability to decomposers (Newman et al., 2015; Ping et al., 2015; Sloan et al., 2014; Vaughn et al., 2016) – may lead to microtopographic differences in the shape of $\Delta^{14}C_{SOC}$ depth profiles and the relative contributions of different carbon source pools to the decomposition flux.

In July and September 2013 and September 2014, ecosystem respiration ranged between 0.32 and $2.5\,\mu mol\,CO_2\,m^{-2}\,s^{-1}$ (Fig. 4). The relationship between the rate and radiocarbon content of ecosystem respiration was highly variable, but with two notable patterns across all data. First, as ecosystem respiration increased, the variance of $\Delta^{14}C_{Reco}$ tended to decrease. Second, there were no negative $\Delta^{14}C_{Reco}$ measurements associated with fluxes above $1\,\mu mol\,CO_2\,m^{-2}\,d^{-1}$ (Fig. 4). These patterns reflect the sensitivity of ecosystem respiration to the most dynamic carbon sources, autotrophic respiration and shallow soil carbon mineralization. Autotrophic respiration can contribute as much as 70 % of ecosystem respiration at the peak of the growing season (Hicks Pries et al., 2013), declining markedly as plants senesce. Similarly, heterotrophic decomposition rates are both highest and most temporally variable in shallow soil layers (Hicks Pries et al., 2013), where soil temperatures exhibit a large seasonal range. Supporting our interpretation, previous studies documenting seasonal variations in CO_2 exchange at this site (Oechel et al., 1995; Vaughn et al., 2016; Zona et al., 2014) indicate that ecosystem respiration, gross primary production, and soil surface temperatures peak in July or August, then decline into the late summer and fall. During the early summer period when respiration fluxes are high, high rates of autotrophic respiration and shallow soil respiration dominate the overall respiration flux. As a result, old, slow-cycling carbon respired from deep soils comprises a large percentage of ecosystem respiration only when respiration rates are low from autotrophic and shallow (fast-cycling) soil sources.

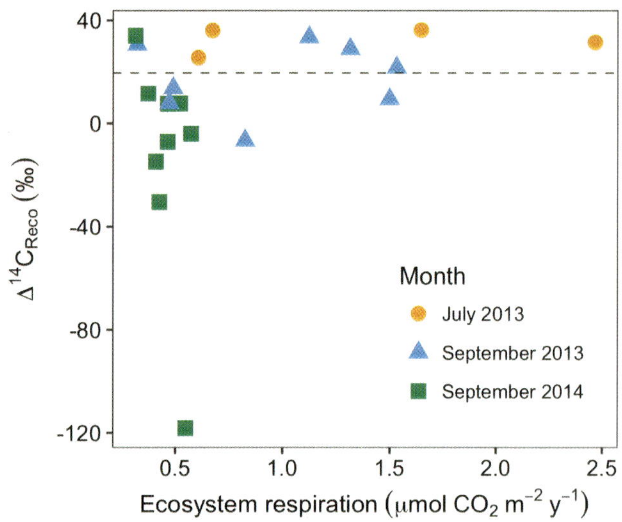

Figure 4. Rates and radiocarbon contents of ecosystem respiration in July and September 2013 and September 2014. The dashed horizontal line indicates the mean $\Delta^{14}C$ of local atmospheric CO_2 during the 2013 and 2014 summer seasons.

5.2 Radiocarbon in soil pore-space CO_2

At a subset of locations and sampling dates, we measured the radiocarbon content of CO_2 in soil pore gas ($\Delta^{14}C_{CO_2p}$). $\Delta^{14}C_{CO_2p}$ varied widely among profiles and sampling dates from -7.1 to $-280\,‰$ (Table 1, Fig. 5). These consistently negative values indicate that pore-space CO_2 was derived primarily from older soil organic matter, with minimal contributions from plant-respired carbon or fast-cycling soil organic carbon. In contrast, ecosystem respiration was generally enriched in radiocarbon relative to soil pore-space CO_2, even at only 10 cm depth (Fig. 5). This observation suggests that although old, slow-cycling carbon dominates soil pore-space CO_2, the shallowest CO_2 sources – autotrophic respiration and/or heterotrophic decomposition of fast-cycling (annual-

Table 1. Isotopic composition of soil profile CO_2.

Profile	Sampling date	Depth (cm)	$\delta^{13}C_{CO_2}$ (‰)	^{14}C analysis year	^{14}C sample size (mg C)	F^{14C}	$\Delta^{14}C_{CO_2}$ (‰)	Mean age of respired C (years)
HC1-center	8/2012	31	–	2013	0.044	0.7464	-259.2 ± 7.5	3050
HC1-center	9/2013	20	−23.7	2013	0.94	0.7270	-278.6 ± 2.6	3350
HC3-center	8/2012	10	−20.1	2013	0.29	0.9473	-59.9 ± 3.0	770
	8/2012	20	–	2013	0.37	0.8864	-120.3 ± 3.2	1320
	8/2012	29	−21.7	2013	0.46	0.8944	-112.4 ± 2.7	1240
HC3-center	7/2013	10	–	2013	0.12	0.9037	-101.3 ± 4.4	1150
	7/2013	20	–	2013	0.14	0.7382	-267.4 ± 2.6	3170
HC3-trough	7/2013	10	–	2013	0.19	0.7898	-216.2 ± 3.3	2440
	7/2013	20	−23.7	2013	0.83	0.8447	-161.7 ± 2.4	1760
FC2-center	8/2012	10	−24.8	2013	1.44	0.9509	-56.3 ± 2.8	740
	8/2012	20	−24.9	2013	1.35	0.9729	-34.5 ± 2.9	580
FC4-center	7/2013	20	−24.8	2013	0.99	0.8034	-202.7 ± 2.1	2260
LC3-trough	7/2013	10	−24.7	2013	0.72	1.0005	-7.1 ± 3.9	410
	7/2013	20	–	2013	0.16	0.8755	-131.2 ± 3.7	1430

decadal) organic carbon – contribute large proportions of the total respiration flux, even late in the season when plants have largely senesced. Detecting and characterizing the decomposition of older, deeper soil organic carbon requires direct measurements of soil pore-space CO_2. Such measurements provide a qualitative indicator of old carbon decomposition; as with surface CO_2 effluxes, proportional or absolute contributions from distinct carbon pools cannot be calculated without well-resolved vertical distributions of ^{14}C and ^{13}C source pools, as well as gas transport rates within the profile.

In most soil profiles, $\Delta^{14}C_{CO_2p}$ became increasingly negative with depth in the soil. This depth trend is similar to that seen in $\Delta^{14}C_{SOC}$ profiles measured from this site (Fig. 3d, Table S2), indicating that the cycling rates of both bulk soil organic matter and the decomposing carbon fraction tend to decrease with depth in the soil. Because frozen or near-freezing temperatures slow decomposition from the deep active layer throughout the majority of summer, this pattern is what would be expected in the absence of vertical mixing or rapid contributions of fast-cycling carbon at depth. In contrast with this general trend, soil pore-space CO_2 from three soil profiles (HC3-trough, HC3-center, and FC2-center) became enriched in radiocarbon near the permafrost table (Fig. 5, Table 1), a pattern that has been previously observed in the Arctic (Lupascu et al., 2014a). In these three profiles, concentrations of CO_2 in deep pore-space samples were 6 to 25 times higher than background atmosphere, so we infer that the higher $\Delta^{14}C_{CO_2p}$ values at depth were not caused by downward transport of atmospheric CO_2. Instead, this observation indicates significant CO_2 production from fast-cycling carbon near the permafrost table. As in other Arctic sites, this depth trend is likely due either to cryoturbation (Bockheim and Tarnocai, 1998) or DOC leaching (Lupascu et al., 2014a), both of which can transport recently fixed, relatively decomposable carbon to the deep section of the active layer.

With $\Delta^{14}C_{CO_2p}$ values, we used a single pool turnover time model to calculate the mean age of pore-space CO_2 (Table 1). This model assumes that CO_2 is respired from a homogeneous pool of decomposing carbon (Trumbore, 2000) such that the turnover time equals the mean age of the respired carbon (Sierra et al., 2017). Across all measurement dates and depths, we found the mean age of respired CO_2 ranged from 410 years in one 10 cm sample to 3350 years at only 20 cm depth, indicating decomposition of old and/or slow-cycling carbon, even in relatively shallow soils. HC polygon centers produced particularly old CO_2, with ages greater than 3000 years from HC3-center in July 2013 and HC1-center in August 2012 and September 2013. This observation aligns with the spatial patterns we observed in surface CO_2 emissions, which were particularly ^{14}C-depleted in HC polygons (Fig. 2, Table S1). Together, these findings suggest that low $\Delta^{14}C_{Reco}$ from HC polygon centers was due to mineralization of very old carbon at depth. Because of unknown diffusion rates within the soil profile, it is challenging to quantitatively use pore-space CO_2 measurements to link soil carbon cycling rates to soil surface $\Delta^{14}C_{Reco}$. Low diffusion rates can lead to the accumulation of high concentrations of ^{14}C-depleted, slow-cycling CO_2 deep in the soil profile (Lee et al., 2010), and vertical mixing can transport

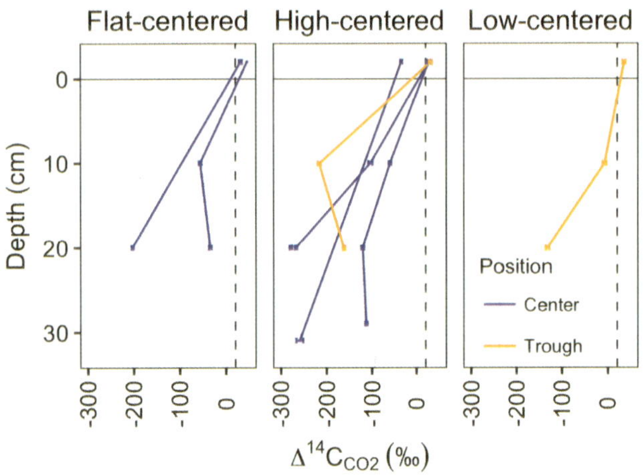

Figure 5. Soil depth profiles of $\Delta^{14}C$-CO_2. Lines connect samples collected in the same vertical profile and month, and error bars represent analytical error. Values at −2 cm depth represent samples collected from surface soil chambers and reflect CO_2 that accumulated after chambers were scrubbed of CO_2. All other samples were collected from subsurface soil probes. Dashed vertical line indicates the mean $\Delta^{14}C$ of local atmospheric CO_2 during the 2012 and 2013 summer seasons.

CO_2 away from the site of production prior to its collection as soil pore gas.

Our findings of consistently negative $\Delta^{14}C$ values from soil pore-space CO_2 are similar to those from the other Arctic sites where it has been measured (Czimczik and Welker, 2010; Lupascu et al., 2014b, a). In contrast, studies in temperate or tropical sites find $\Delta^{14}C_{CO_2p}$ values that are more similar to atmospheric values (Borken et al., 2006; Gaudinski et al., 2000; Trumbore, 2000), where soil respiration is dominated by root respiration and decomposition of rapidly cycling soil organic matter. This difference may be due in part to differences among biomes in rooting distributions. Because of strong vertical gradients in soil temperature and nutrient and water availability, rooting distributions in Arctic tundra are extremely shallow (Iversen et al., 2015), limiting most root respiration to the shallow organic layer. Additionally, such negative $\Delta^{14}C_{CO_2p}$ values suggest that old, slow-cycling organic matter in Arctic soils is readily decomposable under thawed conditions. Cold soil temperatures and frozen conditions limit microbial activity throughout much of the year, allowing otherwise-decomposable soil organic matter to accumulate. During the short season in which Arctic soils are thawed, these large pools of ^{14}C-depleted soil carbon may be largely unprotected against microbial mineralization. Compared with tropical or temperate soils, existing Arctic soil carbon stores may thus be particularly vulnerable to warming.

Due to sampling limitations, ^{14}C data from soil pore-space CO_2 are available from only a subset of polygon types, positions, sampling dates, and depths. No samples were collected from polygon rims, and only few samples were obtained where soils were saturated (Table 1). For this reason, our dataset does not capture the full range of microtopographic variations in deep soil decomposition rates and controls. Soil temperature profiles, soil pore-space oxygen availability, soil organic matter concentration and chemistry, and vegetation composition, rooting distribution, and productivity influence microbial activity and vary among profiles and with time in the thawed season (Lipson et al., 2012; Olivas et al., 2011). In particular, these controls on decomposition dynamics likely differ between saturated and unsaturated soils, due both to the direct effects of saturation and to covarying effects of microtopography. For this reason, radiocarbon values presented in this manuscript cannot be scaled across the full range of environmental variation present at the site. To better characterize the relationships between these variables and soil carbon decomposition rates, further measurements are needed of ^{14}C in soil pore-space DIC and CO_2, across spatial, seasonal, and hydrological gradients.

6 Data availability

Data are stored in the Next-Generation Ecosystem Experiments (NGEE-Arctic) data repository (Vaughn et al., 2018; Vaughn and Torn, 2018) and may be accessed at http://dx.doi.org/10.5440/1364062 and https://doi.org/10.5440/1418853. The code used to generate plots and correct chamber samples for atmospheric contamination can be found at https://github.com/lydiajsvaughn/Radiocarbon_field_Barrow. The code used for turnover time modeling can be found at https://github.com/lydiajsvaughn/Radiocarbon_inc_2012.

7 Conclusions

We measured the radiocarbon contents of ecosystem respiration and soil pore-space CO_2 between 2012 and 2014 in Utqiaġvik, Alaska. As a naturally occurring tracer of soil carbon dynamics, radiocarbon in CO_2 and soil carbon reflects the rates at which carbon cycles through plant and soil pools (Trumbore, 2000). Radiocarbon provides a powerful tool to test and parameterize bottom-up models of the carbon cycle (He et al., 2016), constrain source signatures in inverse modeling studies (Randerson et al., 2002), and evaluate how environmental variables, seasonal changes, and disturbance influence decomposition rates from fast- and slow-cycling carbon pools. In Arctic sites, where cycling rates can be very slow, radiocarbon can be used to detect changes in decomposition from millennial-cycling pools (Hicks Pries et al., 2013; Schuur et al., 2009), which strongly influence long-term carbon dynamics but may be impossible to detect with other methods. Only limited data are available of radiocarbon in ecosystem respiration or soil pore-space CO_2, particularly from Arctic sites where samples can be difficult to obtain due to low rates of CO_2 production. To characterize seasonal and spatial variations in the age of carbon pools contributing to total CO_2 efflux, we measured radiocarbon in ecosystem respiration and soil pore-space CO_2 across three summer seasons in Arctic polygon tundra.

$\Delta^{14}C$ in ecosystem respiration varied between +60.5 and −160 ‰ across the 3 years, with a strong seasonal trend (Fig. 2). July and August $\Delta^{14}C_{Reco}$ measurements were generally close to the $\Delta^{14}CO_2$ values of the local atmosphere, which declined from 22.2 ‰ in 2012 to 17.7 ‰ in 2014. In September, in contrast, over half the $\Delta^{14}CO_2$ measurements differed greatly from the local atmosphere (Fig. 2), contributing ^{14}C-depleted CO_2 to the atmospheric pool. These seasonal variations in $\Delta^{14}C_{Reco}$ and respiration rates contribute to the strong seasonal cycles of atmospheric $\Delta^{14}CO_2$ observed in the high latitudes (Graven et al., 2012). By quantifying these variations – respiration flux rates and radiocarbon values over space and time – this dataset offers useful information for atmospheric budgeting and inversions.

Measurements of radiocarbon in late-season ecosystem respiration indicate that carbon that cycles on millennial timescales contributes substantially to soil respiration. When thaw depth approached its maximum in September and October, highly depleted ^{14}C in respiration indicated that carbon older than 1000 years was a major source of heterotrophic respiration. Decomposition of old, slow-cycling soil carbon, however, may be missed when measuring surface $\Delta^{14}C_{Reco}$ alone. In the soil pore space, $\Delta^{14}C_{CO_2p}$ declined steeply with depth, particularly in high-centered polygons (Fig. 4); at only 20 cm below the soil surface, the mean age of carbon in the decomposition flux was as old as 3000 years (Table 1). Together, these observations suggest that during the short summer thaw season, ancient carbon stores in the deep active layer become available to decomposition, stabilized otherwise by cold, often frozen, and often anaerobic conditions (Mikan et al., 2002; Schmidt et al., 2011; Schuur et al., 2015).

As climate change alters these environmental controls and soils warm and thaw, a key question is how decomposition rates will change. Where permafrost degradation occurs, either through gradual deepening of the active layer or rapid thaw events, old soil carbon that has remained frozen for years to millennia is exposed to thawed conditions. A particularly important – but unknown – factor is the decomposition rate of this carbon released from thawing permafrost (Hicks Pries et al., 2013; Koven et al., 2015; Kuhry et al., 2013). Our measurements cannot differentiate between such newly thawed soil organic matter and carbon that has historically experienced an annual thaw. Our data do show, however, that when it is unfrozen, even very old Arctic soil organic carbon will readily decompose. As thaw depth progressed to the permafrost boundary, we consistently documented CO_2 production via decomposition of centennial carbon pools. As permafrost thaw progresses and Arctic soils warm, further measurements of soil-respired $^{14}CO_2$ across spatial and temporal gradients will provide critical information on soil carbon vulnerability.

Author contributions. LJSV and MST designed the sampling procedure, LJSV collected and analyzed data with assistance and guidance from MST, and LJSV prepared the manuscript with contributions from MST.

Competing interests. The authors declare that they have no conflict of interest.

Acknowledgements. We thank John Bryan Curtis and Oriana Chafe for field sampling assistance, and the Radiocarbon Collaborative for radiocarbon analyses. This research was conducted through the Next Generation Ecosystem Experiments (NGEE-Arctic) project, which is supported by the Office of Biological and Environmental Research in the US Department of Energy Office of Science.

Edited by: Kirsten Elger
Reviewed by: two anonymous referees

References

ACIA: Impacts of a Warming Arctic: Arctic Climate Impact Assessment, Cambridge University Press, Cambridge, UK, available at: https://www.amap.no/documents/doc/impacts-of-a-warming-arctic-2004/786 (last access: 27 September 2018), 2004.

Baldocchi, D.: 'Breathing' of the terrestrial biosphere: lessons learned from a global network of carbon dioxide flux measurement systems, Aust. J. Bot., 56, 1–26, https://doi.org/10.1071/BT07151, 2008.

Billings, W. D. and Peterson, K. M.: Vegetational change and ice-wedge polygons through the thaw-lake cycle in Arctic Alaska, Arctic Alpine Res., 12, 413–432, 1980.

Black, R. F.: Gubik Formation of Quaternary age in northern Alaska, United States Geological Survey, available at: http://pubs.er.usgs.gov/publication/pp302C (last access: 29 July 2014), 1964.

Bockheim, J. G. and Tarnocai, C.: Recognition of cryoturbation for classifying permafrost-affected soils, Geoderma, 81, 281–293, https://doi.org/10.1016/S0016-7061(97)00115-8, 1998.

Bockheim, J. G., Everett, L. R., Hinkel, K. M., Nelson, F. E., and Brown, J.: Soil organic carbon storage and distribution in Arctic tundra, Barrow, Alaska, Soil Sci. Soc. Am. J., 63, 934–940, 1999.

Bond-Lamberty, B. and Thomson, A.: A global database of soil respiration data, Biogeosciences, 7, 1915–1926, https://doi.org/10.5194/bg-7-1915-2010, 2010.

Borken, W., Savage, K., Davidson, E. A., and Trumbore, S. E.: Effects of experimental drought on soil respiration and radiocarbon efflux from a temperate forest soil, Glob. Change Biol., 12, 177–193, https://doi.org/10.1111/j.1365-2486.2005.001058.x, 2006.

Brown, J., Miller, P. C., Tieszen, L. L., and Bunnell, F.: An Arctic ecosystem?: the coastal tundra at Barrow, Alaska, Dowden, Hutchinson and Ross, Inc., Stroudsburg, Pennsylvania, available at: https://darchive.mblwhoilibrary.org/handle/1912/222 (last access: 30 March 2014), 1980.

Bubier, J., Crill, P., and Mosedale, A.: Net ecosystem CO_2 exchange measured by autochambers during the snow-covered season at a temperate peatland, Hydrol. Process., 16, 3667–3682, https://doi.org/10.1002/hyp.1233, 2002.

Chapin III, F. S., Tieszen, L. L., Lewis, M. C., Miller, P. C., and McCown, B. H.: Control of tundra plant allocation patterns and growth, in: An Arctic Ecosystem: The Coastal Tundra at Barrow, Alaska, edited by: Brown, J., Miller, P. C., Tieszen, L. L., and Bunnell, F., Dowden Hutchinson Ross Stroudsburg Penn, 140–185, 1980.

Chasar, L. S., Chanton, J. P., Glaser, P. H., Siegel, D. I., and Rivers, J. S.: Radiocarbon and stable carbon isotopic evidence for transport and transformation of dissolved organic carbon, dissolved inorganic carbon, and CH_4 in a northern Minnesota peatland, Global Biogeochem. Cy., 14, 1095–1108, https://doi.org/10.1029/1999GB001221, 2000.

Czimczik, C. I. and Welker, J. M.: Radiocarbon Content of CO_2 Respired from High Arctic Tundra in Northwest Greenland, Arct. Antarct. Alp. Res., 42, 342–350, https://doi.org/10.1657/1938-4246-42.3.342, 2010.

Davidson, E. A., Savage, K., Verchot, L. V., and Navarro, R.: Minimizing artifacts and biases in chamber-based measurements of soil respiration, Agr. Forest Meteorol., 113, 21–37, https://doi.org/10.1016/S0168-1923(02)00100-4, 2002.

Dennis, J. G.: Distribution Patterns of Belowground Standing Crop in Arctic Tundra at Barrow, Alaska, Arctic Alpine Res., 9, 113–127, https://doi.org/10.2307/1550574, 1977.

Dioumaeva, I., Trumbore, S., Schuur, E. A. G., Goulden, M. L., Litvak, M., and Hirsch, A. I.: Decomposition of peat from upland boreal forest: Temperature dependence and sources of respired carbon, J. Geophys. Res., 107, 8222, https://doi.org/10.1029/2001JD000848, 2002.

Donner, N., Karpov, N. S., de Klerk, P., and Joosten, H.: Distribution, diversity, development and dynamics of polygon mires: examples from Northeast Yakutia (Siberia), available at: http://www.pimdeklerk-palynology.eu/polygon_mires_PI__Minke_et_al_2007_.pdf (last access: 27 September 2018), 2007.

Egan, J., Nickerson, N., Phillips, C., and Risk, D.: A Numerical Examination of $^{14}CO_2$ Chamber Methodologies for Sampling at the Soil Surface, Radiocarbon, 56, 1175–1188, https://doi.org/10.2458/56.17771, 2014.

Elberling, B., Michelsen, A., Schädel, C., Schuur, E. A. G., Christiansen, H. H., Berg, L., Tamstorf, M. P., and Sigsgaard, C.: Long-term CO2 production following permafrost thaw, Nat. Clim. Change, 3, 890–894, https://doi.org/10.1038/nclimate1955, 2013.

Gaudinski, J. B., Trumbore, S. E., Davidson, E. A., and Zheng, S.: Soil carbon cycling in a temperate forest: radiocarbon-based estimates of residence times, sequestration rates and partitioning of fluxes, Biogeochemistry, 51, 33–69, 2000.

Graven, H. D., Guilderson, T. P., and Keeling, R. F.: Observations of radiocarbon in CO_2 at seven global sampling sites in the Scripps flask network: Analysis of spatial gradients and seasonal cycles, J. Geophys. Res., 117, D02303, https://doi.org/10.1029/2011JD016535, 2012.

Hahn, V., Högberg, P., and Buchmann, N.: ^{14}C – a tool for separation of autotrophic and heterotrophic soil respiration, Glob. Change Biol., 12, 972–982, https://doi.org/10.1111/j.1365-2486.2006.001143.x, 2006.

Hardie, S. M. L., Garnett, M. H., Fallick, A. E., Ostle, N. J., and Rowland, A. P.: Bomb-^{14}C analysis of ecosystem respiration reveals that peatland vegetation facilitates release of old carbon, Geoderma, 153, 393–401, https://doi.org/10.1016/j.geoderma.2009.09.002, 2009.

He, Y., Trumbore, S. E., Torn, M. S., Harden, J. W., Vaughn, L. J. S., Allison, S. D., and Randerson, J. T.: Radiocarbon constraints imply reduced carbon uptake by soils during the 21st century, Science, 353, 1419–1424, https://doi.org/10.1126/science.aad4273, 2016.

Hicks Pries, C. E., Schuur, E. A. G., and Crummer, K. G.: Thawing permafrost increases old soil and autotrophic respiration in tundra: Partitioning ecosystem respiration using $\delta^{13}C$ and ^{14}C, Glob. Change Biol., 19, 649–661, https://doi.org/10.1111/gcb.12058, 2013.

Hinkel, K. M. and Nelson, F. E.: Spatial and temporal patterns of active layer thickness at Circumpolar Active Layer Monitoring (CALM) sites in northern Alaska, 1995–2000, J. Geophys. Res., 108, 8168, https://doi.org/10.1029/2001JD000927, 2003.

Hinkel, K. M., Paetzold, F., Nelson, F. E., and Bockheim, J. G.: Patterns of soil temperature and moisture in the active layer and upper permafrost at Barrow, Alaska: 1993–1999, Glob. Planet. Change, 29, 293–309, https://doi.org/10.1016/S0921-8181(01)00096-0, 2001.

Hopkins, F. M., Torn, M. S., and Trumbore, S. E.: Warming accelerates decomposition of decades-old carbon in forest soils, P. Natl. Acad. Sci. USA, 109, E1753–E1761, https://doi.org/10.1073/pnas.1120603109, 2012.

Hubbard, S. S., Gangodagamage, C., Dafflon, B., Wainwright, H., Peterson, J., Gusmeroli, A., Ulrich, C., Wu, Y., Wilson, C., Rowland, J., Tweedie, C., and Wullschleger, S. D.: Quantifying and relating land-surface and subsurface variability in permafrost environments using LiDAR and surface geophysical datasets, Hydrogeol. J., 21, 149–169, https://doi.org/10.1007/s10040-012-0939-y, 2013.

Hugelius, G., Strauss, J., Zubrzycki, S., Harden, J. W., Schuur, E. A. G., Ping, C.-L., Schirrmeister, L., Grosse, G., Michaelson, G. J., Koven, C. D., O'Donnell, J. A., Elberling, B., Mishra, U., Camill, P., Yu, Z., Palmtag, J., and Kuhry, P.: Estimated stocks of circumpolar permafrost carbon with quantified uncertainty ranges and identified data gaps, Biogeosciences, 11, 6573–6593, https://doi.org/10.5194/bg-11-6573-2014, 2014.

Iversen, C. M., Sloan, V. L., Sullivan, P. F., Euskirchen, E. S., McGuire, A. D., Norby, R. J., Walker, A. P., Warren, J. M., and Wullschleger, S. D.: The unseen iceberg: plant roots in arctic tundra, New Phytol., 205, 34–58, https://doi.org/10.1111/nph.13003, 2015.

Kaiser, C., Meyer, H., Biasi, C., Rusalimova, O., Barsukov, P., and Richter, A.: Conservation of soil organic matter through cryoturbation in arctic soils in Siberia, J. Geophys. Res., 112, G02017, https://doi.org/10.1029/2006JG000258, 2007.

Koven, C. D., Lawrence, D. M., and Riley, W. J.: Permafrost carbon-climate feedback is sensitive to deep soil carbon decomposability but not deep soil nitrogen dynamics, P. Natl. Acad. Sci. USA, 112, 3752–3757, https://doi.org/10.1073/pnas.1415123112, 2015.

Kuhry, P., Grosse, G., Harden, J. W., Hugelius, G., Koven, C. D., Ping, C.-L., Schirrmeister, L., and Tarnocai, C.: Characterisation of the Permafrost Carbon Pool, Permafrost Periglac., 24, 146–155, https://doi.org/10.1002/ppp.1782, 2013.

Lara, M. J., McGuire, A. D., Euskirchen, E. S., Tweedie, C. E., Hinkel, K. M., Skurikhin, A. N., Romanovsky, V. E., Grosse, G., Bolton, W. R., and Genet, H.: Polygonal tundra geomorphological change in response to warming alters future CO_2 and CH_4 flux on the Barrow Peninsula, Glob. Change Biol., 21, 1634–1651, https://doi.org/10.1111/gcb.12757, 2014.

Lee, H., Schuur, E. A. G., and Vogel, J. G.: Soil CO_2 production in upland tundra where permafrost is thawing, J. Geophys. Res., 115, G01009, https://doi.org/10.1029/2008JG000906, 2010.

Liljedahl, A. K., Boike, J., Daanen, R. P., Fedorov, A. N., Frost, G. V., Grosse, G., Hinzman, L. D., Iijma, Y., Jorgenson, J. C., Matveyeva, N., Necsoiu, M., Raynolds, M. K., Romanovsky, V. E., Schulla, J., Tape, K. D., Walker, D. A., Wilson, C. J., Yabuki, H., and Zona, D.: Pan-Arctic ice-wedge degradation in warming permafrost and its influence on tundra hydrology, Nat. Geosci., 9, 312–318, https://doi.org/10.1038/ngeo2674, 2016.

Lipson, D. A., Zona, D., Raab, T. K., Bozzolo, F., Mauritz, M., and Oechel, W. C.: Water-table height and microtopography control biogeochemical cycling in an Arctic coastal tundra ecosystem, Biogeosciences, 9, 577–591, https://doi.org/10.5194/bg-9-577-2012, 2012.

Loya, W. M., Johnson, L. C., Kling, G. W., King, J. Y., Reeburgh, W. S., and Nadelhoffer, K. J.: Pulse-labeling studies of carbon cycling in arctic tundra ecosystems: Contribution of photosynthates to soil organic matter, Global Biogeochem. Cy., 16, 1101, https://doi.org/10.1029/2001GB001464, 2002.

Lupascu, M., Welker, J. M., Xu, X., and Czimczik, C. I.: Rates and radiocarbon content of summer ecosystem respiration in response to long-term deeper snow in the High Arctic of NW Greenland, J. Geophys. Res.-Biogeosci., 119, 1180–1194, https://doi.org/10.1002/2013JG002494, 2014a.

Lupascu, M., Welker, J. M., Seibt, U., Xu, X., Velicogna, I., Lindsey, D. S., and Czimczik, C. I.: The amount and timing of precipitation control the magnitude, seasonality and sources (^{14}C) of ecosystem respiration in a polar semi-desert, northwestern Greenland, Biogeosciences, 11, 4289–4304, https://doi.org/10.5194/bg-11-4289-2014, 2014b.

McFarlane, K. J., Torn, M. S., Hanson, P. J., Porras, R. C., Swanston, C. W., Callaham, M. A., and Guilderson, T. P.: Comparison of soil organic matter dynamics at five temperate deciduous forests with physical fractionation and radiocarbon measurements, Biogeochemistry, 112, 457–476, https://doi.org/10.1007/s10533-012-9740-1, 2013.

Meijer, H. J., Pertuisot, M. H., and van der Plicht, J.: High-accuracy 614C measurements for atmospheric CO_2 samples by AMS, Radiocarbon, 48, 355–372, 2008.

Mikan, C. J., Schimel, J. P., and Doyle, A. P.: Temperature controls of microbial respiration in arctic tundra soils above and below freezing, Soil Biol. Biochem., 34, 1785–1795, 2002.

Mueller, C. W., Rethemeyer, J., Kao-Kniffin, J., Löppmann, S., Hinkel, K. M., and G. Bockheim, J.: Large amounts of labile organic carbon in permafrost soils of northern Alaska, Glob. Change Biol., 21, 2804–2817, https://doi.org/10.1111/gcb.12876, 2015.

Newman, B. D., Throckmorton, H. M., Graham, D. E., Gu, B., Hubbard, S. S., Liang, L., Wu, Y., Heikoop, J. M., Herndon, E. M., Phelps, T. J., Wilson, C. J., and Wullschleger, S. D.: Microtopographic and depth controls on active layer chemistry in Arctic polygonal ground, Geophys. Res. Lett., 42, 1808–1817, https://doi.org/10.1002/2014GL062804, 2015.

NOAA: National Weather Service Climate Services, available at: https://w2.weather.gov/climate/xmacis.php?wfo=pafg, last access: 9 October 2018.

Norman, J. M., Kucharik, C. J., Gower, S. T., Baldocchi, D. D., Crill, P. M., Rayment, M., Savage, K., and Striegl, R. G.: A comparison of six methods for measuring soil-surface carbon dioxide fluxes, J. Geophys. Res., 102, 28771–28777, https://doi.org/10.1029/97JD01440, 1997.

Nowinski, N. S., Taneva, L., Trumbore, S. E., and Welker, J. M.: Decomposition of old organic matter as a result of deeper active layers in a snow depth manipulation experiment, Oecologia, 163, 785–792, https://doi.org/10.1007/s00442-009-1556-x, 2010.

Nydal, R. and Lövseth, K.: Carbon-14 measurements in atmospheric CO2 from northern and southern hemisphere sites, 1962–1993, Oak Ridge National Lab., TN (United States), Oak Ridge Inst. for Science and Education, TN, USA, available at: http://www.osti.gov/scitech/biblio/461185 (last access: 24 May 2017), 1996.

Oechel, W. C., Vourlitis, G. L., Hastings, S. J., and Bochkarev, S. A.: Change in Arctic CO_2 Flux Over Two Decades: Effects of Climate Change at Barrow, Alaska, Ecol. Appl., 5, 846–855, https://doi.org/10.2307/1941992, 1995.

Oechel, W. C., Vourlitis, G., and Hastings, S. J.: Cold season CO_2 emission from Arctic soils, Global Biogeochem. Cy., 11, 163–172, https://doi.org/10.1029/96GB03035, 1997.

Olivas, P. C., Oberbauer, S. F., Tweedie, C., Oechel, W. C., Lin, D., and Kuchy, A.: Effects of Fine-Scale Topography on CO_2 Flux Components of Alaskan Coastal Plain Tundra: Response to Contrasting Growing Seasons, Arct. Antarct. Alp. Res., 43, 256–266, https://doi.org/10.1657/1938-4246-43.2.256, 2011.

Phillips, C. L., McFarlane, K. J., LaFranchi, B., Desai, A. R., Miller, J. B., and Lehman, S. J.: Observations of $^{14}CO_2$ in ecosystem respiration from a temperate deciduous forest in Northern Wisconsin, J. Geophys. Res.-Biogeosci., 120, 600–616, https://doi.org/10.1002/2014JG002808, 2015.

Phillips, D. L. and Gregg, J. W.: Uncertainty in source partitioning using stable isotopes, Oecologia, 127, 171–179, https://doi.org/10.1007/s004420000578, 2001.

Ping, C. L., Bockheim, J. G., Kimble, J. M., Michaelson, G. J., and Walker, D. A.: Characteristics of cryogenic soils along a latitudinal transect in Arctic Alaska, J. Geophys. Res., 103, 28917–28928, 1998.

Ping, C. L., Jastrow, J. D., Jorgenson, M. T., Michaelson, G. J., and Shur, Y. L.: Permafrost soils and carbon cycling, SOIL, 1, 147–171, https://doi.org/10.5194/soil-1-147-2015, 2015.

Randerson, J. T., Enting, I. G., Schuur, E. A. G., Caldeira, K., and Fung, I. Y.: Seasonal and latitudinal variability of troposphere $\Delta^{14}CO_2$: Post bomb contributions from fossil fuels, oceans, the stratosphere, and the terrestrial biosphere: seasonal and latitudinal variability of trophosphere $\Delta^{14}CO_2$, Global Biogeochem. Cy., 16, 1112, https://doi.org/10.1029/2002GB001876, 2002.

Raz-Yaseef, N., Torn, M. S., Wu, Y., Billesbach, D. P., Liljedahl, A. K., Kneafsey, T. J., Romanovsky, V. E., Cook, D. R., and Wullschleger, S. D.: Large CO_2 and CH_4 emissions from polygonal tundra during spring thaw in northern Alaska, Geophys. Res. Lett., 44, 504–513, 2017.

Reimer, P. J., Bard, E., Bayliss, A., Beck, J. W., Blackwell, P. G., Ramsey, C. B., Buck, C. E., Cheng, H., Edwards, R. L., Friedrich, M., Grootes, P. M., Guilderson, T. P., Haflidason, H., Hajdas, I., Hatté, C., Heaton, T. J., Hoffmann, D. L., Hogg, A. G., Hughen, K. A., Kaiser, K. F., Kromer, B., Manning, S. W., Niu, M., Reimer, R. W., Richards, D. A., Scott, E. M., Southon, J. R., Staff, R. A., Turney, C. S. M., and van der Plicht, J.: IntCal13 and Marine13 Radiocarbon Age Calibration Curves 0–50,000 Years cal BP, Radiocarbon, 55, 1869–1887, https://doi.org/10.2458/azu_js_rc.55.16947, 2013.

Schmidt, M. W., Torn, M. S., Abiven, S., Dittmar, T., Guggenberger, G., Janssens, I. A., Kleber, M., Kögel-Knabner, I., Lehmann, J., and Manning, D. A.: Persistence of soil organic matter as an ecosystem property, Nature, 478, 49–56, 2011.

Schuur, E. A. G. and Trumbore, S. E.: Partitioning sources of soil respiration in boreal black spruce forest using radiocarbon, Glob. Change Biol., 12, 165–176, https://doi.org/10.1111/j.1365-2486.2005.01066.x, 2006.

Schuur, E. A. G., Vogel, J. G., Crummer, K. G., Lee, H., Sickman, J. O., and Osterkamp, T. E.: The effect of permafrost thaw on old carbon release and net carbon exchange from tundra, Nature, 459, 556–559, 2009.

Schuur, E. A. G., McGuire, A. D., Schädel, C., Grosse, G., Harden, J. W., Hayes, D. J., Hugelius, G., Koven, C. D., Kuhry, P., Lawrence, D. M., Natali, S. M., Olefeldt, D., Romanovsky, V. E., Schaefer, K., Turetsky, M. R., Treat, C. C., and Vonk, J. E.: Climate change and the permafrost carbon feedback, Nature, 520, 171–179, https://doi.org/10.1038/nature14338, 2015.

Shaver, G. R. and Billings, W. D.: Root Production and Root Turnover in a Wet Tundra Ecosystem, Barrow, Alaska, Ecology, 56, 401–409, https://doi.org/10.2307/1934970, 1975.

Shaver, G. R. and Kummerow, J.: Phenology, resource allocation, and growth of arctic vascular plants, in: Arctic Ecosystems in a Changing Climate an Ecophysiological Perspective, edited by: Chapin, F. S., Jefferies, R. L., Reynolds, J. F., and Shaver, G. R., Academic Press, Inc., San Diego, CA, 193–211, 1991.

Sierra, C. A., Müller, M., Metzler, H., Manzoni, S., and Trumbore, S. E.: The muddle of ages, turnover, transit, and residence times in the carbon cycle, Glob. Change Biol., 23, 1763–1773, 2017.

Sloan, V. L., Brooks, J. D., Wood, S. J., Liebig, J. A., Siegrist, J., Iversen, C. M., and Norby, R. J.: Plant community composition and vegetation height, Barrow, Alaska, Ver. 1, Carbon Dioxide Information Analysis Center, Oak Ridge National Laboratory, Oak Ridge, Tennessee, USA, https://doi.org/10.5440/1129476, 2014.

Strauss, J., Schirrmeister, L., Mangelsdorf, K., Eichhorn, L., Wetterich, S., and Herzschuh, U.: Organic-matter quality of deep permafrost carbon – a study from Arctic Siberia, Biogeosciences, 12, 2227–2245, https://doi.org/10.5194/bg-12-2227-2015, 2015.

Stuiver, M. and Polach, H. A.: Discussion reporting of ^{14}C data, Radiocarbon, 19, 355–363, 1977.

Torn, M.: CO2 CH4 flux Air temperature Soil temperature and Soil moisture, Barrow, Alaska 2013 ver. 1, Next Generation Ecosystems Experiment – Arctic, Oak Ridge National Laboratory (ORNL), Oak Ridge, TN, USA, 2015.

Torn, M. S., Lapenis, A. G., Timofeev, A., Fischer, M. L., Babikov, B. V., and Harden, J. W.: Organic carbon and carbon isotopes in modern and 100-year-old-soil archives of the Russian steppe, Glob. Change Biol., 8, 941–953, 2002.

Torn, M. S., Swanston, C. W., Castanha, C., and Trumbore, S. E.: Storage and turnover of organic matter in soil, in: Biophysico-Chemical Processes Involving Natural Nonliving Organic Matter in Environmental Systems, edited by: Senesi, N., Xing, B., and Huang, P. M., 219–272, https://doi.org/10.1002/9780470494950.ch6, 2009.

Trumbore, S.: Age of soil organic matter and soil respiration: radiocarbon constraints on belowground C dynamics, Ecol. Appl., 10, 399–411, 2000.

Trumbore, S.: Radiocarbon and soil carbon dynamics, Annu. Rev. Earth Pl. Sci., 37, 47–66, 2009.

Vaughn, L. J. S. and Torn, M. S: Radiocarbon in soil and CO_2 from laboratory incubation, Barrow, Alaska, 2014, Oak Ridge National Laboratory, Oak Ridge, Tennessee, USA, https://doi.org/10.5440/1418853, 2018.

Vaughn, L. J. S., Conrad, M. E., Bill, M., and Torn, M. S.: Isotopic insights into methane production, oxidation, and emissions in Arctic polygon tundra, Glob. Change Biol., 22, 3487–3502, https://doi.org/10.1111/gcb.13281, 2016.

Vaughn, L. J. S., Torn, M. S., Porras, R. C., Curtis, J. B., and Chafe, O.: Radiocarbon in Ecosystem Respiration and Soil Pore-Space CO_2 with Surface Gas Flux, Air Temperature, and Soil Temperature and Moisture, Barrow, Alaska, 2012–2014, Oak Ridge National Laboratory, U.S. Department of Energy, Oak Ridge, Tennessee, USA, https://doi.org/10.5440/1364062, 2018.

Wainwright, H. M., Dafflon, B., Smith, L. J., Hahn, M. S., Curtis, J. B., Wu, Y., Ulrich, C., Peterson, J. E., Torn, M. S., and Hubbard, S. S.: Identifying multiscale zonation and assessing the relative importance of polygon geomorphology on carbon fluxes in an Arctic Tundra Ecosystem, J. Geophys. Res. Biogeosci., 120, 788–808, https://doi.org/10.1002/2014JG002799, 2015.

Waldrop, M. P., Wickland, K. P., White III, R., Berhe, A. A., Harden, J. W., and Romanovsky, V. E.: Molecular investigations into a globally important carbon pool: Permafrost-protected carbon in Alaskan soils, Glob. Change Biol., 16, 2543–2554, 2010.

Xu, L. and Baldocchi, D. D.: Seasonal variation in carbon dioxide exchange over a Mediterranean annual grassland in California, Agr. Forest. Meteorol., 123, 79–96, https://doi.org/10.1016/j.agrformet.2003.10.004, 2004.

Xu, L., Furtaw, M. D., Madsen, R. A., Garcia, R. L., Anderson, D. J., and McDermitt, D. K.: On maintaining pressure equilibrium between a soil CO_2 flux chamber and the ambient air, J. Geophys. Res., 111, D08S10, https://doi.org/10.1029/2005JD006435, 2006.

Zona, D., Lipson, D. A., Zulueta, R. C., Oberbauer, S. F., and Oechel, W. C.: Microtopographic controls on ecosystem functioning in the Arctic Coastal Plain, J. Geophys. Res., 116, G00I08, https://doi.org/10.1029/2009JG001241, 2011.

Zona, D., Lipson, D. A., Richards, J. H., Phoenix, G. K., Liljedahl, A. K., Ueyama, M., Sturtevant, C. S., and Oechel, W. C.: Delayed responses of an Arctic ecosystem to an extreme summer: impacts on net ecosystem exchange and vegetation functioning, Biogeosciences, 11, 5877–5888, https://doi.org/10.5194/bg-11-5877-2014, 2014.

Zona, D., Gioli, B., Commane, R., Lindaas, J., Wofsy, S. C., Miller, C. E., Dinardo, S. J., Dengel, S., Sweeney, C., Karion, A., Chang, R. Y.-W., Henderson, J. M., Murphy, P. C., Goodrich, J. P., Moreaux, V., Liljedahl, A., Watts, J. D., Kimball, J. S., Lipson, D. A., and Oechel, W. C.: Cold season emissions dominate the Arctic tundra methane budget, P. Natl. Acad. Sci. USA, 113, 40–45, 2016.

Permissions

All chapters in this book were first published in ESSD, by Copernicus Publications; hereby published with permission under the Creative Commons Attribution License or equivalent. Every chapter published in this book has been scrutinized by our experts. Their significance has been extensively debated. The topics covered herein carry significant findings which will fuel the growth of the discipline. They may even be implemented as practical applications or may be referred to as a beginning point for another development.

The contributors of this book come from diverse backgrounds, making this book a truly international effort. This book will bring forth new frontiers with its revolutionizing research information and detailed analysis of the nascent developments around the world.

We would like to thank all the contributing authors for lending their expertise to make the book truly unique. They have played a crucial role in the development of this book. Without their invaluable contributions this book wouldn't have been possible. They have made vital efforts to compile up to date information on the varied aspects of this subject to make this book a valuable addition to the collection of many professionals and students.

This book was conceptualized with the vision of imparting up-to-date information and advanced data in this field. To ensure the same, a matchless editorial board was set up. Every individual on the board went through rigorous rounds of assessment to prove their worth. After which they invested a large part of their time researching and compiling the most relevant data for our readers.

The editorial board has been involved in producing this book since its inception. They have spent rigorous hours researching and exploring the diverse topics which have resulted in the successful publishing of this book. They have passed on their knowledge of decades through this book. To expedite this challenging task, the publisher supported the team at every step. A small team of assistant editors was also appointed to further simplify the editing procedure and attain best results for the readers.

Apart from the editorial board, the designing team has also invested a significant amount of their time in understanding the subject and creating the most relevant covers. They scrutinized every image to scout for the most suitable representation of the subject and create an appropriate cover for the book.

The publishing team has been an ardent support to the editorial, designing and production team. Their endless efforts to recruit the best for this project, has resulted in the accomplishment of this book. They are a veteran in the field of academics and their pool of knowledge is as vast as their experience in printing. Their expertise and guidance has proved useful at every step. Their uncompromising quality standards have made this book an exceptional effort. Their encouragement from time to time has been an inspiration for everyone.

The publisher and the editorial board hope that this book will prove to be a valuable piece of knowledge for researchers, students, practitioners and scholars across the globe.

List of Contributors

Guangsheng Chen, Shufen Pan and Hanqin Tian
International Center for Climate and Global Change Research and School of Forestry and Wildlife Sciences, Auburn University, Auburn, AL, USA

Daniel J. Hayes
School of Forest Resources, University of Maine, Orono, ME, USA

Daniel Odermatt
Odermatt & Brockmann GmbH, Zurich, 8005, Switzerland
Brockmann Consult GmbH, Geesthacht, 20502, Germany
Eawag, Swiss Federal Institute of Aquatic Science and Technology, Dübendorf, 8600, Switzerland

Olaf Danne and Carsten Brockmann
Brockmann Consult GmbH, Geesthacht, 20502, Germany

Petra Philipson
Brockmann Geomatics Sweden AB, Kista, 164 40, Sweden

Fei Lun
College of Resources and Environmental Sciences, China Agricultural University, Beijing 100193, China

Junguo Liu
School of Environmental Science and Engineering, South University of Science and Technology of China, Shenzhen 518055, China

Philippe Ciais, Jinfeng Chang, Rong Wang and Daniel Goll
IPSL - LSCE, CEA CNRS UVSQ, Centre d'Etudes Orme des Merisiers, 91191 Gif-sur-Yvette, France

Thomas Nesme
Bordeaux Sciences Agro, Université de Bordeaux, UMR 1391 ISPA, CS 40201, 33175 Gradignan CEDEX, France

Jordi Sardans
CREAF, Cerdanyola del Vallès 08193, Catalonia, Spain

Josep Peñuelas
CSIC, Global Ecology Unit CREAF-CSIC-UAB, Cerdanyola del Vallès 08193, Catalonia, Spain

Michael Obersteiner
International Institute for Applied Systems Analysis, 2361 Laxenburg, Austria

Heather A. Bouman
Department of Earth Sciences, University of Oxford, Oxford, OX1 3AN, UK

Trevor Platt, Thomas Jackson and Gavin H. Tilstone
Plymouth Marine Laboratory, Prospect Place, The Hoe, PL1 3DH, UK

Martina Doblin and Virginie van Dongen-Vogels
Plant Functional Biology and Climate Change Cluster, Faculty of Science, University of Technology Sydney Broadway, Sydney, NSW 2007, Australia

Francisco G. Figueiras
Instituto de Investigaciones Marinas (CSIC), Eduardo Cabello 6, 36208 Vigo, Spain

Kristinn Gudmundsson and Hafsteinn G. Gudfinnsson
Marine Research Institute, 121 Reykjavík, Iceland

Bangqin Huang
State Key Laboratory of Marine Environmental Science/Key Laboratory of Coastal and Wetland Ecosystems, Ministry of Education, Xiamen University, Xiamen, Fujian 361005, China

Anna Hickman
Ocean and Earth Science, University of Southampton, National Oceanography Centre Southampton, European Way, Southampton, SO14 3ZH, UK

Michael Hiscock
United States Environmental Protection Agency, Ariel Rios Building, 1200 Pennsylvania Avenue, Washington D.C., 20460, USA

Vivian A. Lutz and Valeria Segura
Instituto Nacional de Investigacion y Desarrollo Pesquero, Mar del Plata, Argentina

Frédéric Mélin
European Commission, Joint Research Centre, Ispra 21027, Italy

Francisco Rey
Institute of Marine Research, c/o Department of Biological Sciences, University of Oslo, 0316, Oslo, Norway

Pierre Pepin
Fisheries and Oceans Canada, Northwest Atlantic Fisheries Centre, St John's, Newfoundland, A1C 5X1, Canada

Shubha Sathyendranath
Plymouth Marine Laboratory, Prospect Place, The Hoe, PL1 3DH, UK
National Centre for Earth Observation, Plymouth Marine Laboratory, Prospect Place, Plymouth PL1 3DH, UK

Astrid Cornils, Rainer Sieger, Elke Mizdalski, Stefanie Schumacher, Hannes Grobe and Sigrid B. Schnack-Schiel
Alfred-Wegener-Institut Helmholtz-Zentrum für Polar- und Meeresforschung, Bremerhaven, Germany

Kees Klein Goldewijk
Copernicus Institute of Sustainable Development, Utrecht University, Utrecht, the Netherlands
PBL Netherlands Environmental Assessment Agency, The Hague, the Netherlands

Arthur Beusen
Department of Earth Sciences, Utrecht University, Utrecht, the Netherlands
PBL Netherlands Environmental Assessment Agency, The Hague, the Netherlands

Jonathan Doelman and Elke Stehfest
PBL Netherlands Environmental Assessment Agency, The Hague, the Netherlands

Anbu Aravazhi Arunkumar
Department of Biotechnology, Karpagam Academy of Higher Education, Coimbatore 641021, Tamil Nadu, India

Arunachalam Manimekalan
Biodiversity and Molecular Lab, Department of Environmental Sciences, Bharathiar University, Coimbatore 641046, Tamil Nadu, India

Sviatoslav Levchuk, Marina Zhurba, Valentyn Protsak and Yuri Khomutinin
Ukrainian Institute of Agricultural Radiology of National University of Life and Environmental Sciences of Ukraine, Mashinobudivnykiv str. 7, Chabany, Kiev region, 08162, Ukraine

Nicholas A. Beresford and Jacqueline S. Chaplow
NERC Centre for Ecology and Hydrology, CEH Lancaster, Lancaster Environment Centre, Library Av. Bailrigg, Lancaster, LA1 4AP, UK

Valery Kashparov
Ukrainian Institute of Agricultural Radiology of National University of Life and Environmental Sciences of Ukraine, Mashinobudivnykiv str. 7, Chabany, Kiev region, 08162, Ukraine
CERAD CoE Environmental Radioactivity, Department of Environmental Sciences, Norwegian University of Life Sciences, 1432 Ås, Norway

Oliver Zielinski and Daniela Meier
Institute for Chemistry and Biology of the Marine Environment, University of Oldenburg, Wilhelmshaven, 26382, Germany

Kertu Lõhmus and Michael Kleyer
Institute of Biology and Environmental Sciences, University of Oldenburg, Oldenburg, 26129, Germany

Thorsten Balke
School of Geographical and Earth Sciences, University of Glasgow, Glasgow, G128QQ, UK

Helmut Hillebrand
Institute for Chemistry and Biology of the Marine Environment, University of Oldenburg, Wilhelmshaven, 26382, Germany
Helmholtz-Institute for Functional Marine Biodiversity, University of Oldenburg, Oldenburg, 26129, Germany

Wei Li and Philippe Ciais
Laboratoire des Sciences du Climat et de l'Environnement, LSCE/IPSL, CEA-CNRS-UVSQ, Université Paris-Saclay, 91191 Gif-sur-Yvette, France

Natasha MacBean
School of Natural Resources and the Environment, The University of Arizona, Tucson, AZ 85721, USA

Pierre Defourny, Céline Lamarche and Sophie Bontemps
Earth and Life Institute, Environmental Sciences, Université catholique de Louvain, Louvain-la-Neuve, Belgium

Richard A. Houghton
Woods Hole Research Center, Falmouth, MA, USA

Shushi Peng
Sino-French Institute for Earth System Science, College of Urban and Environmental Sciences, Peking University, Beijing 100871, China

Bowen Zhang, Shree R. S. Dangal and Jia Yang
International Center for Climate and Global Change Research, Auburn University, Auburn, AL 6849, USA

Hanqin Tian and Shufen Pan
International Center for Climate and Global Change Research, Auburn University, Auburn, AL 6849, USA
State Key Laboratory of Urban and Regional Ecology, Research Center for Eco-Environmental Sciences, Chinese Academy of Sciences, Beijing 100085, China

Chaoqun Lu
International Center for Climate and Global Change Research, Auburn University, Auburn, AL 6849, USA
Department of Ecology, Evolution, and Organismal Biology, Iowa State University, Ames, IA 50011, USA

Lydia J. S. Vaughn
Integrative Biology, University of California, Berkeley, Berkeley, CA 94720, USA
Lawrence Berkeley National Laboratory, Berkeley, CA 94720, USA

Margaret S. Torn
Lawrence Berkeley National Laboratory, Berkeley, CA 94720, USA
Energy and Resources Group, University of California, Berkeley, Berkeley, CA 94720, USA

Index

S

T

Z